Multiobjective Genetic Algorithms for Clustering

Ujjwal Maulik · Sanghamitra Bandyopadhyay
Anirban Mukhopadhyay

Multiobjective Genetic Algorithms for Clustering

Applications in Data Mining and Bioinformatics

 Springer

Prof. Ujjwal Maulik
Department of Computer Science
and Engineering
Jadavpur University
700032 Kolkata
West Bengal
India
umaulik@cse.jdvu.ac.in

Prof. Sanghamitra Bandyopadhyay
Machine Intelligence Unit
Indian Statistical Institute
B. T. Road 203
700108 Kolkata
West Bengal
India
sanghami@isical.ac.in

Dr. Anirban Mukhopadhyay
Department of Computer Science
and Engineering
University of Kalyani
741235 Kalyani
Nadia, West Bengal
India
anirban@klyuniv.ac.in

ISBN 978-3-642-16614-3   e-ISBN 978-3-642-16615-0
DOI 10.1007/978-3-642-16615-0
Springer Heidelberg Dordrecht London New York

Library of Congress Control Number: 2011932492

ACM Codes: I.2, J.3, H.2, H.3

Cover design: KünkelLopka

Printed on acid-free paper

Springer is part of Springer Science+Business Media (www.springer.com)

To my parents Manoj Kumar Maulik and Gouri Maulik, my son Utsav, and my students
Ujjwal Maulik

To my parents Satyendra Nath Banerjee and Bandana Banerjee, my son Utsav, and my students
Sanghamitra Bandyopadhyay

To my parents Somnath Mukhopadhyay and Manjusri Mukhopadhyay, my wife Anindita, and my son Agnibho
Anirban Mukhopadhyay

Preface

Clustering is an important unsupervised classification technique where a set of patterns, usually vectors in multidimensional space, are grouped into clusters based on some similarity or dissimilarity criteria. In crisp clustering, each pattern is assigned to exactly one cluster, whereas in fuzzy clustering, each pattern is given a membership degree to each class. Fuzzy clustering is inherently more suitable for handling imprecise and noisy data with overlapping clusters. Clustering techniques aim to find a suitable grouping of the input dataset so that some criteria are optimized. Hence the problem of clustering can be posed as an optimization problem. The objectives to be optimized may represent different characteristics of the clusters, such as compactness, separation, and connectivity. A straightforward way to pose clustering as an optimization problem is to optimize some cluster validity index that reflects the goodness of the clustering solutions. All possible partitionings of the dataset and the corresponding values of the validity index define the complete search space. Traditional partitional clustering techniques, such as K-means and fuzzy C-means, employ greedy search techniques over the search space to optimize the compactness of the clusters. Although these algorithms are computationally efficient, they often get stuck at some local optima depending on the choice of the initial cluster centers. Moreover, they optimize a single cluster validity index (compactness in this case), and therefore do not cover different characteristics of the datasets.

To overcome the problem of local optima, some global optimization tools such as Genetic Algorithms (GAs) have been widely used to reach the global optimum value of the chosen validity measure. GAs are randomized search and optimization techniques guided by the principles of evolution and natural genetics, and have a large amount of implicit parallelism. GAs perform multimodal search in complex landscapes and provide near-optimal solutions for the objective or fitness function of an optimization problem. Conventional GA-based clustering techniques use some validity measure as the fitness value. However, no single validity measure works equally well for different kinds of datasets. Thus it is natural to simultaneously optimize multiple such measures for capturing different characteristics of the data.

Simultaneous optimization of multiple objectives provides improved robustness to different data properties. Hence it is useful to utilize multiobjective GAs (MOGAs) for clustering. In multiobjective optimization (MOO), search is performed over a number of often conflicting objective functions. In contrast to single objective optimization, which yields a single best solution, in MOO the final solution set contains a number of Pareto-optimal solutions, none of which can be further improved with regard to any one objective without degrading another. Multiobjective clustering techniques optimize more than one cluster validity index simultaneously, leading to high-quality results. As the relative importance of different clustering criteria is unknown, it is better to optimize them separately rather than combine them in a single measure to be optimized. The resultant set of near-Pareto-optimal solutions contains a number of non-dominated solutions, which the user can judge relatively to pick the most promising one for the problem requirements.

Data mining involves discovering interesting and potentially useful patterns of different types such as associations, summaries, rules, changes, outliers and significant structures. Clustering is an important unsupervised data mining tool that can be useful in mining different kinds of datasets, including geoscientific data, biological data, World Wide Web data, and multimedia data. In the present book, we have explored the application of multiobjective fuzzy clustering in discovering knowledge from geoscientific data such as satellite images and biological data such as microarray gene expression data.

Remote sensing satellite images have significant applications in different areas such as climate studies, the assessment of forest resources, and the examination of marine environments. For remote sensing applications, classification is an important task where the pixels in the images are classified into homogeneous regions, each of which corresponds to some particular land cover type. The problem of pixel classification is often posed as that of clustering in the intensity space. Clustering techniques used for pixel classification should be able to handle the satellite images that have a large number of classes with overlapping and nonlinear class boundaries. In a satellite image, a pixel represents an area of land space, which may not necessarily belong to a single land cover type. Hence, it is evident that a large amount of imprecision and uncertainty can be associated with the pixels in a remotely sensed image. Therefore, it is appropriate as well as natural to apply the principles of fuzzy set theory in the domain of pixel classification. In this book, we explore the application of multiobjective fuzzy clustering algorithms for classifying remote sensing imagery in order to demonstrate their effectiveness.

In recent years, advances in the field of molecular biology, coupled with advances in genomic technologies, have led to an explosive growth in the biological information generated by the scientific community. Bioinformatics, viewed as the use of computational methods to make biological discoveries, has evolved as a major research direction in response to this deluge of information. The main purpose is to utilize computerized databases to store, organize and index the data, and specialized tools to view and analyze the data. It is an interdisciplinary field involving biology, computer science, mathematics and statistics to analyze biological sequence data and genome content and arrangement, and to predict the function and structure of macromolecules. The ultimate goal of the field is to enable the discovery of new biological insights as well as to create a global perspective from which unifying principles in biology can be derived. Data analysis tools used earlier in bioinformatics were mainly based on statistical techniques like regression and estimation. The landscape of biological and biomedical research is being changed rapidly with the invention of microarrays, which enable simultaneous views of the transcription levels of a large number of genes across different tissue samples or time points. Microarray technology has applications in the areas of medical diagnosis, biomedicine, and gene expression profiling, among others. Clustering is used to identify the sets of genes with similar expression profiles. Clustering algorithms are also used for classifying different tissue samples representing multiple cancer types and for identifying possible gene markers. Another important microarray analysis tool is biclustering, which aims to discover local patterns from microarray datasets that contain a subset of genes co-expressed in a subset of experimental conditions. Biclustering reflects the biological reality better than does the traditional clustering approach. The present volume explores the application of multiobjective GAs in clustering and biclustering in microarray gene expression datasets along with the biological relevance of the experimental results.

Another important area that is addressed in this book is categorical data clustering. Many real-life datasets are categorical in nature, i.e., no natural ordering among the values of a particular attribute can be found. Hence, for datasets having categorical attribute domains, there is no inherent distance measure between any two of their points. Also, the mean of a set of points cannot be computed in such datasets. Thus the conventional clustering algorithms do not work with categorical datasets. With the growing demand for categorical data clustering, a few clustering algorithms focussing on categorical data have been developed in the past few years. This volume addresses the issues in designing multiobjective clustering algorithms for categorical attributes with applications to artificial and real-life categorical datasets.

The present volume is aimed at providing a treatise in a unified framework, with extensive applications to real-life datasets. Chapter 1 introduces the clustering problem and discusses different types of clustering algorithms, cluster validity measures, research issues, challenges and applications, and representation of the clustering problem as an optimization task and the possible use of GAs and MOGAs for this purpose. Chapter 2 describes the basic principles of GAs and MOGAs, their theoretical analysis and their applications to data mining and bioinformatics problems. Chapter 3 gives a broad overview of data mining and knowledge discovery tasks and applications. Chapter 4 provides an introduction to basic molecular biology concepts followed by basic tasks in bioinformatics, with a discussion on some of the existing works. Chapter 5 presents a multiobjective fuzzy clustering algorithm that uses real-valued center-based encoding of chromosomes and simultaneously optimizes two fuzzy cluster validity indices. In Chapter 6, a method based on the

Support Vector Machine (SVM) classifier for obtaining the final solution from the set of non-dominated Pareto-optimal clustering solutions produced by the multiobjective fuzzy clustering method is described. In Chapter 7, a two-stage fuzzy clustering technique is described that utilizes the data points having significant membership to multiple classes (SiMM points). A variable string length genetic fuzzy clustering algorithm and multiobjective clustering algorithm are used for this purpose. Results of all the algorithms described in Chapters 5–7 have been demonstrated on both remote sensing imagery as well as microarray gene expression data. Chapter 8 addresses the problem of multiobjective fuzzy clustering of categorical data. A cluster mode-based encoding technique is used and fuzzy compactness and fuzzy separation have been simultaneously optimized in the context of categorical data. The results have been demonstrated for various synthetic and real-life categorical datasets. Chapter 9 presents an application of the multiobjective clustering technique for unsupervised cancer classification and identifying relevant genetic markers using some statistics followed by multiobjective feature selection. Finally, in Chapter 10, an overview of the biclustering problem and algorithms is presented. Also, application of multiobjective GAs in biclustering has been described. Different characteristics of the biclusters are optimized simultaneously. The effect of incorporation of fuzziness has also been studied. Results have been reported for various artificial and real-life gene expression datasets with biological and statistical significance tests.

The present volume is an attempt dedicated to clustering using multiobjective GAs with extensive real-life application in data mining and bioinformatics. The volume, which is unique in its character, will be useful to graduate students and researchers in bioinformatics, computer science, biotechnology, electrical engineering, system science, and information technology as both a text and a reference book for some parts of the curriculum. Researchers and practitioners in the industry and R & D laboratories in the fields of system design, pattern recognition, data mining, soft computing, geoscience, remote sensing and bioinformatics will also benefit from this volume.

The authors gratefully acknowledge the initiative and the support for the project provided by Mr. Ronan Nugent of Springer. Sanghamitra Bandyopadhyay acknowledges the support provided by the Swarnajayanti Fellowship Project Grant (No. DST/SJF/ET-02/2006-07) of the Department of Science and Technology, Government of India. Ujjwal Maulik and Sanghamitra Bandyopadhyay acknowledge their son Utsav, and their parents for the unwavering support provided by them. Anirban Mukhopadhyay acknowledges the moral support provided by his parents and in-laws, his wife Anindita and his son Agnibho.

Kolkata, India,
April 18, 2011

Ujjwal Maulik
Sanghamitra Bandyopadhyay
Anirban Mukhopadhyay

Contents

Chapter 1
Introduction

1.1 Introduction

Data mining involves discovering patterns of different types such as associations, summaries, rules, changes, outliers and significant structures. The terms *interesting* and *potentially useful* used in the realm of data mining are evidently subjective in nature depending on the problem and the concerned user. Some information that is of immense value to one user may be absolutely useless to another.

In general, data mining techniques comprise three components [155]: a model, a preference criterion and a search algorithm. Association rule mining, classification, clustering, regression, sequence and link analysis and dependency modeling are some of the most common functions in current data mining techniques. Model representation determines both the flexibility of the model for representing the underlying data and the interpretability of the model in human terms. This includes decision trees and rules, linear and nonlinear models, example-based techniques such as the NN-rule and case-based reasoning, probabilistic graphical dependency models (e.g., Bayesian network) and relational attribute models.

The preference criterion is used for determining, depending on the underlying dataset, which model to use for mining, by associating some measure of goodness with the model functions. It tries to avoid overfitting of the underlying data or generating a model function with a large number of degrees of freedom. Once the model and the preference criterion are selected, specification of the search algorithm is defined in terms of these along with the given data.

It is important in the knowledge discovery process to represent the information extracted by mining the data in a way easily understandable by the user. The main components of the knowledge representation step are data and knowledge visualization techniques. A hierarchical presentation of information is often very effective for the user to enable him to focus attention on only the important and interesting concepts. This also provides multiple views of the extracted knowledge to the users at different levels of abstraction. There are several approaches to knowledge representation such as rule generation, summarization using natural languages, tables and cross tabulations, graphical representation in the form of bar charts, pie charts and curves, data cube views and decision trees.

There are mainly two kinds of data mining tasks, descriptive and predictive [197]. The descriptive techniques provide a summary of the data. On the other hand, the predictive techniques learn from the current data in order to make predictions about the behavior of new datasets. Clustering is one of the popular descriptive data mining techniques that is used for partitioning an input dataset into a number of clusters. Each cluster should contain the data points that are similar to each other and dissimilar from the data points belonging to other clusters. Clustering has several applications in different fields and is considered as an important unsupervised learning technique. This book is mainly focussed on the problem of clustering. The next section discusses different issues and algorithms for data clustering.

1.2 Clustering

When the only data available is unlabeled, the classification problems are sometimes referred to as *unsupervised classification*. Clustering [15, 135, 205, 237, 408] is an important unsupervised classification technique where a set of patterns, usually vectors in a multidimensional space, are grouped into clusters in such a way that patterns in the same cluster are similar in some sense and patterns in different clusters are dissimilar in the same sense.

Clustering in d-dimensional Euclidean space $\mathbb{R}^d$ is the process of partitioning a given set of n points into a number, say K, of groups (or clusters) $\{C_1, C_2, \ldots, C_K\}$ based on some similarity/dissimilarity metric. The value of K may or may not be known a priori. The main objective of any clustering technique is to produce a $K \times n$ partition matrix $U(X)$ of the given dataset X consisting of n patterns, $X = \{x_1, x_2, \ldots, x_n\}$. The partition matrix may be represented as $U = [u_{kj}]$, $k = 1, \ldots, K$ and $j = 1, \ldots, n$, where u_{kj} is the membership of pattern x_j to cluster C_k. In the case of hard or crisp partitioning,

$$u_{kj} = \begin{cases} 1 & \text{if } x_j \in C_k, \\ 0 & \text{if } x_j \notin C_k. \end{cases}$$

On the other hand, for probabilistic fuzzy partitioning of the data, the following conditions hold on U (representing non-degenerate clustering):

$$\forall k \in \{1, 2, \ldots, K\}, \ 0 < \sum_{j=1}^{n} u_{kj} < n,$$

$$\forall j \in \{1,2,\ldots,n\}, \quad \sum_{k=1}^{K} u_{kj} = 1, \quad \text{and}$$

$$\sum_{k=1}^{K} \sum_{j=1}^{n} u_{kj} = n.$$

Several clustering methods are available in the literature. These can be broadly categorized into hierarchical, partitional and density-based. In the following subsections, some widely used methods belonging to these categories are discussed. Prior to that some popular distance measures are described that are used to characterize the clusters. Finally, this section contains a detailed discussion on several cluster validity indices popularly used for assigning some means of goodness to the discovered partitions.

1.2.1 Distance Measure

A distance or dissimilarity measure is used to compute the amount of dissimilarity between two data points. The choice of distance function plays an important role in cluster analysis. Suppose the given dataset X contains n points, i.e., $X = \{x_1, x_2, \ldots, x_n\}$. If d is the dimension of the data, then x_{ij}, $i = 1, \ldots, n$, $j = 1, \ldots, d$, denotes the value of the ith point in the jth dimension. The distance between two data points x_i and x_j is denoted by $D(x_i, x_j)$. Any distance function D should satisfy the following properties:

1. $D(x_i, x_j) \geq 0$ for all $x_i, x_j \in X$, and $D(x_i, x_j) = 0$ only if $i = j$.
2. $D(x_i, x_j) = D(x_j, x_i)$ for all $x_i, x_j \in X$; and
3. $D(x_i, x_j) \leq D(x_i, x_l) + D(x_l, x_j)$ for all $x_i, x_j, x_l \in X$.

Two widely used distance measures, viz., Euclidean distance measure [197] and correlation-based distance measure [197], have been used in this book. They are defined as follows.

1.2.1.1 Euclidean Distance Measure

Given two feature vectors x_i and x_j, the Euclidean distance $D(x_i, x_j)$ between them is computed as

$$D(x_i, x_j) = \sqrt{\sum_{l=1}^{d} (x_{il} - x_{jl})^2}. \tag{1.1}$$

The Euclidean distance measure is a special case of the Minkowski distance metric, which is defined as follows:

$$D_p(x_i, x_j) = \left(\sum_{l=1}^{d} (x_{il} - x_{jl})^p \right)^{\frac{1}{p}}. \tag{1.2}$$

For Euclidean distance, $p = 2$. When $p = 1$, the distance is called Manhattan distance.

1.2.1.2 Pearson Correlation-Based Distance Measure

Given two feature vectors x_i and x_j, the Pearson correlation coefficient $Cor(x_i, x_j)$ between them is computed as

$$Cor(x_i, x_j) = \frac{\sum_{l=1}^{d}(x_{il} - \mu_{x_i})(x_{jl} - \mu_{x_j})}{\sqrt{\sum_{l=1}^{d}(x_{il} - \mu_{x_i})^2}\sqrt{\sum_{l=1}^{d}(x_{jl} - \mu_{x_j})^2}}. \tag{1.3}$$

Here μ_{x_i} and μ_{x_j} represent the arithmetic means of the components of the feature vectors x_i and x_j, respectively. The Pearson correlation coefficient, defined in Equation 1.3, is a measure of similarity between two points in the feature space. The distance between two points x_i and x_j is computed as $D(x_i, x_j) = 1 - Cor(x_i, x_j)$, which represents the dissimilarity between those two points.

There are mainly three categories of clustering algorithms [237, 238, 408], viz., *hierarchical*, *partitional* and *density-based*. Some basic clustering techniques from each category are described below.

1.2.2 Hierarchical Clustering

In hierarchical clustering, the clusters are generated in a hierarchy, where every level of the hierarchy provides a particular clustering of the data, ranging from a single cluster (where all the points are put in the same cluster) to n clusters (where each point comprises a cluster). Hierarchical clustering may be either *agglomerative* or *divisive*.

Agglomerative clustering techniques begin with singleton clusters, and combine the two least distant clusters at every iteration. Thus in each iteration two clusters are merged, and hence the number of clusters reduces by 1. This proceeds iteratively in a hierarchy, providing a possible partitioning of the data at every level. When the target number of clusters (K) is achieved, the algorithms terminate. *Single*, *average* and *complete linkage* agglomerative algorithms differ only in the linkage metric used. The distance $\mathscr{D}(C_i, C_j)$ between two clusters C_i and C_j for the single, average and complete linkage algorithms are defined as follows.

For single linkage,

$$\mathscr{D}(C_i, C_j) = \min_{x \in C_i, y \in C_j} \{D(x,y)\}.$$

For average linkage,

$$\mathscr{D}(C_i, C_j) = \frac{1}{|C_i||C_j|} \sum_{x \in C_i, y \in C_j} D(x,y).$$

For complete linkage,

$$\mathscr{D}(C_i, C_j) = \max_{x \in C_i, y \in C_j} \{D(x,y)\}.$$

Here $D(x,y)$ denotes the distance between the data points x and y, and $|C_i|$ denotes the number points in cluster C_i.

Divisive clustering just follows the reverse process, i.e., it starts from a single cluster containing all the points. At each step, the biggest cluster is divided into two clusters until the target number of clusters is achieved. In general, the hierarchical clustering methods generate a *dendrogram*, which is graphical representation of the arrangement of the clusters as provided by the clustering method. This is a binary tree-like representation, where the leaves represent the data points and each internal node represents a cluster formed by merging two of its children (agglomerative case). Thus each internal node represents a clustering step.

One main advantage of hierarchical clustering is that it offers flexibility regarding the levels of granularity. Moreover it can adopt any kind of distance metric easily. However, this clustering technique is criticized for its inability to improve already constructed clusters. Also, the problem of choosing a suitable stopping criteria in the process of agglomeration or division is another important drawback of hierarchical clustering.

1.2.3 Partitional Clustering

The partitional clustering algorithms obtain a single clustering solution for a dataset instead of a hierarchical clustering structure, such as the dendrogram produced by a hierarchical clustering method. Partitional methods are more computationally efficient compared to hierarchical techniques. The two widely used partitional clustering algorithms are *K-means* [408] and its fuzzy version, called *Fuzzy C-means (FCM)* [62].

1.2.3.1 K-Means Clustering

The K-means clustering algorithm consists of the following steps:

1. Choose K initial cluster centers $z_1, z_2, \ldots, z_K$ randomly from the n points $\{x_1, x_2, \ldots, x_n\}$.
2. Assign point x_i, $i = 1, 2, \ldots, n$ to cluster C_j, $j \in \{1, 2, \ldots, K\}$, iff

$$D(z_j, x_i) < D(z_p, x_i), \quad p = 1, 2, \ldots, K, \text{ and } j \neq p.$$

Ties are resolved arbitrarily.
3. Compute new cluster centers $z_1^*, z_2^*, \ldots, z_K^*$ as follows :

$$z_i^* = \frac{1}{n_i} \sum_{x_j \in C_i} x_j, \quad i = 1, 2, \ldots, K,$$

where n_i is the number of elements belonging to cluster C_i.
4. If $z_i^* = z_i$, $i = 1, 2, \ldots, K$, then terminate. Otherwise continue from Step 2.

K-means clustering minimizes the global cluster variance J to maximize the compactness of the clusters. The global cluster variance J can be represented as

$$J = \sum_{k=1}^{K} \sum_{x_i \in C_k} D^2(z_k, x_i). \tag{1.4}$$

Note that if the process does not terminate at Step 4 normally, then it is executed for a maximum fixed number of iterations. It has been shown in [379] that the K-means algorithm may converge to values that are not optimal. Also, global solutions of large problems cannot be found with a reasonable amount of computation effort [391].

1.2.3.2 Fuzzy C-Means Clustering

Fuzzy C-means (FCM) [62, 335] is a widely used technique that uses the principles of fuzzy sets to evolve a partition matrix $U(X)$ while minimizing the measure

$$J_m = \sum_{k=1}^{K} \sum_{i=1}^{n} u_{ki}^m D^2(z_k, x_i), \tag{1.5}$$

where u is the fuzzy membership matrix (partition matrix) and m denotes the fuzzy exponent.

The FCM algorithm is based on an alternating optimizing strategy. This involves iteratively estimating the partition matrix followed by computation of new cluster centers. It starts with random K initial cluster centers, and then at every iteration it finds the fuzzy membership of each data point to every cluster using the following equation [62]:

$$u_{ki} = \frac{1}{\sum_{j=1}^{K} \left(\frac{D(z_k,x_i)}{D(z_j,x_i)} \right)^{\frac{2}{m-1}}}, \quad \text{for } 1 \leq k \leq K, \ 1 \leq i \leq n. \tag{1.6}$$

Note that while computing u_{ki} using Equation 1.6, if $D(z_j,x_i)$ is equal to zero for some j, then u_{ki} is set to zero for all $k = 1,\ldots,K, \ k \neq j$, while u_{ji} is set equal to 1.

Based on the membership values, the cluster centers are recomputed using the following equation [62]:

$$z_k = \frac{\sum_{i=1}^{n} (u_{ki})^m x_i}{\sum_{i=1}^{n} (u_{ki})^m}, \quad 1 \leq k \leq K. \tag{1.7}$$

The algorithm terminates when there is no more movement in the cluster centers. Finally, each data point is assigned to the cluster to which it has maximum membership. FCM clustering usually performs better than K-means clustering for overlapping clusters and noisy data. However, this algorithm may also get stuck at local optima [187].

Both the K-means and FCM algorithms are known to be sensitive to outliers, since such points can significantly affect the computation of the centroid, and hence the resultant partitioning. K-medoids and fuzzy K-medoids attempt to alleviate this problem by using the medoid, the most centrally located object, as the representative of the cluster. Partitioning Around Medoids (PAM) [243] was one of the earliest K-medoid algorithms introduced. PAM finds K clusters by first finding a representative object for each cluster, the medoid. The algorithm then repeatedly tries to make a better choice of medoids by analyzing all possible pairs of objects such that one object is a medoid and the other is not. PAM is computationally quite inefficient for large datasets and a large number of clusters. The CLARA algorithm was proposed by the same authors [243] to tackle this problem. CLARA is based on data sampling, where only a small portion of the real data is chosen as a representative of the data and medoids are chosen from this sample using PAM. CLARA draws multiple samples and outputs the best clustering from these samples. As expected, CLARA can deal with larger datasets than PAM. However, if the best set of medoids is never chosen in any of the data samples, CLARA will never find the best clustering.

Ng and Han [325] proposed the CLARANS algorithm, which tries to mix both PAM and CLARA by searching only the subset of the dataset. However, unlike CLARA, CLARANS does not confine itself to any sample at any given time, but draws it randomly at each step of the algorithm. In order to make CLARANS applicable to large datasets, use of efficient spatial access methods, such as R*-tree, was proposed [153]. CLARANS had that limitation that it could provide good clustering only when the clusters were mostly equisized and convex.

Another extension of the K-means is the *K-modes* algorithm [222, 223]. Here the cluster centroids are replaced by *cluster modes*. A cluster mode is a d-dimensional vector (where d is the data dimension) where each component of the vector is computed as the most frequently occurring value in the corresponding attribute domain.

A fuzzy version of the K-modes algorithm, i.e., *fuzzy K-modes*, is also proposed in [224]. K-modes and fuzzy K-modes algorithms are usually suitable for categorical datasets [315].

1.2.4 Density-Based Clustering

In density-based clustering approaches, clusters are considered as regions in the data space in which the points are densely situated, and which are separated by regions of low point density (noise). These regions may have arbitrary shapes and the points inside a region may be arbitrarily distributed.

DBSCAN (Density-Based Spatial Clustering of Applications with Noise) [152] is a popularly used density-based clustering technique. In DBSCAN, a cluster is defined using the notion of density-reachability. A point y is said to be directly density-reachable from another point x if it is within a given distance ε (i.e., y is a part of x's ε-neighborhood), and if x is surrounded by sufficiently many points in its ε-neighborhood such that x and y may be considered as a part of a cluster. Again, y is called indirectly density-reachable from x if there is a sequence $x_1, x_2, \ldots, x_p$ of points with and $x_1 = x$ and $x_p = y$ in which each x_{i+1} is directly density-reachable from x_i. Two parameters are required as the input of DBSCAN: ε (radius of the neighborhood of a point) and the minimum number of points (*minpts*) required to form a cluster. It starts with an arbitrary starting data point that has not been visited yet. This point's ε-neighborhood is found, and if it contains at least *Minpts* points, a cluster formation is started. Otherwise, the point is considered as noise. It should be noted that this point may later be found within the ε-neighborhood of another point that contains at least *Minpts* points. Hence the point can be assigned to some cluster at a later stage. DBSCAN clusters a dataset by a single scan over all the data points. DBSCAN can handle nonconvex and non-uniformly sized clusters. Moreover, DBSCAN can find the number of clusters automatically and it is robust in the presence of noise. However, performance of DBSCAN highly depends on its input parameters.

Two other widely used density-based clustering algorithms are GDBSCAN (Generalized DBSCAN) [376] and OPTICS (Ordering Points To Identify the Clustering Structure) [25]. GDBSCAN is an attempt to generalize the DBSCAN algorithm so that it can cluster point objects as well as spatially extended objects depending on their spatial and their nonspatial attributes. OPTICS extends the DBSCAN algorithm to handle different local densities by building an augmented ordering of the data points. In general, density-based algorithms are suitable for detecting outliers. However their performance degrades in the presence of overlapping clusters.

1.2.5 Other Clustering Methods

In addition to the above-mentioned clustering techniques, several other clustering algorithms are available in the literature [40]. As the K-means and FCM algorithms often get stuck at local optima, several approximate methods, including Genetic Algorithms and simulated annealing [38, 39, 42, 298], are developed to solve the underlying optimization problem. These methods have also been extended to the case where the number of clusters is variable [36, 300, 333]. Balanced Iterative Reducing and Clustering using Hierarchies (BIRCH), proposed by Zhang et al. [463], is another algorithm for clustering large datasets. It uses two concepts, the clustering feature and the clustering feature tree, to summarize cluster representations which help the method achieve good speed and scalability in large databases. Among the other clustering approaches, a widely used technique of statistical clustering is the expectation maximization (EM) algorithm [133] and its variants [338]. Other clustering techniques include spanning tree-based graph theoretic clustering [443], spectral clustering [324], and Self Organizing Map (SOM) clustering [396]. A good review of several other clustering algorithms may be found in [197, 238].

1.2.6 Cluster Validity Indices

The result of one clustering algorithm can be very different from that of another for the same input dataset as the other input parameters of an algorithm can substantially affect the behavior and execution of the algorithm. The main objective of a *cluster validity index* [65, 195, 266, 299, 334, 335, 427, 438] is to validate a clustering solution, i.e., to find how good the clustering is. Validity measures can be used to find the partitioning that best fits the underlying data. Beyond three-dimensional space, it is not possible to visualize the clustering result in the feature space. Therefore cluster validity measures can effectively be used to compare the performance of several clustering techniques, specially for high-dimensional data. There are mainly two types of cluster validity indices: *external* and *internal*.

1.2.6.1 External Cluster Validity Indices

External validity measures are used to compare the resultant clustering solution with the true clustering of data. These indices are very useful for comparing the performance of different clustering techniques when the true clustering is known. Some widely used external cluster validity indices are the *Jaccard index* [194], the *Minkowski index* [61], the *Rand index* [421], the *percentage of correctly classified pairs* [41], and the *adjusted Rand index* [449].

Suppose T is the true clustering of a dataset and C is a clustering result given by some clustering algorithm. Let a, b, c and d respectively denote the number of pairs of points belonging to the same cluster in both T and C, the number of pairs belonging to the same cluster in T but to different clusters in C, the number of pairs belonging to different clusters in T but to the same cluster in C, and the number of pairs belonging to different clusters in both T and C. Then the external validity indices are defined as follows:

Jaccard Index

The Jaccard index $J(T,C)$ is defined as

$$J(T,C) = \frac{a}{a+b+c}. \tag{1.8}$$

The value of the Jaccard index varies between 0 and 1, with a higher value indicating a better match between T and C. Obviously, $J(T,T) = 1$.

Minkowski Index

The Minkowski index $M(T,C)$ is defined as

$$M(T,C) = \sqrt{\frac{b+c}{a+b}}. \tag{1.9}$$

Lower values of the Minkowski index indicate better matching between T and C, where the minimum value $M(T,T) = 0$.

Rand Index

The Rand index $R(T,C)$ is defined as

$$R(T,C) = \frac{a+d}{a+b+c+d}. \tag{1.10}$$

The value of the Rand index also lies between 0 and 1, with a higher value indicating a better match between T and C, and $R(T,T) = 1$. Rand index is also known as the percentage of correctly classified pairs $\%CP$ [41] when multiplied by 100, since it gives the percentage of the pairs of points which are correctly classified among all the pairs of points.

Adjusted Rand Index

The adjusted Rand index $ARI(T,C)$ is then defined as

$$ARI(T,C) = \frac{2(ad-bc)}{(a+b)(b+d)+(a+c)(c+d)}. \tag{1.11}$$

The value of $ARI(T,C)$ also lies between 0 and 1; a higher value indicates that C is more similar to T. Also, $ARI(T,T) = 1$.

1.2.6.2 Internal Cluster Validity Indices

Internal validity indices evaluate the quality of a clustering solution in terms of the geometrical properties of the clusters, such as *compactness, separation* and *connectedness*. Some widely used internal cluster validity indices are J_m [62], the *Davies-Bouldin (DB) index* [118], the *Dunn index* [148], *Partition Coefficient (PC)* [63], *Partition Entropy (PE)* [194], the *Xie-Beni (XB) index* [442], the *Fukuyama-Sugeno (FS) index* [194], the $\mathscr{I}$ *index* [299], the *Silhouette index* [369], and the K-index [266]. Internal validity indices can act as objective functions to be optimized in order to determine the cluster structure of the underlying dataset. Moreover, the validity measures that are functions of cluster compactness and separation, such as DB, *Dunn*, XB, $\mathscr{I}$, the Silhouette index, and the K-index, can be used to determine the number of clusters also.

J_m Index

The J_m index is minimized by fuzzy C-means clustering. It is defined as follows:

$$J_m = \sum_{k=1}^{K} \sum_{i=1}^{n} u_{ki}^m D^2(z_k, x_i), \tag{1.12}$$

where u is the fuzzy membership matrix (partition matrix) and m denotes the fuzzy exponent. $D(z_k, x_i)$ denotes the distance between the kth cluster center z_k and the ith data point x_i. J_m can be considered as the global fuzzy cluster variance. A lower value of J_m index indicates more compact clusters. However, the J_m value is not independent of the number of clusters K, i.e., as the value of K increases, the J_m value gradually decreases and it takes the minimum value 0 when $K = n$. It is possible to have a crisp version of J_m when the partition matrix u has only binary values.

Davies-Bouldin Index

The Davies-Bouldin (DB) index is a function of the ratio of the sum of within-cluster scatter to between-cluster separation. The scatter within the ith cluster, S_i, is computed as

$$S_i = \frac{1}{|C_i|} \sum_{x \in C_i} D^2(z_i, x). \tag{1.13}$$

Here $|C_i|$ denotes the number of data points belonging to cluster C_i. The distance between two clusters C_i and C_j, d_{ij} is defined as the distance between the centers.

$$d_{ij} = D^2(z_i, z_j). \tag{1.14}$$

The *DB* index is then defined as

$$DB = \frac{1}{K} \sum_{i=1}^{K} R_i, \tag{1.15}$$

where

$$R_i = \max_{j, j \neq i} \left\{ \frac{S_i + S_j}{d_{ij}} \right\}. \tag{1.16}$$

The value of *DB* index is to be minimized in order to achieve proper clustering.

Dunn Index Family

Suppose $\delta(C_i, C_j)$ denotes the distance between two clusters C_i and C_j, and $\Delta(C_i)$ denotes the diameter of cluster C_i; then any index of the following form falls under Dunn family of indices:

$$DN = \min_{1 \leq i \leq K} \left\{ \min_{1 \leq j \leq K, j \neq i} \left\{ \frac{\delta(C_i, C_j)}{\max_{1 \leq k \leq K} \{\Delta(C_k)\}} \right\} \right\}. \tag{1.17}$$

Originally Dunn used the following forms of δ and Δ:

$$\delta(C_i, C_j) = \min_{x \in C_i, y \in C_j} \{D(x, y)\}, \tag{1.18}$$

and

$$\Delta(C_i) = \max_{x, y \in C_i} \{D(x, y)\}. \tag{1.19}$$

Here $D(x, y)$ denotes the distance between the data points x and y. A larger value of the Dunn index implies compact and well-separated clusters. Hence the objective is to maximize the Dunn index.

Partition Coefficient and Partition Entropy

Partition Coefficient (*PC*) and Partition Entropy (*PE*) are two fuzzy cluster validity indices defined on the evolved fuzzy partition matrix $U = u_{ij}$, $1 \leq i \leq K$, $1 \leq j \leq n$, where u_{ij} denotes the degree of membership of point j to cluster i. K and n are the number of clusters and number of data points, respectively. The *PC* index is defined as

$$PC = \frac{1}{n} \sum_{i=1}^{K} \sum_{j=1}^{n} u_{ij}^2.$$ (1.20)

PC index value ranges between $\frac{1}{K}$ and 1, and the larger the value, the crisper is the clustering. A value close to $\frac{1}{K}$ indicates absence of clustering structure in the dataset. The PE index is defined as follows:

$$PE = -\frac{1}{n} \sum_{i=1}^{K} \sum_{j=1}^{n} u_{ij} \log_b u_{ij},$$ (1.21)

where b is the base of the logarithm. The value of PE lies between 0 and $\log_b K$. When the PE value is closer to 0, the clustering is crisper. A value close to $\log_b K$ indicates absence of proper clustering structure in the dataset.

Two problems with the PC and PE indices are that they do not have any direct relationship with the geometric properties of the dataset and that they have monotonic dependency on the number of clusters K.

Xie-Beni Index

The Xie-Beni (XB) index is defined as a function of the ratio of the total fuzzy cluster variance σ to the minimum separation sep of the clusters. Here σ and sep can be written as

$$\sigma = \sum_{k=1}^{K} \sum_{i=1}^{n} u_{ki}^2 D^2(z_k, x_i)$$ (1.22)

and

$$sep = \min_{k \neq l} \{ D^2(z_k, z_l) \},$$ (1.23)

The XB index is then written as

$$XB = \frac{\sigma}{n \times sep} = \frac{\sum_{k=1}^{K} \sum_{i=1}^{n} u_{ki}^2 D^2(z_k, x_i)}{n \times (\min_{k \neq l} \{ D^2(z_k, z_l) \})}.$$ (1.24)

Note that when the partitioning is compact and good, the value of σ should be low while sep should be high, thereby yielding lower values of the XB index. The objective is therefore to minimize the XB index for achieving proper clustering.

Fukuyama-Sugeno Index

The fuzzy version of the Fukuyama-Sugeno (FS) index is defined as follows:

$$FS = \sum_{i=1}^{n} \sum_{k=1}^{K} u_{ki}^m \left[D^2(z_k, x_i) - D^2(z_k, \bar{z}) \right],$$ (1.25)

where $\bar{z}$ represents the mean of all the data points. It is evident that to obtain compact and well-separated clusters, the objective is to have a lower value of the FS index.

$\mathscr{I}$ Index

The $\mathscr{I}$ index is defined as follows:

$$\mathscr{I} = (\frac{1}{K} \times \frac{E_1}{E_K} \times D_K)^p, \tag{1.26}$$

where

$$E_K = \sum_{k=1}^{K} \sum_{j=1}^{n} u_{kj} D(z_k, x_j) \tag{1.27}$$

and

$$D_K = \max_{i,j=1}^{K} \{D(z_i, z_j)\}. \tag{1.28}$$

The different symbols used are as discussed earlier. The index $\mathscr{I}$ is a composition of three factors, namely, $\frac{1}{k}$, $\frac{E_1}{E_K}$ and D_K. The first factor will try to reduce index $\mathscr{I}$ as K is increased. The second factor consists of the ratio of E_1, which is constant for a given dataset, to E_K, which decreases with increase in K. Hence, because of this term, index $\mathscr{I}$ increases as E_K decreases. This, in turn, indicates that formation of more clusters that are compact in nature would be encouraged. Finally, the third factor, D_K (which measures the maximum separation between two clusters over all possible pairs of clusters), will increase with the value of K. However, note that this value is upper bounded by the maximum separation between two points in the dataset. Thus, the three factors are found to compete with and balance each other critically. The power p is used to control the contrast between the different cluster configurations. Larger value of the $\mathscr{I}$ index indicates better clustering.

Silhouette Index

Suppose a_i represents the average distance of a point x_i from the other points of the cluster to which the point is assigned, and b_i represents the minimum of the average distances of the point from the points of the other clusters. Then the silhouette width s_i of the point is defined as

$$s_i = \frac{b_i - a_i}{max\{a_i, b_i\}}. \tag{1.29}$$

Silhouette index S is the average silhouette width of all the data points, i.e.,

$$S = \frac{1}{n} \sum_{i=1}^{n} s_i. \tag{1.30}$$

The value of the Silhouette index varies from -1 to 1 and a higher value indicates a better clustering result.

K-index

The K index is defined as follows:

$$K(U,Z;X) = \frac{\sum_{k=1}^{K} \sum_{i=1}^{n} u_{ki}^2 D^2(z_k, x_i) + \frac{1}{K} \sum_{k=1}^{K} D^2(z_k, \bar{z})}{\min_{k \neq l}\{D^2(z_k, z_l)\}}, \qquad (1.31)$$

where $\bar{z} = \frac{1}{n} \sum_{i=1}^{n} x_i$, i.e., $\bar{z}$ is the center of all the points. The different symbols are as described earlier.

The first term of the numerator in Equation 1.31 measures the intraclass similarity, which is low for compact clusters. The second term in the numerator is an ad hoc penalizing function used to eliminate the decreasing tendency as the number of clusters K tends to the number of data points n. The denominator in Equation 1.31 measures the inter-class difference and its larger value indicates that the clusters are well separated. Hence, a lower value of the K-index means compact and well-separated clusters.

1.3 Clustering as an Optimization Problem

As stated earlier, the aim of a clustering technique is to find a suitable grouping of the input dataset so that some criteria are optimized. Hence the problem of clustering can be posed as an optimization problem. The objective to be optimized may represent different characteristics of the clusters, such as compactness, separation, and connectivity. A straightforward way to pose clustering as an optimization problem is to optimize some cluster validity index that reflects the goodness of the clustering solutions. All possible partitionings of a dataset and the corresponding values of the validity index define the complete search space.

Traditional partitional clustering techniques such as K-means and FCM employ a greedy search technique over the search space in order to optimize the compactness of the clusters. Although these algorithms are computationally efficient, they suffer from the following drawbacks:

1. They often get stuck at some local optimum depending on the choice of the initial cluster centers.
2. They optimize a single cluster validity index (compactness in this case), and therefore may not cover the different characteristics of the datasets.
3. The number of clusters has to be specified a priori.

To overcome the problem of local optima, some global optimization tools, such as Genetic Algorithms (GAs) [120, 181], Simulated Annealing (SA) [252, 415], and Particle Swarm Optimization (PSO) [104], can be used to reach the global optimum value of the validity measure chosen. In this book, Genetic Algorithms are used for this purpose. The second issue, i.e., *multiobjective clustering*, is the main focus of this book. By multiobjective clustering, more than one cluster validity index can be optimized simultaneously. The third issue, i.e., prior specification of the number of clusters, is also tackled by evolving the number of clusters automatically using variable string-length encoding in GAs.

1.3.1 Genetic Algorithms for Clustering

In this section, first we provide a short introduction to Genetic Algorithms (GAs). Subsequently we discuss how GAs can be used for clustering purposes.

1.3.1.1 Genetic Algorithms (GAs)

Genetic Algorithms (GAs) [120, 181, 308], introduced by John Holland, are efficient methods for the solution of many search and optimization problems. Genetic Algorithms are robust and stochastic search procedures with a large amount of implicit parallelism. GAs are based on the principle of natural genetics and the evolutionary theory of genes. They follow the evolutionary process as stated by Charles Darwin. The algorithm starts by initializing a population of potential solutions encoded into strings called *chromosomes*. Each solution has some *fitness* value based on which the fittest parents that would be used for reproduction are found (*survival of the fittest*). The new generation is created by applying genetic operators such as *selection* (based on natural selection to create the mating pool), *crossover* (exchange of information among parents) and *mutation* (sudden small change in a parent) on selected parents. Thus the quality of the population is improved as the number of generations increases. The process continues until some specific criterion is met or the solution converges to some optimized value.

GAs are efficient in solving complex problems where the main focus is on obtaining good, not necessarily the best, solutions quickly. Moreover, they are well suited for applications involving search and optimization, where the space is huge, complex and multimodal. For example, for the problem of clustering n points into K clusters such that some cluster validity measure is optimized, the number of possibilities is $\frac{1}{K!}\sum_{j=0}^{K}(-1)^{K-j}\binom{K}{j}j^n$. The number becomes immensely large even for small problems. It is expected that application of an effective search strategy like GAs is likely to provide significantly superior performance compared to that of schemes that are ad hoc or mostly local in nature.

Applications of Genetic Algorithms and related techniques in data mining include extraction of association rules [326], predictive rules [161, 162, 326], clustering [36, 38, 39, 298, 300], program evolution [361, 404] and Web mining [330, 331, 349–351].

1.3.1.2 GA-Based Clustering

In order to overcome the difficulties in using greedy search techniques, global optimization tools such as evolutionary techniques, specially Genetic Algorithms, are popularly used for clustering. For using GAs for clustering purposes, one must first choose a suitable way of encoding to represent a possible clustering solution as a chromosome. Among the different encoding approaches, two are widely used: point-based encoding and center-based encoding.

Point-Based Encoding:

In point-based encoding techniques [260, 263, 284, 285, 322], the length of a chromosome is the same as the number of points, and the value assigned to each gene (corresponding to each point) is drawn from $\{1, \ldots, K\}$, K being the number of clusters. K may be fixed or variable. If gene i is assigned a value j, then the ith point is assigned to the jth cluster. Point-based encoding techniques are straightforward, but suffer from large chromosome lengths and hence slow rates of convergence.

Center-Based Encoding:

In center-based encoding [36, 38, 39, 298, 300], cluster centers are encoded into chromosomes. Hence each chromosome is of length $K \times d$, where d is the dimension of the data. Here also, K may be varied, resulting in variable length chromosomes. The advantage of center-based encoding is that the chromosome length is not very large, and thus it usually has a faster convergence rate than point-based encoding techniques.

As a fitness function, GA-based clustering techniques usually use a cluster validity index, such as Jm [62], XB [442], DB [118], $Dunn$ [147, 148], $Silhouette$ [369], and $\mathscr{I}$ [299]. The aim is to obtain the best possible value of the index, and the corresponding chromosome represents the final clustering.

1.3.2 Multiobjective Evolutionary Algorithms for Clustering

This section first briefly introduces the concept of multiobjective optimization. Thereafter, how the clustering problem is posed as a multiobjective optimization problem is discussed, and the literature on some existing multiobjective clustering techniques is surveyed.

1.3.2.1 Multiobjective Optimization and Evolutionary Algorithms

Conventional GAs are traditionally single objective in nature, i.e., they optimize a single criterion in the searching process. Thus, finally, a single solution is generated that corresponds to the optimized value of the chosen optimization criterion. However, in many real-world situations there may be several objectives that must be optimized simultaneously in order to solve a certain problem. This is in contrast to the problems tackled by conventional GAs, which involve optimization of just a single criterion. The main difficulty in considering multiobjective optimization (MOO) is that there is no accepted definition of optimum in this case, and therefore it is difficult to compare one solution with another one. In general, these problems admit multiple solutions, each of which is considered acceptable and all of which are equivalent when the relative importance of the objectives is unknown. The best solution is subjective and depends on the need of the designer or decision maker.

Traditional search and optimization methods such as gradient descent search, and other non-conventional ones such as simulated annealing, are difficult to extend to the multiobjective case, since their basic design precludes the consideration of multiple solutions. In contrast, population-based methods like Genetic Algorithms are well suited for handling such situations. There are different approaches to solving multiobjective optimization problems [106–108, 125], such as aggregating approaches, population-based non-Pareto approaches, Pareto-based non-elitist approaches and Pareto-based elitist approaches.

Besides GA and SA, other evolutionary optimization tools such as tabu search [55], particle swarm optimization [365], ant colony optimization [169], and differential evolution [444] have also been modified to cope with multiobjective optimization. An important aspect of multiobjective optimization is that they provide a set of non-dominated alternative solutions to the decision maker who can choose one or more solutions from the set depending on the problem requirements.

1.3.2.2 Multiobjective Clustering

Conventional Genetic Algorithms-based clustering techniques use some validity measure as the fitness value. However, there is no single validity measure that works equally well for different kinds of datasets. Thus it is natural to simultaneously optimize a number of such measures that can capture different characteristics of the data. Simultaneous optimization of multiple objectives provides improved robustness to different data properties. Hence it is useful to utilize multiobjective Genetic Algorithms [106–108, 125] for clustering. Multiobjective clustering techniques optimize more than one cluster validity index simultaneously, leading to high-quality results. As the relative importance of different clustering criteria is unknown, it is better to optimize compactness and separation separately rather than combine them in a single measure to be optimized.

There are some instances in literature that apply multiobjective techniques for data clustering. One of the earliest approaches in this field is found in [131], where objective functions representing compactness and separation of the clusters were optimized in a crisp clustering context and with a deterministic method. In [79], a tabu search-based multiobjective clustering technique was proposed, where the partitioning criteria are chosen as the within-cluster similarity and between-cluster dissimilarity. This technique uses solution representation based on cluster centers as in [298]. However, experiments are mainly based on artificial distance matrices. A series of works on multiobjective clustering has been proposed in [199, 200, 202], where the authors have adopted chromosome encoding of length equal to the number of data points. The two objectives that are optimized are overall deviation (compactness) and connectivity. The algorithm in [199] is capable of handling categorical data, whereas the other two papers deal with numeric and continuous datasets. These methods have the advantages that they can automatically evolve the number of clusters and also be used to find non-convex-shaped clusters. It may be noted that the chromosome length in these works is equal to the number of points to be clustered. Hence, as discussed in [38], when the length of the chromosomes becomes equal to the number of points, n, to be clustered, the convergence becomes slower for the large values of n. This is due to the fact that the chromosomes, and hence the search space, in such cases become large. However, in [200], a special mutation operator is used to reduce the effective search space by maintaining a list of L nearest neighbors for each data point, where L is a user-defined parameter. This allows faster convergence of the algorithm towards the global Pareto-optimal front, making it scalable for larger datasets. However, this algorithm uses special initialization routines based on the minimum spanning tree method and is intended for crisp clustering of continuous data.

All the above multiobjective clustering algorithms work in the crisp/hard clustering context. As fuzzy clustering is better equipped to handle overlapping clusters [62, 427], there is need to develop some multiobjective fuzzy clustering techniques. Also, as stated earlier, a multiobjective clustering algorithm produces a set of nondominated solutions in the final generation. Therefore it is a challenge to devise some techniques to obtain a final solution from the set of alternative solutions. Some studies have been made in this book to address these challenges.

1.4 Applications of Clustering

Clustering algorithms have their applications in many real-life domains such as medical diagnosis, medical image segmentation, disease classification, computational biology and bioinformatics, natural resource study, VLSI design, remote sensing applications, and investment analysis. In this book, the clustering algorithms have been applied to two real-life fields, viz., unsupervised pixel classification of remote sensing imagery and clustering of microarray gene expression data.

1.4.1 Unsupervised Pixel Classification in Remote Sensing Imagery

Remote sensing satellite images have significant applications in different areas such as climate studies, the assessment of forest resources, and the examination of marine environments. For remote sensing applications, classification is an important task where the pixels in the images are classified into homogeneous regions, each of which corresponds to some particular land cover type. The problem of pixel classification is often posed as that of clustering in the intensity space [41, 49, 300, 338]. Clustering techniques used for pixel classification should be able to handle satellite images that have a large number of classes with overlapping and nonlinear class boundaries.

In a satellite image, a pixel represents an area of land space, which may not necessarily belong to a single land cover type. Hence, it is evident that a large amount of imprecision and uncertainty can be associated with the pixels in a remotely sensed image. Therefore, it is appropriate as well as natural to apply the principles of fuzzy set theory in the domain of pixel classification. Multiobjective clustering algorithms have been applied to classify remote sensing imagery in order to demonstrate their effectiveness.

1.4.2 Microarray Gene Expression Data Clustering

The landscape of biological and biomedical research is being changed rapidly by the invention of microarrays which enable simultaneous views of the transcription levels of a huge number of genes across different tissue samples or time points. Microarray technology has applications in the areas of medical diagnosis, bio-medicine, gene expression profiling, and so on [13, 43, 151, 282]. Usually, the gene expression values during some biological experiment are measured along time series.

A microarray gene expression dataset consisting of $\mathscr{G}$ genes and $\mathscr{C}$ experimental conditions is typically organized in a 2D matrix $E = [e_{ij}]$ of size $\mathscr{G} \times \mathscr{C}$. Each element e_{ij} gives the expression level of the ith gene at the jth time point. Clustering is used to identify the sets of genes with similar expression profiles. Clustering algorithms are also used for classifying different tissue samples representing multiple cancer types and for identifying possible gene markers. Recently, biclustering algorithms have been applied to find subsets of genes that are co-expressed in a subset of experimental conditions. Biclusters have been found to be more biologically relevant than clusters. In this book, the multiobjective clustering and biclustering algorithms have been applied to real-life gene expression datasets to show their effectiveness.

1.5 Summary and Scope of the Book

This book focuses on multiobjective GA-based fuzzy clustering and its application to real-life data mining and bioinformatics problems. As the application areas, two important fields, viz., remote sensing and microarray gene expression data analysis, have been chosen. The book is organized into ten chapters, including this introductory chapter.

In this chapter, we have presented an introduction to the clustering problem. Various popular clustering algorithms and different cluster validity measures have been discussed. Subsequently, the clustering problem has been posed as an optimization problem. A brief introduction to GAs and multiobjective optimization has also been provided. Moreover, how multiobjective Genetic Algorithms can be applied to clustering problems has been discussed. Finally, two possible application fields of clustering, viz., remote sensing and bioinformatics, have been introduced.

The next chapter discusses various aspects of GAs in detail. The different steps of GAs and their various operators, such as selection, crossover and mutation, are described with examples. Thereafter, the fundamental concepts of multiobjective optimization are provided and several multiobjective optimization algorithms are discussed. Discussions on theoretical studies on both traditional GAs and multiobjective evolutionary algorithms are provided. Finally, the chapter ends with a literature review of the different applications of GAs in data mining and bioinformatics.

In Chapter 3, we discuss in detail the process of knowledge discovery and data mining. Different data mining tasks are discussed with examples of existing algorithms. Moreover, major issues and challenges in data mining are discussed. Furthermore, a discussion on the recent trends and applications of the data mining methodologies is provided.

The application of computer science in solving biological problems is becoming more and more popular. In Chapter 4, an overview of different issues of bioinformatics and computational biology is provided. In this context, we also discuss the basic concepts of molecular biology. Subsequently, a discussion on different bioinformatics tasks with a brief review of related existing literature is presented.

In Chapter 5, the clustering problem is posed as a multiobjective optimization problem. In this regard, first a popular multiobjective crisp clustering method (MOCK) is described. Thereafter, multiobjective Genetic Algorithm is applied to the fuzzy clustering problem, and a multiobjective fuzzy clustering algorithm has been discussed in detail. This algorithm uses real-valued center-based encoding of chromosomes. The fuzzy validity indices, viz., J_m [62] and XB [442], are simultaneously optimized. The final solution is chosen from the resultant set of non-dominated solutions using a third validity index $\mathscr{I}$ [299]. The performance of the multiobjective fuzzy clustering algorithm is demonstrated on artificial and real-life datasets. Moreover, the multiobjective fuzzy clustering technique is applied to remote sensing image segmentation and microarray gene expression data clustering to demonstrate its effectiveness.

In Chapter 6, a method for obtaining the final solution from the set of non-dominated Pareto-optimal clustering solutions produced by the multiobjective fuzzy clustering method is described. In this regard, the support vector machine (SVM) classifier is efficiently integrated with the multiobjective clustering technique through a fuzzy voting algorithm. The application of the integrated method (MOGA-SVM) is demonstrated on remote sensing and gene expression data analysis.

In Chapter 7, a two-stage fuzzy clustering technique (SiMM-TS) is described that utilizes the data points having significant membership to multiple classes (SiMM points). These points are first identified and the remaining points are reclustered. Finally the SiMM points are assigned class memberships by an SVM classifier trained with the remaining points. A variable string-length genetic fuzzy clustering algorithm and a multiobjective fuzzy clustering algorithm are used for this purpose. The results of the SiMM-TS clustering algorithm are demonstrated on both remote sensing imagery as well as gene expression data.

Chapter 8 addresses the problem of multiobjective fuzzy clustering of categorical data. A cluster mode-based encoding technique is used and fuzzy compactness and fuzzy separation are simultaneously optimized in the context of categorical data. The results are demonstrated for various synthetic and real-life categorical datasets.

In Chapter 9, we present an application of the multiobjective fuzzy clustering technique for unsupervised cancer classification. The results are demonstrated on some publicly available real-life microarray cancer tumor datasets. Moreover, how relevant genetic markers are identified from the clustering results by using some statistic followed by multiobjective feature selection, is demonstrated.

Finally, Chapter 10 presents the notion of biclustering the microarray gene expression data, followed by a comprehensive literature review of biclustering algorithms. Subsequently, the need for multiobjective optimization for biclustering problem is discussed and in this regard a multiobjective GA-based biclustering (MO-GAB) technique is described. Results demonstrating the effectiveness of the multiobjective biclustering algorithm are reported for various artificial and real-life gene expression datasets and compared with other well-known biclustering techniques. Biological and statistical significance tests are also carried out. Finally we discuss how fuzziness can be incorporated into the multiobjective biclustering method and demonstrate its usefulness.

Chapter 2
Genetic Algorithms and Multiobjective Optimization

2.1 Introduction

There are several powerful optimization techniques that are inspired by nature. Evolutionary Algorithms (EAs) were developed based on the principle of natural genetics for performing search and optimization in complex landscapes. The important components of EAs are Genetic Algorithms (GAs), Genetic Programming (GP) and Evolutionary Strategies (ESs). Also, various other bio-inspired search and optimization techniques, such as Particle Swarm Optimization (PSO), Ant Colony Optimization (ACO), Differential Evolution (DE) and Simulated Annealing (SA), are also broadly classified under evolutionary computation.

All the search and optimization algorithms mentioned above, in their traditional form, are single objective by nature. That is, they are designed for optimizing a single criterion. However, there are many real-life problems that require optimization of more than one criterion, which are often contradictory in nature, simultaneously. This necessitates the need for Multiobjective Optimization (MOO) techniques and the definition of optimality in the multiobjective context. There are many works in the literature that explore this problem and extend the traditional evolutionary algorithms to the multiobjective case. In this book, we have used multiobjective Genetic Algorithms mainly for clustering. Therefore, in this chapter, we start with a gentle introduction to GAs and related issues. Next, a discussion on multiobjective evolutionary techniques is provided. Finally, we provide a brief literature survey on the application of GAs in data mining and bioinformatics.

2.2 Genetic Algorithms

Genetic Algorithms (GAs) [120, 181, 308] are parallel, efficient and robust search and optimization techniques which get inspiration from the principles of natural genetics and evolution. GAs are very effective when the search space is large, complex and multimodal. These algorithms encode a potential solution to a problem on a simple string-like data structure called *chromosome*. An implementation of a Genetic Algorithm starts with a population of (typically random) chromosomes or individuals. Each chromosome is then evaluated based on some *fitness value* that measures the goodness of the solution encoded in the chromosome. Thereafter, the chromosomes are given the opportunities for reproduction by *selection*. During selection, the chromosomes having better fitness are given more chance to reproduce than the others, which have worse fitness values. Thereafter, some recombination/reproduction operators, such as *crossover* and *mutation*, are applied on the selected chromosomes to preserve critical information and to produce potentially better solutions for the next generation. This process continues for a number of generations or until some termination criterion is met.

2.2.1 Characteristics of GAs

Genetic Algorithms are different from most of the traditional search and optimization techniques in several ways. GAs work with the coding of the problem variables, not with the variables themselves. They search simultaneously from multiple points, not from a single point. This involves parallelism in the searching process. GAs are blind search techniques that search via sampling using only payoff information. Moreover, while searching, they use stochastic operators, not deterministic rules. GAs work simultaneously on a set of encoded solutions. Therefore there is very little chance of getting stuck at a local optimum when GAs are used as an optimization technique. Furthermore, the search space need not to be continuous, and no auxiliary information, such as the derivative of the optimization function, is required. Moreover, the resolution of the possible search space is increased by operating on encoded potential solutions and not on the solutions themselves. Usually GAs perform best when potential solutions can be represented in a way which exposes important components of possible solutions, and operators to mutate and hybridize these components are available. In contrast, GAs are hampered when the chosen encoding technique does not represent the key features of the potential solutions, or the operators do not generate interesting new candidates. The following are essential components of GAs [50, 181]:

- An encoding strategy that determines the way in which potential solutions will be represented to form the chromosomes.
- A population of chromosomes or individuals.
- Mechanism (fitness function) for evaluating each chromosome.
- Selection/reproduction procedure.
- Genetic operators like crossover and mutation.
- Probabilities of performing genetic operators.
- Some termination criterion.

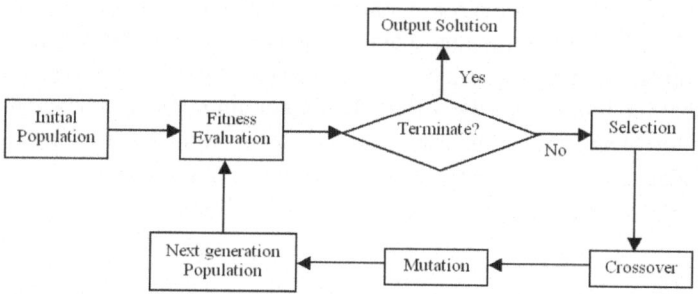

Fig. 2.1 Steps of a Genetic Algorithm

GAs operate through a cycle of evaluation of each chromosome in the population to get the fitness value, the selection of chromosomes, and the genetic manipulation to create a new population of chromosomes over a number of iterations (generations) until some termination criterion is satisfied. The termination criterion may be one of the following [50]:

- The average fitness value of a population becomes more or less constant over a specified number of generations.
- A desired objective function value is attained by at least one chromosome in the population.
- The number of generations is grater than some predefined threshold.

Figure 2.1 shows a schematic diagram of the basic steps of a Genetic Algorithm. The steps are described in detail below.

2.2.2 Solution Encoding and Population Initialization

In order to solve an optimization problem using GAs, one has to encode the variables of the problem into chromosomes. The variables are to be encoded as a finite length string over an alphabet of finite length. Usually, the chromosomes are represented as binary strings, i.e., strings with only 0s and 1s. For example, the string $< 1\ 0\ 0\ 1\ 0\ 1\ 0\ 1\ 1\ 1 >$ is a binary string of length 10. If we choose the string length to be l, it is obvious that number of possible combinations, and hence number of possible encoded solutions, is 2^l. A set of chromosomes is called a population, the size of which may be constant or may vary from generation to generation. It is general practice to choose the initial population randomly. However, one can generate the chromosomes of the initial population using some domain knowledge about the problem.

A commonly used principle for encoding is known as the *principle of minimum alphabet* [214]. The principle states that for efficient encoding, the smallest alphabet set that permits a natural expression of the problem should be chosen. It has been observed that the binary alphabet offers the maximum number of schemata per bit of information of any encoding. Therefore binary encoding is one of the commonly used strategies, although other types of encoding techniques such as integer value encoding and floating point encoding are also used widely.

2.2.3 Fitness Evaluation

The next step is to evaluate the fitness of the encoded solutions. The fitness function represents the goodness of a chromosome and it is dependent on the problem at hand. This is also called as the objective function. The fitness/objective function should be chosen in such a way that a chromosome that is closer to the optimal solution in the search space should have a higher fitness value. The fitness function is the only information (also called the payoff information) that GAs use while searching for possible solutions.

2.2.4 Selection

The *selection* operator mimics the process of *natural selection* and the *survival of the fittest* of Darwinian evolution theory. In this process, an intermediate population, called *mating pool*, is generated by copying the chromosomes from the parent population. Usually, the number of copies a chromosome receives in the mating pool is taken to be directly proportional to its fitness value. Only the selected chromosomes in the mating pool take part in the subsequent genetic operations like crossover and mutation. Among the several available selection methods, *roulette wheel selection*, *stochastic universal sampling* and *binary tournament selection* are three widely used techniques.

- In *roulette wheel selection*, the wheel has as many slots as the population size P, where the area of the slot i is proportional to the relative fitness of the ith chromosome in the population. A chromosome is selected by spinning the wheel and noting the position of a marker when the wheel stops. Therefore, the number of times a chromosome will be selected is proportional to its fitness (or the area of the slot) in the population. In Figure 2.2(a), the roulette selection procedure is illustrated.

- In *stochastic universal sampling*, P equidistant markers are placed on the wheel. All the P individuals are selected by spinning the wheel, the number of copies that an individual gets being equal to the number of markers that lie within the

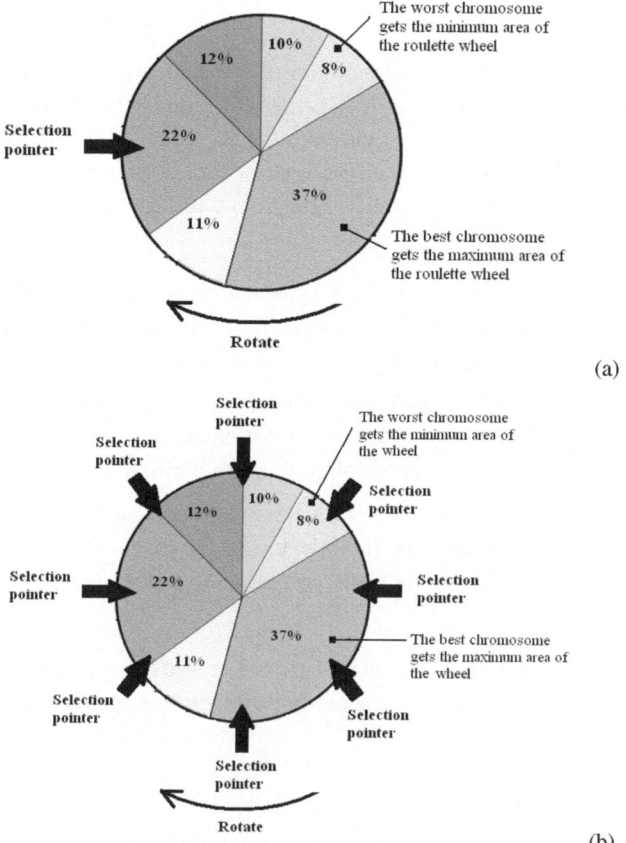

Fig. 2.2 Different selection operations: (a) Roulette wheel selection, (b) Stochastic universal sampling

corresponding slot. The stochastic universal sampling process is illustrated in Figure 2.2(b).

- In *binary tournament selection*, two chromosomes are taken at random, and the better chromosome is put into the mating pool. The process is repeated until the mating pool becomes full. The binary tournament selection may be implemented with and without replacement of the competing individuals. The tournament size also can be varied, i.e., instead of a binary tournament, it can be a tertiary tournament. It should be noted that in binary tournament, the worst chromosome will never be selected in the mating pool.

2.2.5 Crossover

Crossover is a genetic operator that combines (mates) two chromosomes (parents) to produce two new chromosomes (offspring). The idea behind crossover is that the new chromosomes may be better than both the parents if they take the best characteristics from each of the parents. Crossover occurs during evolution according to a user-definable crossover probability. Some popular crossover methods are *single-point*, *two-point* and *uniform* crossover.

- A crossover operator that randomly selects a crossover point within a chromosome and then interchanges the two parent chromosomes at this point to produce two new offspring is called single-point crossover.

- A crossover operator that randomly selects two crossover points within a chromosome and then interchanges the two parent chromosomes between these points to produce two new offspring is called two-point crossover.

- In uniform crossover, a binary mask having same length as that of the parent chromosomes is generated and then the bits of the parent chromosomes are interchanged in the positions where the corresponding positions in the binary mask are '1'.

Figure 2.3 illustrates the above three types of crossover operation for binary chromosomes. One can define crossover methods depending on the specific problem at hand.

2.2.6 Mutation

Mutation is a genetic operator that alters one or more gene values in a chromosome from its initial state. This can result in entirely new gene values being added to the gene pool. With these new gene values, the Genetic Algorithm may be able to arrive at a better solution than was previously possible. Mutation is an important part of the genetic search as it helps to prevent the population from stagnating at any local optimum. Mutation occurs during evolution according to a user-definable mutation probability. This probability should usually be set fairly low (0.01 is a good first choice). If it is set too high, the search will turn into a primitive random search. A commonly used mutation operator for binary chromosomes is *bit-flip* mutation, where each bit of a chromosome is subjected to mutation with the mutation probability and if the bit is selected to be mutated, it is just flipped. In Figure 2.4, the bit-flip mutation operation is demonstrated.

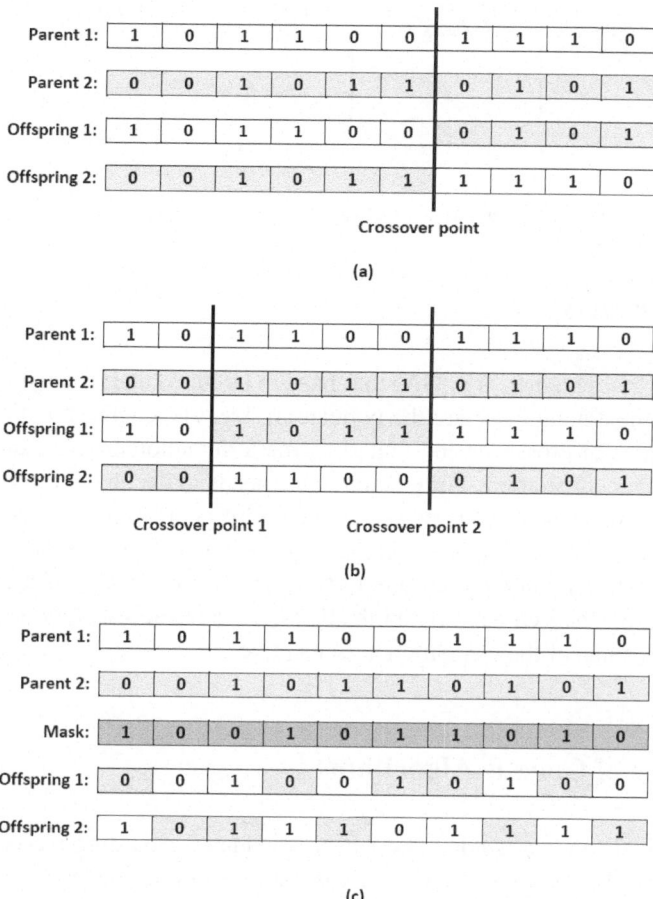

Fig. 2.3 Illustration of different crossover operations on binary chromosomes: (a) Single-point crossover, (b) Two-point crossover, (c) Uniform crossover

2.2.7 Elitism

To ensure that the best chromosome found so far is not lost due to randomized operators in GAs, *elitist* models of GAs are usually used. In the elitist models of GAs, the best chromosome seen up to the current generation is retained either in the population or in a location outside it. Sometimes elitism is performed by replacing the worst chromosome of the current generation by the best chromosome of the previous generation, provided that the latter has better fitness than the former.

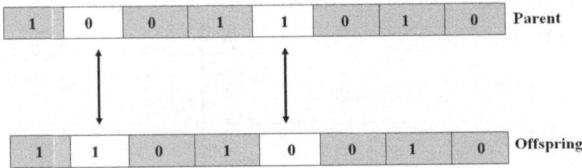

Fig. 2.4 Illustration of bit-flip mutation operation

2.2.8 Parameters

There are several parameters in GAs that have to be tuned and/or fixed by the programmer. Some among these are the population size, the length of a chromosome, the probabilities of crossover and mutation, the termination criteria, and the population replacement strategy. For example, one needs to decide whether to use the generational replacement strategy, where the entire population is replaced by the new population, or the steady state replacement policy, where only the less fit individuals are replaced. Such parameters in GAs are mostly problem-dependent, and no guidelines for their choice exist in the literature. Several researchers have therefore also kept some of the GA parameters adaptive.

2.3 Theory of Genetic Algorithms

A number of theoretical studies have been conducted regarding the convergence properties of GAs. In this section, we discuss some of these concepts.

2.3.1 Schema Theorem

The *schema theorem* is one of the most fundamental theoretical results of GAs. In the following discussion, available in detail in [181], the schema theorem for conventional GAs is derived. A *schema h* is a similarity template describing a subset of strings with similarities at certain string positions. For binary representation, it is composed of 0, 1 and # (don't care) symbols. For example, # 1 # 1 # # 0 # # # is a schema of length 10. The schema will be subsequently referred to as h'. A schema indicates the set of all strings that match it in the positions where the schema has either a 0 or a 1.

The *defining positions* of a schema are the positions in it that have either a 1 or a 0. The *defining length* of a schema h, denoted by $\delta(h)$, is defined as the distance between the last defining position and the first defining position of the schema, and is obtained by subtracting the first defining position from the last defining position. For the schema h' given above, the first defining position (counted from the left) is 2 and the last defining position is 7. Hence $\delta(h') = 7 - 2 = 5$. The *order* of a schema h, denoted by $O(h)$, is the number of defining positions in the schema. For the schema h', $O(h') = 3$.

A schema h_1 is said to be contained in another schema h_2 if for each defining position in h_2, the position is defined in h_1, the defining bit being the same. For example, let $h_1 = \#\,1\,1\,1\,0\,1\,0\,\#\,\#$; then h_1 is contained in h'. Note that if h_1 is contained in h_2, then $m(h_2,t) \geq m(h_1,t)$, where $m(h,t)$ represents the number of instances of h in the population at time t.

The *schema theorem* [181] estimates the lower bound of the number of instances of different schemata at any point of time. According to this theorem, a short-length, low-order, above-average schema will receive an exponentially increasing number of instances in subsequent generations at the expense of below-average ones. This is now discussed in detail. The following notations are used in this discussion:

h	: a short-length, low-order, above-average schema
$m(h,t)$	: number of instances of schema h in a population at generation t
$\delta(h)$	: the defining length of schema h
$O(h)$	: order of schema h
l	: length of a chromosome
$\overline{f_h}$	: average fitness value of schema h
$\overline{f}$	: average fitness value of the population

As already mentioned, the number of copies of each string that go into the mating pool is proportional to its fitness in the population. Accordingly, the expected number of instances of the schema h that go into the mating pool after selection, $m(h,t+1)$, is given by

$$m(h,t+1) = m(h,t) \times \frac{\overline{f_h}}{\overline{f}}. \tag{2.1}$$

Hence the number of above-average schemata will grow exponentially, and below-average ones will receive decreasing numbers of copies.

If a crossover site is selected uniformly at random from among $(l-1)$ possible sites, a schema h may be destroyed when the crossover site falls within its defining length $\delta(h)$. Therefore, the probability (p_d) of schema disruption due to crossover is

$$p_d = \frac{\delta(h)}{(l-1)}. \tag{2.2}$$

Hence the survival probability (p_s) (when the crossover site falls outside the defining length) is

$$p_s = 1 - p_d = 1 - \frac{\delta(h)}{(l-1)}. \tag{2.3}$$

If μ_c is the crossover probability, then

$$p_s \geq 1 - \mu_c \times \frac{\delta(h)}{(l-1)}. \tag{2.4}$$

Again, for the survival of a schema h, all the fixed positions of h, $(O(h))$, should remain unaltered. If μ_m is the mutation probability, the probability that one fixed position of the schema will remain unaltered is $(1 - \mu_m)$. Hence for $O(h)$ fixed positions of a schema h to survive, the survival probability is

$$(1 - \mu_m)^{O(h)}.$$

If $\mu_m << 1$, the above value becomes $(1 - O(h) \times \mu_m)$. Therefore,

$$m(h,t+1) \geq m(h,t) \times \frac{\overline{f_h}}{\overline{f}} \times \left(1 - \mu_c \times \frac{\delta(h)}{(l-1)}\right) \times (1 - \mu_m \times O(h)). \tag{2.5}$$

Neglecting the smaller term, we have

$$m(h,t+1) \geq m(h,t) \times \frac{\overline{f_h}}{\overline{f}} \times \left(1 - \mu_c \times \frac{\delta(h)}{(l-1)} - \mu_m \times O(h)\right). \tag{2.6}$$

Equation 2.6 indicates that short-length (small $\delta(h)$), low-order (small $O(h)$) and above-average $(\overline{f_h} > \overline{f})$ schemata will receive an exponentially increasing number of instances in subsequent generations [181, 214]. Note that the schema theorem does not guarantee the convergence of the process to the global optimal solution. Vose extended this work and interpreted GAs as constrained random walks and generalized the concept of schemata [419]. Some results of the analysis of schema distribution and deceptive problems, i.e., the class of problems which mislead the GA, are available in [117, 225, 277, 358, 435].

2.3.2 Markov Chain Modeling of GAs

Several researchers have analyzed GAs by modeling them as states of a Markov chain [66, 183, 216, 395, 419]. The states of the chain are denoted by the populations of GAs. A population Q is a collection of strings of length l generated over a finite alphabet A and is defined as follows:

$$Q = \{S_1, S_1, \ldots, (\sigma_1 \text{ times}), S_2, S_2, \ldots, (\sigma_2 \text{ times}), \ldots,$$
$$S_\xi, S_\xi, \ldots, (\sigma_\xi \text{ times}) : S_i \in \Sigma;$$
$$\sigma_i \geq 1 \text{ for } i = 1, 2, \ldots, \xi; \ S_i \neq S_j \ \forall i \neq j \text{ and } \sum_{i=1}^{\xi} = P\}.$$

Here Σ denotes the search space of the GA consisting of 2^l strings. Let Q denotes the set of all populations of size P. The number of populations or states in a Markov chain is finite and is given by

$$N = \binom{2^l + P - 1}{P}. \tag{2.7}$$

In [183], proportional selection has been assumed without mutation. Since single-bit chromosomes are considered in [183], for a population size P, there are $P +$ 1 possible states i, where i is the population with exactly i 1s and $(P - i)$ 0s. A $(P+1) \times (P+1)$ transition matrix $\tau[i, j]$ is defined that maps the current state i to the next state j. The major focus has been the study of the pure genetic drift of proportional selection using the transition probability expression, where the authors have investigated the expected times to absorption for the drift case. In [216], this work has been extended by modeling a niched GA using finite, discrete time Markov chains, limiting the niching operator to *fitness sharing*, introduced in [182].

In [419], the transition and asymptotic behavior of the GAs have been studied by modeling them as Markov chains. In [121], an attempt has been made to provide a Markov chain model and an accompanying theory for the simple GA by extrapolating the existing theoretical foundation of the simulated annealing algorithm [119, 252, 415]. The asymptotic behavior of elitist GAs with regard to the optimal string can also be found in [372]. A Markov chain analysis of GAs using a modified elitist strategy has been presented in [395] by considering selection, crossover and mutation for generational change. By evaluating the eigenvalues of the transition matrix of the Markov chain, the convergence rate of GAs has been computed in terms of the mutation probability μ_m. It is shown that the probability of the population including the individual with the highest fitness value is lower-bounded by $1 - O(|\lambda^*|^n)$, $|\lambda^*| < 1$, where n is the number of generation changes and λ^* is a specified eigenvalue of the transition matrix.

A survey of a number of recent developments concerning the simple GAs, its formalism, and the application of the Walsh transformation to the theory of GAs can be found in [420]. The issue of asymptotic behavior of GAs has also been pursued in [66], where GAs are again modeled as Markov chains having a finite number of states. A state is represented by a population together with a potential string. Irrespective of choice of the initial population, GAs have been proved to approach the optimal string after infinitely many iterations, provided the conventional mutation operation is incorporated. In [321], a stopping criterion called ε-optimal stopping time is provided for the elitist model of the GAs. Subsequently, the ε-optimal stopping time has been derived for GAs with elitism under a "practically valid assumption". Two approaches, mainly pessimistic and optimistic, have been considered to find out the ε-optimal stopping time. It has been found that the total number of strings to be searched in the optimistic approach to obtain an ε-optimal string is less than the number of all possible strings for a sufficiently large string length. This observation validates the use of GAs in solving complex optimization problems.

2.4 Evolutionary Multiobjective Optimization

Conventional GAs are traditionally single objective in nature, i.e., they optimize a single criterion in the searching process. Thus, finally, a single solution is generated that corresponds to the optimized value of the chosen optimization criterion. However, in many real-world situations there may be several objectives that must be optimized simultaneously in order to solve a certain problem. This is in contrast to the problems tackled by conventional GAs, which involve optimization of just a single criterion. The main difficulty in considering multiobjective optimization is that there is no accepted definition of optimum in this case, and therefore it is difficult to compare one solution with another one. In general, these problems admit multiple solutions, each of which is considered acceptable and equivalent when the relative importance of the objectives is unknown. The best solution is subjective and depends on the need of the designer or decision maker [50].

The multiobjective optimization can formally be stated as follows [106–108, 125]: Find the vector $\bar{x}^* = [x_1^*, x_2^*, \ldots, x_n^*]^T$ of decision variables which will satisfy the m inequality constraints

$$g_i(\bar{x}) \geq 0, \quad i = 1, 2, \ldots, m, \tag{2.8}$$

and the p equality constraints

$$h_i(\bar{x}) = 0, \quad i = 1, 2, \ldots, p, \tag{2.9}$$

and optimize the vector function

$$\bar{f}(\bar{x}) = [f_1(\bar{x}), f_2(\bar{x}), \ldots, f_k(\bar{x})]^T. \tag{2.10}$$

The constraints given in Equations 2.8 and 2.9 define the feasible region $\mathscr{F}$ which contains all the admissible solutions. Any solution outside this region is inadmissible since it violates one or more constraints. The vector $\bar{x}^*$ denotes an optimal solution in $\mathscr{F}$. In the context of multiobjective optimization, the difficulty lies in the definition of optimality, since it is rare that we will find a situation where a single vector $\bar{x}^*$ represents the optimum solution to all the objective functions.

The concept of *Pareto-optimality* comes handy in the domain of multiobjective optimization. A formal definition of Pareto-optimality from the viewpoint of the minimization problem may be given as follows: A decision vector $\bar{x}^*$ is called Pareto-optimal if and only if there is no $\bar{x}$ that dominates $\bar{x}^*$, i.e., there is no $\bar{x}$ such that

$$\forall i \in 1, 2, \ldots, k, f_i(\bar{x}) \leq f_i(\bar{x}^*)$$

and

$$\exists i \in 1, 2, \ldots, k, f_i(\bar{x}) < f_i(\bar{x}^*).$$

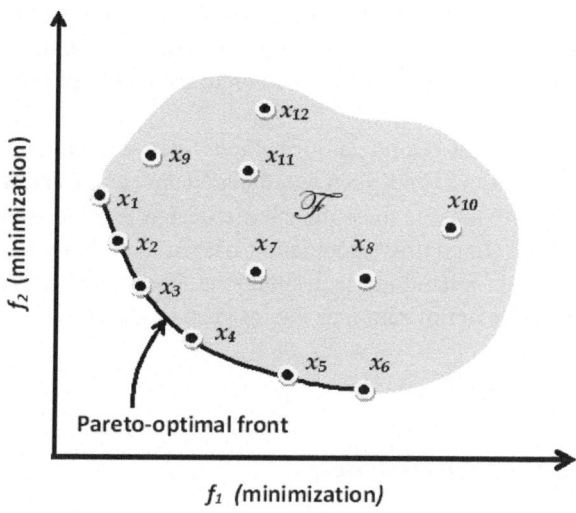

Fig. 2.5 Example of non-domination and Pareto-optimality: Two objectives f_1 and f_2 are minimized simultaneously. The shaded region represents the complete set $\mathscr{F}$ of feasible solutions. The x_is represent solutions in the objective space. Solutions x_1, x_2, x_3, x_4, x_5 and x_6 are not dominated by any other solution in $\mathscr{F}$. Hence the set $\{x_i\}$, $i = 1, 2, \ldots, 6$, represents the Pareto-optimal set

In other words, $\bar{x}^*$ is Pareto-optimal if there exists no feasible vector $\bar{x}$ which causes a reduction in some criterion without a simultaneous increase in at least one other (Figure 2.5). In this context, two other notions, *weakly non-dominated* and *strongly non-dominated* solutions, are defined [106, 125]. A point $\bar{x}^*$ is a weakly non-dominated solution if there exists no $\bar{x}$ such that $f_i(\bar{x}) < f_i(\bar{x}^*)$ for $i = 1, 2, \ldots, k$. A point $\bar{x}^*$ is a strongly non-dominated solution if there exists no $\bar{x}$ such that $f_i(\bar{x}) \leq f_i(\bar{x}^*)$ for $i = 1, 2, \ldots, k$, and for at least one i, $f_i(\bar{x}) < f_i(\bar{x}^*)$. In general, a Pareto-optimum usually admits a set of solutions called *non-dominated* solutions.

A multiobjective optimization approach should achieve the following three conflicting goals [106–108, 125]:

1. The best-known Pareto front should be as close as possible to the true Pareto front. Ideally, the best-known Pareto set should be a subset of the Pareto-optimal set.
2. Solutions in the best-known Pareto set should be uniformly distributed and diverse over of the Pareto front in order to provide the decision maker a true picture of trade-offs.
3. The best-known Pareto front should capture the whole spectrum of the Pareto front. This requires investigating solutions at the extreme ends of the objective function space.

For a given computational time limit, the first goal is best served by focusing (intensifying) the search on a particular region of the Pareto front. In contrast, the second goal demands the search effort to be uniformly distributed over the Pareto front. The third goal aims at extending the Pareto front at both ends, exploring new extreme solutions.

Traditional search and optimization methods such as gradient descent search, and other non-conventional ones such as simulated annealing, are difficult to extend as in the multiobjective case, since their basic design precludes the consideration of multiple solutions. In contrast, population-based methods such as Genetic Algorithms are well suited for handling such situations. There are different approaches to solving multiobjective optimization problems [106–108, 125]. They are categorized as follows.

2.4.1 Aggregating Approaches

- *Weighted sum approach:* Here, different objectives are combined using some weights w_i, $i = 1, 2, \ldots, k$ (k is the number of objectives). The objective to be optimized is $\sum_{i=1}^{k} w_i f_i(\overline{x})$.

- *Goal programming-based approach:* Here, the users are required to assign targets T_i, $i = 1, 2, \ldots, k$, or goals for each objective f_i. The aim is then to minimize the deviation from the targets to the objectives, i.e., $\sum_{i=1}^{k} |f_i(\overline{x}) - T_i|$.

- *Goal attainment-based approach:* Here, the users are required to provide, along with the target vector, a weight vector w_i, $i = 1, 2, \ldots, k$, relating the relative under- or over-attainment of the desired goals.

- *ε-Constraint approach:* Here, the primary objective function is to be optimized considering the other objective functions as constraints bound by some allowable levels ε_i.

2.4.2 Population-Based Non-Pareto Approaches

- *Vector Evaluated Genetic Algorithm (VEGA) [377]:* This algorithm incorporates a special selection operator where a number of subpopulations are generated by applying proportional selection based on each objective function in turn.

- *Lexicographic ordering [265]:* In this approach the objectives are ranked in order of importance by the user. The optimization is performed on these objectives according to this order.

- *Game theory-based approach:* This approach assumes that a player is associated with each objective.

- *Use of gender for identifying the objectives:* In this approach, panmitic reproduction where several parents combine to produce a single offspring is allowed.

- *Use of contact theorem:* Here, the fitness of an individual is set according to its relative distance from the Pareto front.

- *Use of non-generational GA:* In this strategy, a multiobjective problem is transformed into a single objective one through a set of appropriate transformations. The fitness of an individual is calculated incrementally. Genetic operators are applied to yield a single individual that replaces the worst individual in the population.

2.4.3 Pareto-Based Non-Elitist Approaches

- *Multiple Objective GA (MOGA) [163]:* In this approach, an individual is assigned a rank corresponding to the number of individuals in the current population by which it is dominated plus 1. All non-dominated individuals are ranked 1. The fitness values of individuals with the same rank are averaged so that all of them are sampled at the same rate. A niche formation method is used to distribute the population over the Pareto-optimal region.

- *Niched Pareto GA (NPGA) [217]:* Here, a Pareto dominance-based tournament selection with a sample of population is used to determine the winner from two candidate solutions. Around ten individuals are used to determine dominance and the non-dominated individual is selected. If both the individuals are either dominated or non-dominated, then the result of the tournament is decided through fitness sharing.

- *Non-dominated Sorting GA (NSGA) [393]:* In this approach, all non-dominated individuals are classified into one category, with a dummy fitness value proportional to the population size. This group is then removed and the remaining population is reclassified. The process is repeated until all the individuals in the entire population are classified. Stochastic remainder proportionate selection is used here.

2.4.4 Pareto-Based Elitist Approaches

- *Strength Pareto Evolutionary Algorithm (SPEA) [467]:* This algorithm implements elitism explicitly by maintaining an external population called an archive. Hence, at any generation t both the main population P_t of constant size N and the archive population P_t' (external) of maximum size N' exist. All non-dominated solutions of P_t are stored in P_t'. For each individual in the archive P_t', a strength value similar to the ranking value in MOGA is computed. The strength of an individual is proportional to the number of individuals dominated by it. The fitness of each individual in the current population P_t is computed according to the strength of all the archived non-dominated solutions that dominate it. Moreover, average linkage clustering is used to maintain diversity in the population.

- *Strength Pareto Evolutionary Algorithm 2 (SPEA2) [466]:* The SPEA algorithm has two potential weaknesses. Firstly, the fitness assignment is entirely based on the strengths of the archive members. This results in individuals having the same fitness value in P_t if the corresponding set of non-dominated members in the archive P_t' is same. In the worst case, if the archive contains a single member, then all the members of P_t will have same rank. In SPEA2, a technique is developed to avoid the situation where individuals dominated by the same archive members have the same fitness values. Here both the main population and the archive are considered to determine the fitness values of the individuals.
  Secondly, the clustering technique used in SPEA to maintain the diversity may lose outer and boundary solutions, which should be kept in the archive in order to obtain a good spread of the non-dominated solutions. In SPEA2, a different scheme has been adopted that prevents the loss of the boundary solutions during archive updating.

- *Pareto Archived Evolutionary Strategy (PAES) [255]:* It uses a (1+1) evolution strategy (i.e., a single parent generates a single offspring) together with an external archive that records all the non-dominated vectors previously found. To maintain the population diversity, an adaptive grid is used. Each solution is placed in a certain grid location based on its objective values. During selection, the solution from the grid location containing a lower number of solutions is selected.

- *Pareto Envelope-Based Selection Algorithm (PESA) [114]:* In this approach, a smaller internal (or primary) population and a larger external (or secondary) population are used. PESA uses the same hyper-grid division of objective space adopted by PAES to maintain the diversity. Its selection mechanism is based on the crowding measure used by the hyper-grid. This same crowding measure is used to decide which solutions to introduce into the external population (i.e., the archive of non-dominated individuals found during the evolutionary process).

- *Pareto Envelope-Based Selection Algorithm-II (PESA-II) [113]:* PESA-II is the revised version of PESA in which *region-based selection* is proposed. In the case of region-based selection, the unit of selection is a hyperbox instead of an individual. The procedure of selection is to select (using any of the traditional selection techniques) a hyperbox and then randomly select an individual within such a hyperbox.

- *Elitist Non-dominated Sorting GA (NSGA-II) [130]:* NSGA-II was proposed to eliminate the weaknesses of NSGA, specially its non-elitist nature and specification of the sharing parameter. Here, the individuals in a population undergo *non-dominated sorting* as in NSGA and individuals are given ranks based on this. A new selection technique, called *crowded tournament selection*, is proposed where selection is done based on *crowding distance* (representing the neighbourhood density of a solution). For implementing elitism, the parent and child populations are combined and the non-dominated individuals from the combined population are propagated to the next generation.

Most of the multiobjective clustering algorithms described in this book use NSGA-II as the underlying multiobjective optimization strategy. Therefore, here we describe the NSGA-II algorithm in detail.

The non-dominated sorting method is an important characteristic of NSGA-II. This is performed as follows: Given a set of solutions $\mathscr{S}$, the non-dominated set of solutions $\mathscr{N} \subseteq \mathscr{S}$ is composed of the those solutions of $\mathscr{S}$ which are not dominated by any other solutions of $\mathscr{S}$. To find the non-dominated set, the following are the necessary steps.

1. Set $i = 1$ and initialize the non-dominated set $\mathscr{N} = \phi$.
2. Set $j = 1$.
3. While $j \leq |\mathscr{S}|$ do
   a. If solution j ($j \neq i$) dominates solution i then go to Step 5.
   b. Set $j = j + 1$.
4. Set $\mathscr{N} = \mathscr{N} \cup \{i\}$.
5. Set $i = i + 1$.
6. If $i \leq |\mathscr{S}|$ then go to Step 2.
7. Output $\mathscr{N}$ as the non-dominated set.

The non-dominated sorting procedure first finds the non-dominated set $\mathscr{N}$ from the given set of solutions $\mathscr{S}$. Each solution belonging to $\mathscr{N}$ is given the rank 1. Next, the same process is repeated on the set $\mathscr{S} = \mathscr{S} - \mathscr{N}$ and the next set of non-dominated solutions $\mathscr{N}'$ is found. Each solution of the set $\mathscr{N}'$ is given the rank 2. This procedure goes on until all the solutions in the initial set are given some rank, i.e., $\mathscr{S}$ becomes empty.

A measure called crowding distance has been defined on the solutions of the non-dominated front for diversity maintenance. The crowding distances for the boundary solutions are set to maximum values (logically infinite). For each solution i among the remaining solutions, the crowding distance is computed as the average distance of the $(i+1)$th and $(i-1)$th solutions along all the objectives. The following are the steps for computing the crowding distance d_i of each point i in the non-dominated front $\mathscr{N}$.

1. For $i = 1, 2, \ldots, |\mathscr{N}|$, initialize $d_i = 0$.
2. For each objective function f_k, $k = 1, 2, \ldots, M$, do the following:
   a. Sort the set $\mathscr{N}$ according to f_k in ascending order.
   b. Set $d_1 = d_{|\mathscr{N}|} = \infty$.
   c. For $j = 2$ to $(|\mathscr{N}| - 1)$, set $d_j = d_j + \frac{(f_k^{j+1} - f_k^{j-1})}{(f_k^{max} - f_k^{min})}$.

In NSGA-II, a binary tournament selection operator is used based on crowding distance. If two solutions a and b are compared during a tournament, then solution a wins the tournament if either

1. The rank of a is better (less) than the rank of b, i.e., a and b belong to two different non-dominated fronts, or
2. The ranks of a and b are same (i.e., they belong to same non-dominated front) and a has higher crowding distance than b. This means that if two solutions belong to the same non-dominated front, the solution situated in the less crowded region is selected.

The following are the steps of the NSGA-II algorithm.

1. Initialize the population.
2. While the termination criterion is not met repeat the following:
   a. Evaluate each solution in the population by computing M objective function values.
   b. Rank the solutions in the population using non-dominated sorting.
   c. Perform selection using the crowded binary tournament selection operator.
   d. Perform crossover and mutation (as in conventional GAs) to generate the offspring population.
   e. Combine the parent and child populations.
   f. Replace the parent population by the best members (selected using non-dominated sorting and the crowded comparison operator) of the combined population.
3. Output the first non-dominated front of the final population.

- *Archived Multiobjective Simulated Annealing (AMOSA) [52]:* AMOSA is a recently proposed simulated annealing-based multiobjective optimization technique. This technique incorporates the concept of an archive that stores the non-dominated solutions found so far. As the size of the archive is finite, two limits, *hard limit* and *soft limit*, are kept on the size of the archive. During the process, the non-dominated solutions are stored in the archive as and when they are generated until the size of the archive increases to the soft limit. Thereafter, if more non-dominated solutions are generated, the size of the archive is first reduced to the hard limit by applying clustering, after which new non-dominated elements can be introduced. The algorithm starts with the initialization of a number (soft limit) of solutions. Each of these solutions is refined by using simple hill-climbing and the domination relation for a number of iterations. Thereafter, the non-dominated solutions are stored in the archive, up to a maximum of the hard limit. Clustering is used if required. Then, one of the solutions is randomly selected from the archive. This is taken as the *current-pt*, or the initial solution, at temperature $T = T_{max}$. The is perturbed to generate a new solution named *new-pt*. The domination status of the new-pt is checked with respect to the current-pt and the solutions in the archive. Based on domination status, different cases may arise viz., accept (i) the new-pt, (ii) the current-pt or (iii) a solution from the archive. The inverse of the amount of domination is used in determining the probability of this acceptance. The process is repeated *iter* times for each temperature that is annealed with a cooling rate till the minimum temperature T_{min} is attained.

Besides the above multiobjective techniques, other optimization tools, such as tabu search [55], particle swarm optimization [365], ant colony optimization [169], and differential evolution [444], have also been modified to cope with multiobjective optimization. A comparison of the performance of multiobjective optimization techniques with their single objective counterparts has been made in [240]. An important aspect of multiobjective optimization is that it provides a set of non-dominated alternative solutions to the decision maker who can choose one or more solutions from the set depending on the problem requirements. There is numerous scope for the application of multiobjective optimization, ranging from data mining, medicine, biology, and remote sensing to scheduling, VLSI designing, and control theory.

2.5 Theory of Multiobjective Evolutionary Algorithms

Unlike with traditional single objective GAs, there is not much work available in the literature on the theoretical analysis of multiobjective evolutionary algorithms. Some attempts in this regard can be found in [71, 204, 269, 371, 372, 418]. However, none of the multiobjective evolutionary algorithms discussed in the previous section has a proof of convergence towards the Pareto-optimal set having wide diversity among the solutions. In this section, we discuss some of the theoretical concepts of multiobjective evolutionary algorithms and their convergence.

2.5.1 ε-Dominance

In [269], some archiving/selection strategies have been proposed that guarantee the progress towards the Pareto-optimal set and the coverage of the whole range of the non-dominated solutions simultaneously. The finite-sized archives containing the non-dominated solutions are iteratively updated using the concept of ε-dominance. The ε-dominance concept helps to control the resolution of the approximated Pareto front.

Let $a, b \in \mathbb{R}^{+^k}$ be two solutions. The solution a is said to ε-dominate solution b for some $\varepsilon > 0$, denoted by $a \succ_\varepsilon b$ (maximization problem), iff [269]

$$\forall i \in \{1, \ldots, k\} \quad (1 + \varepsilon).f_i(a) \geq f_i(b). \tag{2.11}$$

Consider a set of solutions $F \subseteq \mathbb{R}^{+^k}$ and $\varepsilon > 0$. A set F_ε is called an ε-approximate Pareto set of F if any solution $b \in F$ is ε-dominated by at least one solution $a \in F_\varepsilon$ [269], i.e.,

$$\forall b \in F \ : \ \exists a \in F_\varepsilon \text{ such that } a \succ_\varepsilon b. \tag{2.12}$$

The set of all ε-approximate Pareto sets of F is denoted by $P_\varepsilon(F)$. Note that the set F_ε is not unique. A survey of different concepts of ε-efficiency and the corresponding Pareto set approximation is available in [210]. The concept of approximation can also be used with additive approximation as follows [269]:

$$\forall i \in 1, 2, \ldots, k \quad \varepsilon_i + f_i(a) \geq f_i(b), \tag{2.13}$$

where the ε_is are constants defined separately for each objective function f_i. The concept of ε-approximate Pareto set is further refined to obtain the following definition [269]: Suppose $F \subseteq \mathbb{R}^{+^k}$ is a set of solutions and $\varepsilon > 0$. Then a set $F_\varepsilon^* \subseteq F$ is called an ε-Pareto set of F if F_ε^* is an ε-approximate Pareto set of F, i.e., $F_\varepsilon^* \in P_\varepsilon(F)$, and F_ε^* contains Pareto points of F only, i.e., $F_\varepsilon^* \subseteq F^*$. The set of all ε-Pareto sets of F is denoted by $P_\varepsilon^*(F)$.

Using the above Pareto front approximation concepts, in [269] a two-level selection strategy has been proposed that guarantees that the optimal solution is never lost and no deterioration occurs. In the coarse level, the search space is discretized by dividing it into boxes, where each solution uniquely belongs to a box. The algorithm maintains a set of non-dominated boxes using the ε-approximation property. In the fine level, a maximum of one solution is kept in a box. The solution in a box can be replaced by a dominating solution only, thus ensuring the convergence [269]. There have been some other studies using the ε-dominance concept and it has been found to be an efficient mechanism for maintaining diversity in multiobjective optimization problems without losing convergence properties towards the Pareto-optimal set [128, 129, 364].

2.5.2 Markov Chain Modeling of Multiobjective EAs

In [418] an asymptotic convergence analysis of the metaheuristic algorithms for multiobjective optimization that use a uniform mutation rule is performed. The multiobjective EAs have been modeled as Markov chains as in the case of traditional GAs described in Section 2.3.2. The matatheuristic algorithms have been assumed to operate on binary strings with a population-based approach. Moreover, they are assumed to be memoryless, i.e., they do not have any way of remembering the past through the searching process. This is required to model them as Markov chains since they only use the current state to determine the next action to be performed. Furthermore, the algorithms are assumed to be operated only by a uniform mutation operator (no crossover operation). Finally, the algorithms may adopt some form of elitism.

The concerned algorithm has been modeled as a Markov chain $\{X_k : k \geq 0\}$. The state space S of the Markov chain is the set of all possible populations of n individual solutions. Each solution is represented by a binary string of length l. A state i can be denoted by $i = \{i_1, i_2, \ldots, i_n\}$, where each i_s is a binary string of length l. The chain's transition probability from state i to state j is given by [418]:

$$P_{ij} = \mathbb{P}(X_{k+1} = j | X_k = i). \tag{2.14}$$

Therefore the transition matrix is given as $P = (P_{ij}) = LM$. Here M denotes the transition matrix corresponding to the mutation operation and L denotes the transition matrix representing some other operation. For L and M, the following properties hold [418]: $\forall i, j \; L_{ij} \geq 0, M_{ij} \geq 0$ and

$$\forall i \in S, \quad \sum_{x \in S} L_{ix} = 1 \quad \text{and} \quad \sum_{x \in S} M_{ix} = 1. \tag{2.15}$$

In order to compute the mutation probability for each individual from the state i to state j, the individual i_s is assumed to be transformed into the individual j_s by applying uniform mutation. Thus each entry of i_s of i is transformed into the corresponding one of j_s of j with probability p_m. Therefore, the mutation probability for each individual solution in the population is given by [418]

$$\forall s \in \{1, 2, \ldots, n\}, \quad p_m^{H(i_s, j_s)} (1 - p_m)^{1 - H(i_s, j_s)}, \tag{2.16}$$

where $H(i_s, j_s)$ is the Hamming distance between i_s and j_s. Now the mutation probability from state i to j is computed as [418]

$$M_{ij} = \prod_{s=1}^{n} p_m^{H(i_s, j_s)} (1 - p_m)^{1 - H(i_s, j_s)}. \tag{2.17}$$

When elitism is incorporated, the states are represented as [418]

$$\hat{i} = (i^e; i) = (i_1^e, \ldots, i_r^e; i_1, \ldots, i_n),$$

where $i_1^e, \ldots, i_r^e$ are the members of the elite set (with size r) of the state. It is assumed that the cardinality of the Pareto-optimal set P^* is greater than or equal to r and $r \leq n$. If every individual in the elite set of a state is Pareto-optimal, then any state containing an individual in the elite set that is not a Pareto-optimal is not accepted [418], i.e.,

$$\text{if } \{i_1^e, \ldots, i_r^e\} \subset P^* \text{ and } \{j_1^e, \ldots, j_r^e\} \not\subset P^* \text{ then } \hat{P}_{ij} = 0, \tag{2.18}$$

where $\hat{P}$ is the transition matrix associated with the new states.

The authors of [418] showed that the associated Markov chain converges geo-metrically to its stationary distribution, but not necessarily to the optimal solution set of the multiobjective problem. Convergence to the Pareto-optimal solution set is ensured only when elitism is incorporated.

In addition to the above two theoretical analyses, some more theoretical stud-ies on multiobjective optimization can be found in [71, 204, 371, 372], but still it is far from complete. There is more scope for research in the area of multiobjec-tive optimization, including for convergence criteria, elitism and effective constraint handling. Extensive studies related to ways of measuring the quality of solutions as well as devising standard benchmark problems need to be undertaken. Moreover, a theoretical analysis of multiobjective GAs should be conducted so that the working principles of such techniques may be formalized and the effects of different pa-rameter settings as well as operators may be explained. In this regard, appropriate stopping times also need to be devised.

2.6 Genetic Algorithms for Data Mining and Bioinformatics

Both traditional and multiobjective GAs have been used in solving different prob-lems in data mining and bioinformatics. This section briefly discusses some of them.

2.6.1 Genetic Algorithms in Data Mining

Data mining refers to the task of extracting meaningful information from complex data. GAs have been applied to perform different data mining tasks such as classifi-cation, clustering, feature selection, and association rule mining.

Supervised classification is an important data mining task. GAs have been used to determine class boundaries in the form of hyperplanes. The objective is to place a number of hyperplanes in the feature space such that the decision boundary of the training dataset can be closely approximated. Here, the chromosomes are used to encode the set of hyperplanes and are evaluated on the basis of classification accuracy. Studies in this direction, with several modifications and improvements, can be found in [45, 46, 48, 49]. Another approach for direct employment of GAs in classification is to use it for extracting meaningful *if-then* classification rules [123, 186, 292, 388]. In these approaches, chromosomes encode possible classification rules and the objective is to minimize the misclassification rate.

Genetic and other evolutionary algorithms have been extensively used in data clustering problems. These algorithms pose the clustering problem as an optimiza-tion one and try to optimize some cluster validity measures. There are mainly two types of encoding strategies for clustering using GAs. In point-based encoding, where the chromosome lengths are equal to the number of points in the dataset, each position represents the class label of the corresponding data point. This strategy has been adopted in [260, 263, 284, 285, 322]. The second approach is center-based encoding, where the chromosomes encode the coordinates of the cluster centers. The algorithms in [36, 38, 39, 298, 300] have used this approach. The application of multiobjective GAs and other evolutionary techniques have gained popularity recently. In this book, we mainly focus on this area. Some recent multiobjective clustering techniques can be found in [41, 44, 200, 302, 314, 315].

GAs have also found application in feature selection, an important problem in data mining. Both supervised and unsupervised approaches are there. GAs have mainly been used in wrapper-based feature selection approaches. Here the chromosomes are encoded as binary strings of length equal to the total number of features, and the bits '1' and '0' indicate selected and ignored features, respectively. Thus each chromosome encodes a subset of features. As the fitness function for the supervised case, the classification accuracy is computed based on some classifier trained on only selected features, and the objective is to maximize the classification accuracy. GAs have been used to search for feature subsets in conjunction with various classification algorithms such as decision trees [32], neural networks [74], k-nearest neighbors [245], Naïve Bayes [82, 235] and support vector machines [465]. For the unsupervised case, instead of a classification method, a clustering technique is used to cluster the input dataset on the selected features and some internal cluster validity index is computed as the fitness function [201, 250, 311].

Another important data mining task is association rule mining, which deals with discovering interesting relationships among the attributes of large databases. GAs have widely been used in association rule mining. The main approach in this regard deals with chromosomes that encode possible rules, and the fitness functions are used to evaluate the rules [373, 387]. Furthermore, some multiobjective approaches for association rule mining, where more than one rule evaluation criterion have been optimized simultaneously, can be found in recent literature [220, 422].

Genetic Algorithms have several applications in various fields of data mining, such as remote sensing [41, 49, 314], Web mining [330, 331, 350, 351], multimedia data mining [77, 97], text mining [28, 134], and biological data mining [43, 402]. The strength of GAs is continuously being explored in these fields.

2.6.2 Genetic Algorithms in Bioinformatics

Different bioinformatics tasks such as gene sequence analysis, gene mapping, deoxyribonucleic acid (DNA) fragment assembly, gene finding, microarray analysis, gene regulatory network analysis, phylogenetic tree construction, structure prediction and analysis of DNA, ribonucleic acid (RNA) and protein, and molecular docking with ligand design have been performed using GAs [337]. Several tasks in bioinformatics involve optimization of different criteria (such as energy, alignment score, and overlap strength), thereby making the application of GAs more natural and appropriate. Moreover, the problems in bioinformatics seldom need the exact optimum solution; rather, they require robust, fast, and close approximate solutions, which GAs are known to provide effectively.

The task of sequence alignment, which involves mutual placement of two or more DNA, RNA or protein sequences in order to find similarity among them, has been solved using GAs. It was first described in the Sequence Alignment by Genetic Algorithm (SAGA) [327] how to use GA to deal with sequence alignments in a general manner. In SAGA, the population is made of alignments (encoded in chromosomes), and the mutations are processing programs that shuffle the gaps using complex methods. There has been a similar approach proposed in [459]. Other recent approaches for sequence alignment can be found in [328, 451], where more efficient mutation operators have been designed.

Gene mapping is defined as the determination of relative positions of genes on a chromosome and the distance between them. The method of genetic mapping described in [171] is embodied in a hybrid framework that relies on statistical optimization algorithms (e.g., expectation maximization) to handle the continuous variables (recombination probabilities), while GAs handle the ordering problem of genes. In canonical GAs with a fixed map it is difficult to design the map without a prior knowledge of the solution space. This is overcome in [320], where GAs using a coevolutionary approach are utilized for exploring not only within a part of the solution space defined by the genotype-phenotype map, but also within the map itself. A comparison of GA- and simulated annealing-based gene mapping can be found in [191].

Automatic identification of the genes from the large DNA sequences is an important problem in bioinformatics. In [244], a GA has been used to design sets of appropriate oligonucleotide probes capable of identifying new genes belonging to a defined gene family within a cDNA or genomic library. In [275], a GA-based method for recognizing promoter regions of eukaryotic genes with an application to Drosophila melanogaster has been described. The method of prediction of eukaryotic Pol II promoters from a DNA sequence [256] takes advantage of a combination of elements similar to those of neural networks and GAs to recognize a set of discrete subpatterns with variable separation as one pattern, a promoter.

In interpretation and analysis of microarray gene expression data, GAs have been used actively. In the context of microarray gene expression analysis, GAs have been used in optimal ordering of genes [273, 409], classification of tumor samples and gene selection [11, 264, 309], clustering of genes and samples [44, 203, 302, 313, 317, 339, 400], biclustering of gene expression data [69, 88, 140, 141, 303, 304], and inferring of gene regulatory networks [8, 18–20].

GAs have also been employed in the construction of phylogenetic trees. In [296], each chromosome in a GA is encoded as a permutation of 15 taxas (objects) and selection, crossover, and mutation operations are performed to minimize the distance among the taxas. GAs have also been used [389] for automatic self-adjustment of the parameters of the optimization algorithm of phylogenetic trees.

Prediction of three-dimensional structures of DNA, RNA and proteins from their one-dimensional sequence is an important task in bioinformatics. GAs have been widely utilized in structure prediction of DNA [58, 341], RNA [54, 190, 382, 383] and proteins [111, 231, 332, 344, 412–414].

Molecular design and docking, which has immense application in drug design, is a difficult optimization problem, requiring efficient sampling across the entire range of positional, orientational, and conformational possibilities. GA is becoming a popular choice for the heuristic search method in molecular design and docking applications. A novel and robust automated docking method that predicts the bound conformations (structures) of flexible ligands to macromolecular targets has been developed [312]. The method combines GAs with a scoring function that estimates the free energy change upon binding. In [30, 35], a variable string length GA is presented for designing a ligand molecule that can bind to the active site of a target protein. In [91], a population-based annealing Genetic Algorithm (PAG) using GAs and simulated annealing (SA) is proposed for finding binding structures for three drug protein molecular pairs, including the anti-cancer drug methotrexate (MTX). In [368], the use of GAs with local search in molecular docking has been evaluated. A survey on the application of GAs for molecular modeling and docking of flexible ligands into protein active sites, can be found in [437].

There are several other potential bioinformatics tasks where GAs can be used effectively. This includes characterization of protein content and metabolic pathways between different genomes, identification of interacting proteins, assignment and prediction of gene products, large-scale analysis of gene expression levels, and mapping expression data to sequence, structural and biochemical data.

2.7 Summary

Genetic Algorithms are randomized search and optimization algorithms that are inspired by natural genetics and evolution. In this chapter, a gentle overview of Genetic Algorithms is provided. Different characteristics of GAs and various operations such as encoding strategy, selection, crossover, mutation and elitism have been discussed. Moreover, we have also discussed about some theoretical studies on GAs.

Traditionally GAs have been developed as single objective optimization methods. As many real-life applications are multiobjective in nature, multiobjective versions of GAs and other evolutionary optimization techniques have become popular. In this chapter, we have discussed the different concepts and important definitions of multiobjective optimization, with examples. Moreover, a discussion on various types of multiobjective evolutionary optimization algorithms such as aggregating, non-

Pareto and Pareto-based approaches is provided. Furthermore, a discussion on some theoretical studies made in the domain of multiobjective evolutionary algorithms has been presented. Finally, the chapter is concluded with a survey of existing literature on the application of GAs in various fields of data mining and bioinformatics.

Chapter 3
Data Mining Fundamentals

3.1 Introduction

To extract useful information from different types of vast data repositories, knowledge discovery and data mining have recently emerged as a significant research direction. A huge amount of data is nowadays being routinely collected as a consequence of widespread automation, which may be considered to be a major advantage of advanced data collection and storage technologies. However it creates the imperative need for developing methods to integrate and make sense out of this immense volume of data. Knowledge discovery from databases (KDD) evolved as a major research area where various fields of computer science such as databases, machine learning, pattern recognition, statistics, artificial intelligence, reasoning with uncertainty, expert systems, information retrieval, signal processing, high-performance computing and networking are involved. Typical examples of domains requiring the use of data mining techniques are the World Wide Web, credit card data, biological data, geoscientific data, mobile and sensor networks, VLSI chip layout and routing, multimedia, and time series data as in financial markets and meteorology. This chapter provides an overview of the area with the basic concepts and principles of data mining and knowledge discovery, the tasks involved, the research issues and the recent trends in this domain.

3.2 Knowledge Discovery Process

The knowledge discovery task can be classified into data preparation, including data cleaning and integration, data mining and knowledge representation. [40] In the data mining step, the techniques for extracting useful and interesting patterns are applied on the cleaned data. Data preparation and knowledge presentation can therefore be considered as the preprocessing and post-processing steps of data mining, respectively. A schematic diagram of the steps involved in the process of knowledge discovery is presented in Figure 3.1. The different issues pertaining to KDD are described below.

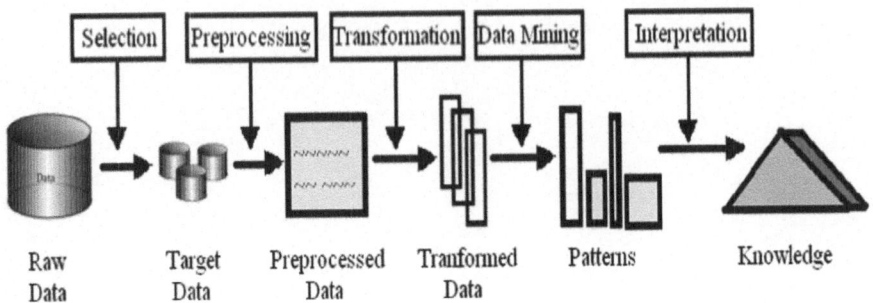

Fig. 3.1 The process of knowledge discovery.

3.2.1 Data Warehousing

Data warehousing [40, 155, 234] can be considered to comprise the tasks of collecting and cleaning transactional data to make it available for analytical processing. The different components of a data warehouse are cleaned and integrated data, detailed and summarized data, historical data, and metadata. The use of data warehouses greatly improves the efficiency of data mining.

For the purpose of knowledge discovery, it is necessary to develop efficient techniques based on machine learning tools, statistical and mathematical models on top of the data warehouses. As an example, the implementation of classification algorithms such as support vector machines (SVM), nearest neighbor, C4.5 and neural networks on top of a large database requires tighter coupling with the database system and intelligent use of coupling techniques [196, 232].

Apart from developing algorithms that can work on top of existing DBMSs, it is also necessary to develop new knowledge and data discovery management systems (KDDMSs) to manage KDD systems [232]. For this purpose it is necessary to define KDD objects that may be far more complex than database objects (records or tuples), and queries that are more general than SQL and that can operate on the complex objects. Here, KDD objects may be rules, classes or clusters [232]. The KDD objects may be pre-generated (e.g., as a set of rules) or may be generated at runtime (e.g., a clustering of the data objects). Also, KDD queries should be able to operate on both KDD objects and database objects. Some attempts in this direction may be found in [233, 385].

3.2.2 Data Mining

Data mining involves discovering *interesting* and *potentially useful* patterns of different types such as associations, summaries, rules, changes, outliers and significant structures. Commonly, data mining and knowledge discovery are treated as synonymous, although some scientists consider data mining to be an integral step in the knowledge discovery process. In general, data mining techniques comprise of three components [155]: a model, a preference criterion and a search algorithm. Association rule mining, classification, clustering, regression, sequence and link analysis and dependency modeling are some of the most common functions in current data mining techniques. Model representation determines both the flexibility of the model for representing the underlying data and the interpretability of the model in human terms. This includes decision trees and rules, linear and nonlinear models, example-based techniques such as NN rule and case-based reasoning, probabilistic graphical dependency models (e.g., Bayesian network) and relational attribute models.

Depending on the underlying dataset, the preference criterion determines which model to use for mining by associating some measure of goodness with the model functions [40]. It tries to avoid overfitting of the underlying data or generating a model function with a large number of degrees of freedom. The specification of the search algorithm is defined in terms of the selected model and the preference criterion along with the given data.

3.2.3 Knowledge Representation

Representing the information extracted by mining the data so that it is easily understood by the user is an important issue in knowledge discovery. This module acts as an interface between the users and the knowledge discovery step. Therefore it makes the entire process more useful and effective. Important components of the knowledge representation step are data and knowledge visualization techniques. Often, presenting the information in a hierarchical manner is considered to be very useful for the user to enable him to focus attention on only the important and interesting concepts. This also provides multiple views of the extracted knowledge to the users at different levels of abstraction. Some knowledge representation approaches include [40] rule generation, summarization using natural languages, tables and cross tabulations, graphical representation in the form of bar charts, pie charts and curves, data cube views, and decision trees.

3.3 Data Mining Tasks

Data mining tasks can be classified into two categories, descriptive and predictive [197]. While the descriptive techniques provide a summary of the data, the predictive techniques learn from the current data in order to make predictions about the behavior of new datasets. The following are the commonly used tasks in the domain of data mining.

3.3.1 Classification

The problem of classification is basically one of partitioning the feature space into regions, one region for each category of input [50]. Thus it attempts to assign every data point in the entire feature space to one of the possible (say, k) classes. Classifiers are usually, but not always, designed with labeled data, in which case these problems are sometimes referred to as *supervised classification* (where the parameters of a classifier function D are learned). Some common examples of the *supervised* pattern classification techniques are the nearest neighbor (NN) rule, the Bayes maximum likelihood classifier, the perceptron rule [17,21,135,146,165,166,172,185,346,408], and Support Vector Machines (SVM) [115,417]. Some of the related classification techniques are described below.

3.3.1.1 NN Rule

Let us consider a set of n pattern points of known classification $\{\mathbf{x}_1, \mathbf{x}_2, \ldots, \mathbf{x}_n\}$, where it is assumed that each pattern belongs to one of the classes $C_1, C_2, \ldots, C_k$. The NN classification rule then assigns a pattern $\mathbf{x}$ of unknown classification to the class of its nearest neighbor, where $\mathbf{x}_i \in \{\mathbf{x}_1, \mathbf{x}_2, \ldots, \mathbf{x}_n\}$ is defined to be the nearest neighbor of $\mathbf{x}$ if

$$D(\mathbf{x}_i, \mathbf{x}) = \min_l \{D(\mathbf{x}_l, \mathbf{x})\}, \quad l = 1, 2, \ldots, n, \tag{3.1}$$

where D is any distance measure definable over the pattern space. Since the aforesaid scheme employs the class label of only the nearest neighbor to $\mathbf{x}$, it is known as the 1-NN rule. If k neighbors are considered for classification, then the scheme is called the k-NN rule [146, 166, 408]. The k-NN rule assigns a pattern $\mathbf{x}$ of unknown classification to class C_i if the majority of the k nearest neighbors belongs to class C_i. Details of the k-NN rule along with the probability of error are available in [146, 166, 408]. The k-NN rule suffers from two severe limitations. Firstly, all the n training points need to be stored for classification and, secondly, n distance computations are required for computing the nearest neighbors. Some attempts at alleviating the problem may be found in [37].

3.3.1.2 Bayes Maximum Likelihood Classifier

In most of the practical problems, the features are usually noisy and the classes in the feature space are overlapping. In order to model such systems, the feature values $x_1, x_2, \ldots, x_j, \ldots, x_N$ are considered as random values in the probabilistic approach. The most commonly used classifier in such probabilistic systems is the Bayes maximum likelihood classifier [17, 408], which is now described. Let P_i denotes the a priori probability and $p_i(\mathbf{x})$ denotes the class conditional density corresponding to the class C_i ($i = 1, 2, \ldots, k$). If the classifier decides that $\mathbf{x}$ is from class C_i, when it actually comes from C_l, it incurs a loss equal to L_{li}. The expected loss (also called the conditional average loss or risk) incurred in assigning an observation $\mathbf{x}$ to class C_i is given by

$$r_i(\mathbf{x}) = \sum_{l=1}^{k} L_{li}\, p(C_l/\mathbf{x}), \tag{3.2}$$

where $p(C_l/\mathbf{x})$ represents the probability that $\mathbf{x}$ is from C_l. Using Bayes's formula, Equation 3.2 can be written as,

$$r_i(\mathbf{x}) = \frac{1}{p(\mathbf{x})} \sum_{l=1}^{k} L_{li}\, p_l(\mathbf{x})P_l, \tag{3.3}$$

where

$$p(\mathbf{x}) = \sum_{l=1}^{k} p_l(\mathbf{x})P_l. \tag{3.4}$$

The pattern $\mathbf{x}$ is assigned to the class with the smallest expected loss. The classifier which minimizes the total expected loss is called the *Bayes classifier*.

Let us assume that the loss (L_{li}) is zero for the correct decision and greater than zero but the same for all erroneous decisions. In such situations, the expected loss, Equation 3.3, becomes

$$r_i(\mathbf{x}) = 1 - \frac{P_i p_i(\mathbf{x})}{p(\mathbf{x})}. \tag{3.5}$$

Since $p(\mathbf{x})$ is not dependent on the class, the Bayes decision rule is nothing but the implementation of the decision functions

$$D_i(\mathbf{x}) = P_i p_i(\mathbf{x}), \qquad i = 1, 2, \ldots, k, \tag{3.6}$$

where a pattern $\mathbf{x}$ is assigned to class C_i if $D_i(\mathbf{x}) > D_l(\mathbf{x})$, $\forall l \neq i$. This decision rule provides the minimum probability of error.

It is to be noted that if the a priori probabilities and the class conditional densities are estimated from a given dataset, and the Bayes decision rule is implemented using these estimated values (which may be different from the actual values), then the resulting classifier is called the *Bayes maximum likelihood classifier*.

Assuming normal (Gaussian) distribution of patterns, with mean vector μ_i and covariance matrix Σ_i, the Gaussian density $p_i(\mathbf{x})$ may be written as

$$p_i(\mathbf{x}) = \frac{1}{(2\pi)^{\frac{N}{2}}|\Sigma_i|^{\frac{1}{2}}} \exp[-\frac{1}{2}(\mathbf{x}-\mu_i)'\Sigma_i^{-1}(\mathbf{x}-\mu_i)], \qquad (3.7)$$
$$i = 1, 2, \ldots, k.$$

Then, $D_i(\mathbf{x})$ becomes (taking the log)

$$D_i(\mathbf{x}) = \ln P_i - \frac{1}{2}\ln|\Sigma_i| - \frac{1}{2}(\mathbf{x}-\mu_i)'\Sigma_i^{-1}(\mathbf{x}-\mu_i), \qquad (3.8)$$
$$i = 1, 2, \ldots, k.$$

Note that the decision functions in Equation 3.9 are hyperquadrics, since no terms higher than the second degree in the components of $\mathbf{x}$ appear in it. It can thus be stated that the Bayes maximum likelihood classifier for a normal distribution of patterns provides a second-order decision surface between each pair of pattern classes. An important point to be mentioned here is that if the pattern classes are truly characterized by normal densities, then, on average, no other surface can yield better results. In fact, the Bayes classifier designed over known probability distribution functions provides, on average, the best performance for datasets which are drawn according to the distribution. In such cases, no other classifier can provide better performance, on average, because the Bayes classifier gives the minimum probability of misclassification over all decision rules.

3.3.1.3 Decision Trees

A decision tree is an acyclic graph, of which each internal node, branch and leaf node represents a test on a feature, an outcome of the test and a class or class distribution, respectively. It is easy to convert any decision tree into classification rules. Once the training data points are available, a decision tree can be constructed from them from top to bottom using a recursive divide and conquer algorithm. This process is also known as decision tree induction. A version of ID3 [356], a well-known decision tree induction algorithm, is described below.

Decision_tree_induction (training data points, features)

1. Create a node N.

2. If all training data points belong to the same class (C) then return N as a leaf node labeled with class C.
3. If cardinality (features) is NULL then return N as a leaf node with the class label of the majority of the points in the training dataset.
4. Select a feature (F) corresponding to the highest information gain.
5. Label node N with F.
6. For each known value f_i of F, partition the data points as s_i.
7. Generate a branch from node N with the condition feature $= f_i$.
8. If s_i is empty then attach a leaf labeled with the most common class in the data points.
9. Else attach the node returned by Decision_tree_induction(s_i,(features-F)).

The information gain of a feature is measured in the following way. Let the training dataset (D) has n points with k distinct class labels. Moreover, let n_i be the number of data points belonging to class C_i (for $i = 1,2,\ldots,k$). The expected information needed to classify the training dataset is

$$I(n_1,n_2,\ldots,n_k) = -\sum_{i=1}^{k} p_i \log_b(p_i), \tag{3.9}$$

where $p_i \ (= \frac{n_i}{n})$ is the probability that a randomly selected data point belongs to class C_i. In the case where the information is encoded in binary, the base b of the log function is set to 2. Let the feature space be d-dimensional, i.e., F has d distance values $\{f_1, f_2, \ldots, f_d\}$, and these are used to partition the data points D into s subsets $\{D_1, D_2, \ldots, D_s\}$. Moreover, let n_{ij} be the number of data points of class C_i in a subset D_j. The entropy or expected information based on the partition by F is given by

$$E(A) = \sum_{j=1}^{s} \left(\frac{n_{1j}, n_{2j} \ldots n_{kj}}{n}\right) I(n_{1j}, n_{2j} \ldots n_{kj}), \tag{3.10}$$

where

$$I(n_{1j}, n_{2j} \ldots n_{kj}) = -\sum_{j=1}^{k} p_{ij} \log_b(p_{ij}). \tag{3.11}$$

Here, p_{ij} is the probability that a data point in D_i belongs to class C_i. The corresponding information gain by branching on F is given by

$$Gain(F) = I(n_1, n_2, \ldots, n_k) - E(A). \tag{3.12}$$

The ID3 algorithm finds the feature corresponding to the highest information gain and chooses it as the test feature. Subsequently a node labeled with this feature is created. For each value of the attribute, branches are generated and accordingly the data points are partitioned.

Due to the presence of noise or outliers some of the branches of the decision tree may reflect anomalies causing the overfitting of the data. In these circumstances tree-pruning techniques are used to remove the least reliable branches, which allows better classification accuracy as well as convergence.

For classifying unknown data, the feature values of the data point are tested against the constructed decision tree. Consequently, a path is traced from the root to the leaf node that holds the class prediction for the test data.

3.3.1.4 Support Vector Machine Classifier

Viewing the input data as two sets of vectors in d-dimensional space, a Support Vector Machine (SVM) classifier [115, 417] constructs a maximally separating hyperplane to separate the two classes of points in that space. On each side of the separating hyperplane, two parallel hyperplanes are constructed that are pushed up against the two classes of points. A good separation is achieved by the hyperplane that has the largest distance from the neighboring data points of both the classes. A larger distance between these parallel hyperplanes indicates better generalization error of the classifier. Fundamentally the SVM classifier is designed for two-class problems. It can be extended for multi-class problems by designing a number of two-class SVMs.

Suppose a dataset contains n feature vectors $< x_i, y_i >$, $i = 1, 2, \ldots, n$, where $y_i \in \{+1, -1\}$ denotes the class label for the data point x_i. The problem of finding the weight vector w can be formulated as minimizing the following function

$$L(w) = \frac{1}{2}||w||^2 + C \sum_{i=1}^{n} \xi_i \qquad (3.13)$$

subject to $y_i[w.\phi(x_i) + b] \geq 1 - \xi_i$, $i = 1, \ldots, n$, $\xi_i \geq 0$. Here, b is the bias and the function $\phi(x)$ maps the input vector to the feature vector. The dual formulation is given by maximizing the following

$$Q(\lambda) = \sum_{i=1}^{n} \lambda_i - \frac{1}{2} \sum_{i=1}^{n} \sum_{j=1}^{n} y_i y_j \lambda_i \lambda_j \kappa(x_i, x_j) \qquad (3.14)$$

subject to $\sum_{i=1}^{n} y_i \lambda_i = 0$ and $0 \leq \lambda_i \leq C$, $i = 1, \ldots, n$. The parameter C, the regularization parameter, controls the trade-off between the complexity of the SVM and the misclassification rate. Only a small fraction of the λ_i coefficients are nonzero. The corresponding pairs of x_i entries are known as support vectors and they fully define the decision function. Geometrically, the support vectors are the points lying near the separating hyperplane. Here $\kappa(x_i, x_j) = \phi(x_i).\phi(x_j)$ is the *kernel function*.

Kernel functions map the input space into higher-dimensional space. Linear, polynomial, sigmoidal, and radial basis functions (RBF) are examples of kernel functions. These are defined as follows:

Linear: $\kappa(x_i, x_j) = x_i^T x_j$.

Polynomial: $\kappa(x_i, x_j) = (\gamma x_i^T x_j + r)^d$.

Sigmoidal: $\kappa(x_i, x_j) = \tanh(\kappa(x_i^T x_j) + \theta)$.

Radial Basis Function (RBF): $\kappa(x_i, x_j) = e^{-\gamma |x_i - x_j|^2}$.

The two-class version of SVM can be extended to deal with the multi-class classification problem by designing a number of one-against-all two-class SVMs [115, 219]. For example, a K-class problem is handled with K two-class SVMs, each of which is used to separate a class of points from all the remaining points.

Some other classification approaches are based on learning classification rules, Bayesian belief networks [93], neural networks [122, 208, 340], genetic algorithms [45–48, 51, 336], and so on.

3.3.2 Regression

Regression is a technique used to learn the relationship between one or more independent (or *predictor*) variables and a dependent (or *criterion*) variable. The simplest form of regression is linear regression, where the relationship is modeled with a straight line learned using the training data points as follows.

Let us assume that for the input vector X $(x_1, x_2, \ldots, x_n)$ (known as the predictor variable), the value of the vector Y (known as the response variable) $(y_1, y_2, \ldots, y_n)$ is known. A straight line through the vectors X, Y can be modeled as $Y = \alpha + \beta X$ where α and β are the regression coefficients and slope of the line, computed as

$$\beta = \frac{\sum_{i=1}^{n}(x_i - x^*)(y_i - y^*)}{\sum_{i=1}^{n}(x_i - x^*)^2}, \tag{3.15}$$

$$\alpha = y^* - \beta x^*, \tag{3.16}$$

where x^* and y^* are the averages of $(x_1, x_2, \ldots, x_n)$ and $(y_1, y_2, \ldots, y_n)$, respectively.

An extension of linear regression which involves more than one predictor variable is multiple regression. Here, a response variable can be modeled as a linear function of a multidimensional feature vector. For example,

$$Y = \alpha + \beta_1 X_i + \beta_2 X_2 + \ldots + \beta_n X_n \tag{3.17}$$

is a multiple regression model based on n predictor variables $(X_1, X_2, \ldots X_n)$. For evaluating α, β_1 and β_2, the least square method can be applied.

Data having nonlinear dependence may be modeled using polynomial regression. This is done by adding polynomial terms to the basic linear model. Transformation can be applied to the variable to convert the nonlinear model into a linear one. Subsequently, it can be solved using the method of least squares. For example, consider the following polynomial

$$Y = \alpha + \beta_1 X + \beta_2 X^2 + \ldots + \beta_n X^n. \tag{3.18}$$

The above polynomial can be converted to the following linear form by defining the new variables $X_1 = X, X_2 = X^2, \ldots, X_n = X^n$ and can be solved using the method of least squares:

$$Y = \alpha + \beta_1 X_1 + \beta_2 X_2 + \ldots + \beta_n X_n. \tag{3.19}$$

3.3.3 Association Rule Mining

The principle of the association rule mining problem lies in market basket or transaction data analysis. Most information about the day to day transactions taking place in supermarkets is hidden. For example, a customer who is buying diapers also like to purchase baby food at the same time. Association analysis is the discovery of rules showing attribute–value associations that occur frequently.

Let $I = \{i_1, i_2, \ldots, i_n\}$ be a set of n items and X be an itemset where $X \subset I$. A k-itemset is a set of k items. Let $T = \{(t_1, X_1), (t_2, X_2) \ldots, (t_m, X_m)\}$ be a set of m transactions, where t_i and X_i, $i = 1, 2, \ldots, m$, are the transaction identifier and the associated itemset respectively. The *cover* of an itemset X in T is defined as follows:

$$cover(X, T) = \{t_i | (t_i, X_i) \in T, X \subset X_i\}. \tag{3.20}$$

The *support* of an itemset X in T is

$$support(X, T) = |cover(X, T)| \tag{3.21}$$

and the *frequency* of an itemset is

$$frequency(X, T) = \frac{support(X, T)}{|T|}. \tag{3.22}$$

Thus, support of an itemset X is the number of transactions where all the items in X appear in each transaction. The frequency of an itemset is the probability of its occurrence in a transaction in T. An itemset is called frequent if its support in T is greater than some threshold min_sup. The collection of frequent itemsets with respect to a minimum support min_sup in T, denoted by $\mathscr{F}(T, min\_sup)$, is defined as

$$\mathscr{F}(T, min\_sup) = \{X \subseteq I, support(X, T) > min\_sup\}. \tag{3.23}$$

The objective in association rule mining is to find all rules of the form $X \Rightarrow Y$, $X \cap Y = \emptyset$ with probability c, indicating that if itemset X occurs in a transaction, the itemset Y also occurs with probability c. X and Y are called the *antecedent* and the *consequent* of the rule respectively. Support of a rule denotes the percentage of transactions in T that contains both X and Y. This is taken to be the probability $P(X \cup Y)$. An association rule is called *frequent* if its support exceeds a minimum value *min_sup*.

The confidence of a rule $X \Rightarrow Y$ in T denotes the percentage of the transactions in T containing X that also contains Y. It is taken to be the conditional probability $P(X|Y)$. In other words,

$$confidence(X \Rightarrow Y, T) = \frac{support(X \cup Y, T)}{support(X, T)}. \tag{3.24}$$

A rule is called *confident* if its confidence value exceeds a threshold *min_conf*. Formally the association rule mining problem can be defined as follows: Find the set of all rules R of the form $X \Rightarrow Y$ such that

$$R = \{ X \Rightarrow Y | X, Y \subset I, X \cap Y = \emptyset, X \cup Y \in \mathscr{F}(T, min\_sup),$$
$$confidence(X \Rightarrow Y, T) > min\_conf \}. \tag{3.25}$$

Other than support and confidence measures, there are other measures of interestingness associated with association rules [397].

Generally the association rule mining process consists of the following two steps [180, 212]:

1. Find all frequent itemsets,
2. Generate strong association rules from the frequent itemsets.

Apart from the above-mentioned general framework adopted in most of the research in association rule mining, there is another approach for immediately generating a large subset of all association rules [431].

The number of itemsets grows exponentially with the number of items $|I|$. A commonly used algorithm for generating frequent itemsets is the *Apriori* algorithm [4, 5]. It is based on the idea that if even one subset of an itemset X is not frequent, then X cannot be frequent. It starts from all itemsets of size 1, and proceeds in a recursive fashion. If any itemset X is not frequent, then that branch of the tree is pruned, since any possible superset of X can never be frequent.

3.3.4 Clustering

Clustering [15, 135, 205, 237, 408] is an important unsupervised classification technique where a set of patterns, usually vectors in multidimensional space, are grouped into clusters in such a way that patterns in the same cluster are similar in some sense and patterns in different clusters are dissimilar in the same sense. There are different clustering techniques available in the literature, such as hierarchical, partitional, and density-based. Various clustering algorithms and different issues regarding clustering have already been discussed in Chapter 1.

3.3.5 Deviation Detection

Deviation detection, an inseparably important part of KDD, deals with identifying if and when the present data changes significantly from previously measured or normative data. This is also known as the process of detection of outliers. Outliers are those patterns that are distinctly different from the normal, frequently occurring patterns, based on some measurement. Such deviations are generally infrequent or rare. Depending on the domain, deviations may be just some noisy observations that often mislead the standard classification or clustering algorithms, and hence should be eliminated. Alternatively, they may become more valuable than the average dataset because they contain useful information on the abnormal behavior of the system, described by the dataset.

The wide range of applications of outlier detection includes fraud detection, customized marketing, detection of criminal activity in e-commerce, network intrusion detection, and weather prediction. The different approaches for outlier detection can be broadly categorized into three types [197]:

- Statistical approach: Here, the data distribution or the probability model of the dataset is considered as the primary factor.
- Distance-based approach: The classical definition of an outlier in this context is the following: An object O in a dataset T is a $DB(p, D)$ outlier if at least fraction p of the objects in T lies greater than distance D from O [254].
- Deviation-based approach: Deviations from the main characteristics of the objects are basically considered here. Objects that "deviate" from the description are treated as outliers.

Some algorithms for outlier detection in data mining applications may be found in [3, 359].

3.3.6 Feature Selection

Feature selection [280] refers to the selection of an optimum relevant set of features or attributes that are necessary for the recognition process (classification or clustering) and reduce the dimensionality of the measurement space. The objective of feature selection is mainly three fold. Firstly, it is practically and computationally infeasible to process all the features if the number of features is too high. Secondly, many of the given features may be noisy, redundant and irrelevant to the classification or clustering task at hand. Finally, it is a problem when the number of features becomes much larger than the number of input data points. For such cases, reduction in dimensionality is necessary to allow meaningful data analysis [201]. Feature selection facilitates the use of easily computable algorithms for efficient classification or clustering.

The general feature selection problem (Ω, P) is defined formally as an optimization problem: determine the feature set F^* for which

$$P(F^*) = \min_{F \in \Omega}\{P(F,X)\}, \tag{3.26}$$

where Ω is the set of possible feature subsets, F denotes a feature subset and $P : \Omega \times \psi \rightarrow (R)$ denotes a criterion to judge the quality of a feature subset with respect to its utility in classifying the set of points $X \in \psi$. The elements of X, which are vectors in d-dimensional space are projected into the subspace of dimension $d_F = |F| \leq d$ defined by F. P is used to measure the quality of this subspace.

Feature selection can be either *supervised* or *unsupervised*. For the supervised case, the actual class labels of the data points are known. In *filter* approaches to supervised feature selection, features are selected based on their discriminatory power with regard to the target classes. In *wrapper* approaches to supervised feature selection, the utility of F is usually measured in terms of the performance of a classifier by comparing the class labels predicted by the classifier for feature space F with the actual class labels. For the unsupervised case, actual class labels are not available. Hence, in filter approaches to unsupervised feature selection, features are selected based on the distribution of their values across the set of point vectors available. In wrapper-based unsupervised feature selection, the utility of a feature subset F is generally computed in terms of the performance of a clustering algorithm when applied to the input dataset in the feature space F. Unsupervised feature selection is more challenging than supervised feature selection, since actual class memberships are not known. Therefore it is not possible to directly compare the class labels. It is usual to compute some internal cluster validity measures [238] to assess the quality of the chosen feature subset.

3.4 Mining Techniques Based on Soft Computing Approaches

Soft computing [457] is a consortium of methodologies that work synergistically and provides, in one form or another, flexible information processing capabilities for handling real-life ambiguous situations. Its aim is to exploit the tolerance for imprecision, uncertainty, approximate reasoning, and partial truth in order to achieve tractability, robustness, and low-cost solutions. The guiding principle is to devise methods of computation that lead to an *acceptable solution at low cost* by seeking an approximate solution to an imprecisely or precisely formulated problem. In data mining, it is often impractical to expect the optimal or exact solution. Moreover, in order for the mining algorithms to be useful, they must be able to provide good solutions reasonably fast. As such, the requirements of a data mining algorithm are often found to be the same as the guiding principle of soft computing, thereby making the application of soft computing in data mining natural and appropriate.

Some of the main components of soft computing include fuzzy logic, neural networks and probabilistic reasoning, with the latter subsuming belief networks, evolutionary computation and genetic algorithms, chaos theory and parts of learning theory [239,390]. Rough sets, wavelets and other optimization methods such as tabu search, simulated annealing, ant colony optimization, particle swarm optimization and differential evolution are also considered to be components of soft computing. In the following subsections, some of the major components in the soft computing paradigm, viz., fuzzy sets, genetic algorithms and neural networks, are discussed followed by a brief description of their applications in data mining.

3.4.1 Fuzzy Sets

Fuzzy set theory was developed in order to handle uncertainties arising from vague, incomplete, linguistic or overlapping patterns in various problem-solving systems. This approach is developed based on the realization that an object may belong to more than one class, with varying degrees of class membership. Uncertainty can result from the incomplete or ambiguous input information, the imprecision in the problem definition, ill-defined and/or overlapping boundaries among the classes or regions, and the indefiniteness in defining or extracting features and relations among them.

Fuzzy sets were introduced in 1965 by Lotfi A. Zadeh [454–456,458] as a way to represent vagueness in everyday life. We almost always speak in fuzzy terms, e.g., *he is more or less tall, she is very beautiful*. The concepts of *tall* and *beautiful* are fuzzy, and the *gentleman* and *lady* have membership values to these fuzzy concepts indicating their degree of belongingness. Since this theory is a generalization of the classical set theory, it has greater flexibility in capturing various aspects of incompleteness, imprecision or imperfection in information about a situation. It has been applied successfully to computing with words and the matching of linguistic terms for reasoning.

Fuzzy set theory has found a lot of applications in data mining [34, 347, 445]. Examples of such applications may be found in clustering [86,261,348], association rules [89,432], time series [95], and image retrieval [164,305].

3.4.2 Evolutionary Computation

Evolutionary computation (EC) is a computing paradigm comprising problem-solving techniques that are based on the principles of biological evolution. The essential components of EC are a strategy for representing or encoding a solution to the problem under consideration, a criterion for evaluating the fitness or goodness of an encoded solution, and a set of biologically inspired operators applied on the encoded solutions. Because of the robustness and effectiveness of the techniques in the EC family, they have widespread application in various engineering and scientific circles such as pattern recognition, image processing, VLSI design, and embedded and real-time systems. The commonly known techniques in EC are Genetic Algorithms (GAs) [181], evolutionary strategies [29] and genetic programming [258]. Of these, GAs appear to be the most well-known and widely used technique in this computing paradigm.

GAs, which are efficient, adaptive and robust search and optimization processes, use biologically inspired operators to guide the search in very large, complex and multimodal search spaces. As discussed in Chaper 2, in GAs, the genetic information of each individual or potential solution is encoded in structures called *chromosomes*. They use some domain or problem-dependent knowledge for directing the search into more promising areas; this is known as the *fitness function*. Each individual or chromosome has an associated fitness function, which indicates its degree of goodness with respect to the solution it represents. Various biologically inspired operators such as *selection, crossover* and *mutation* are applied on the chromosomes to yield potentially better solutions. GAs represent a form of multi-point, stochastic search in complex landscapes. Applications of genetic algorithms and related techniques in data mining include extraction of association rules [326], predictive rules [161,162,326], clustering [36,38,39,41,44,298,300,302,314], program evolution [361,404] and web mining [330,331,349–351].

3.4.3 Neural Networks

Neural networks can be formally defined as massively parallel interconnections of simple (usually adaptive) processing elements that interact with objects of the real world in a manner similar to that in biological systems. Their origin can be traced to the work of Hebb [209], where a local learning rule is proposed. The benefit of neural nets lies in the high computation rate provided by their inherent massive parallelism. This allows real-time processing of huge datasets with proper hardware backing. All information is stored distributed among the various connection weights. The redundancy of interconnections produces a high degree of robustness resulting in a *graceful degradation* of performance in the case of noise or damage to a few nodes or links.

Neural network models have been studied for many years with the hope of achieving human-like performance (artificially), particularly in the field of pattern recognition, by capturing the key ingredients responsible for the remarkable capabilities of the human nervous system. Note that these models are extreme simplifications of the actual human nervous system. Some commonly used neural networks are the multi-layer perceptron, the Hopfield network, Kohonen's self-organizing maps and the radial basis function network [208].

Neural networks have been widely used in searching for patterns in data [68] because they appear to bridge the gap between the generalization capability of human beings and the deterministic nature of computers. The more important of these applications are rule generation and classification [283], clustering [10], data modeling [278], time series analysis [144, 177, 226] and visualization [257]. Neural networks may be used as a direct substitute for autocorrelation, multivariable regression, linear regression, and trigonometric and other regression techniques [218, 392]. Apart from data mining tasks, neural networks have also been used for data preprocessing, such as for data cleaning and handling missing values. Various applications of supervised and unsupervised neural networks to the analysis of the gene expression profiles produced using DNA microarrays have been studied in [293]. A hybridization of genetic algorithms and perceptrons has been used in [242] for supervised classification in microarray data. Issues involved in research on the use of neural networks for data mining include model selection, determination of an appropriate architecture and training algorithm, network pruning, convergence and training time analysis, data representation and tackling of missing values. Hybridization of neural networks with other soft computing tools such as fuzzy logic, genetic algorithms, rough sets and wavelets has proved to be effective for solving complex problems.

3.5 Major Issues and Challenges in Data Mining

In this section major issues and challenges in data mining regarding underlying data types, mining techniques, user interaction and performance are described [40, 197].

3.5.1 Issues Related to the Underlying Data Types

- *Complex and high-dimensional data:*
  Databases with a very large number of records having high dimensionality (large numbers of attributes) are quite common. Moreover, these databases may contain complex data objects such as hypertext and multimedia, graphical data, transaction data, and spatial and temporal data. Consequently mining these data may re-

quire exploring combinatorially explosive search spaces and may sometimes result in spurious patterns. Therefore, it is important that the algorithms developed for data mining tasks be very efficient and able to also exploit the advantages of techniques such as dimensionality reduction, sampling, approximation methods, incorporation of domain specific prior knowledge, and so on. Moreover, it is essential to develop different techniques for mining different databases, given the diversity of data types and goals.

- *Missing, incomplete and noisy data:*
  Sometime data stored in a database either may not have a few important attributes or may have noisy values. These can result from operator error, actual system and measurement failure, or a revision of the data collection process. These incomplete or noisy objects may confuse the mining process causing the model to overfit or underfit the data. As a result, the accuracy of the discovered patterns can be poor. Data cleaning techniques, more sophisticated statistical methods to identify hidden attributes and their dependencies, and techniques for identifying outliers are therefore required.

- *Handling changing data and knowledge:*
  Situations where the dataset is changing rapidly (e.g., time series data or data obtained from sensors deployed in real-life situations) may make previously discovered patterns invalid. Moreover, the variables measured in a given application database may be modified, deleted or augmented with time. Incremental learning techniques are required to handle these types of data.

3.5.2 Issues Related to Data Mining Techniques

- *Parallel and distributed algorithms:*
  The very large size of the underlying databases, and the complex nature of the data and their distribution motivated researchers to develop parallel and distributed data mining algorithms.

- *Problem characteristics:*
  Though a number of data mining algorithms have been developed, there is none that is equally applicable to a wide variety of datasets and can be called the universally best data mining technique. For example, there exist a number of classification algorithms such as decision tree classifiers, nearest neighbor classifiers, and neural networks. When the data is high dimensional with a mixture of continuous and categorical attributes, decision tree-based classifiers may be a good choice. However they may not be suitable when the true decision boundaries are nonlinear multivariate functions. In such cases, neural networks and probabilistic

models may be a better choice. Thus, the particular data mining algorithm chosen is critically dependent on the problem domain.

3.5.3 Issues Related to Extracted Knowledge

- *Mining different types of knowledge:*
  Different users may be interested in different kinds of knowledge from the same underlying database. Therefore, it is essential that the data mining methods allow a wide range of data analysis and knowledge discovery tasks such as data characterization, classification and clustering.

- *Understandability of the discovered patterns:*
  In most of the applications, it is important to represent the discovered patterns in more human understandable form such as natural language, visual representation, graphical representation, and rule structuring. This requires the mining techniques to adopt more sophisticated knowledge representation techniques such as rules, trees, tables, and graphs.

3.5.4 Issues Related to User Interaction and Prior Knowledge

- *User interaction:*
  The knowledge discovery process is interactive and iterative in nature as sometimes it is difficult to estimate exactly what can be discovered from a database. User interaction helps the mining process to focus the search patterns, appropriately sampling and refining the data. This in turn results in better performance of the data mining algorithm in terms of discovered knowledge as well as convergence.

- *Incorporation of a priori knowledge:*
  Incorporation of a priori domain-specific knowledge is important in all phases of a knowledge discovery process. This knowledge includes integrity constraints, rules for deduction, probabilities over data and distribution, number of classes, and so on. This a priori knowledge helps with better convergence of the data mining search as well as with the quality of the discovered patterns.

3.5.5 Issues Related to Performance of the Data Mining Techniques

- *Scalability:*
  Data mining algorithms must be scalable in the size of the underlying data, meaning both the number of patterns and the number of attributes. The size of datasets to be mined is usually huge, and hence it is necessary either to design faster algorithms or to partition the data into several subsets, executing the algorithms on the smaller subsets and possibly combining the results [353].

- *Efficiency and accuracy:*
  Efficiency and accuracy of a data mining technique are key issues. Data mining algorithms must be very efficient such that the time required to extract the knowledge from even a very large database is predictable and acceptable. Moreover, the accuracy of the mining system needs to be better than or as good as the acceptable range.

- *Ability to deal with minority classes:*
  Data mining techniques should have the capability to deal with minority or low-probability classes whose occurrence in the data may be rare.

3.6 Recent Trends in Knowledge Discovery

Data mining is widely used in different application domains, where the data is not necessarily restricted to conventional structured types, e.g., those found in relational databases, transactional databases and data warehouses. Complex data that are nowadays widely collected and routinely analyzed include the following.

- *Spatial data:*
  This type of data is often stored in Geographical Information Systems (GIS), where the spatial coordinates constitute an integral part of the data. Some examples of spatial data are maps, preprocessed remote sensing and medical image data, and VLSI chip layouts. Clustering of geographical points into different regions characterized by the presence of different types of land cover, such as lakes, mountains, forests, residential and business areas, and agricultural land, is an example of spatial data mining.

- *Biological data:*
  DNA, RNA and proteins are the most widely studied molecules in biology. A large number of databases store biological data in different forms, such as sequences (of nucleotides and amino acids), atomic coordinates and microarray data (that measure the levels of gene expression). Finding homologous se-

quences, identifying the evolutionary relationships of proteins and clustering gene microarray data are some examples of biological data mining.

- *Multimedia data:*
  This type of data may contain text, image, graphics, video clips, music and voice. Summarizing an article, identifying the content of an image using features such as shape, size, texture and color, summarizing the melody and style of a piece of music, are some examples of multimedia data mining.

- *Time series data:*
  This consists of data that is temporally varying. Examples of such data include financial/stock market data. Typical applications of mining time series data involve prediction of the time series at some future time point given its past history.

- *Web data:*
  The World Wide Web is a vast repository of unstructured information distributed over wide geographical regions. Web data can typically be categorized into those that constitute *Web content* (e.g., text, images, sound clips), those that define *Web structure* (e.g., hyperlinks, tags) and those that monitor *Web usage* (e.g., http logs, application server logs). Accordingly, Web mining can also be classified into Web content mining, Web structure mining and Web usage mining.

In order to deal with different types of complex problem domains, specialized algorithms have been developed that are best suited to the particular problem that they are designed for.

3.7 Summary

This chapter presented the basic concepts and issues in KDD, and also discussed the challenges that data mining researchers are facing. Such challenges arise due to different reasons, such as very high-dimensional and extremely large datasets, unstructured and semi-structured data, temporal and spatial patterns and heterogeneous data. Here, different data mining tasks, such as classification, regression, association rule mining, clustering, deviation detection and feature selection have been described. Also, discussions on existing methods have been provided. Subsequently, a brief introduction of soft computing techniques including fuzzy sets, evolutionary algorithms and neural networks has been given. Moreover, some literature references have been provided on the use of soft computing techniques in data mining problems. Thereafter, various major issues and challenges in data mining, such as data types, methods, extracted knowledge, user interaction, prior knowledge, and performance measures, have been discussed. Finally, recent trends in knowledge discovery from different types of datasets, such as spatial, biological, multimedia, the World Wide Web, have been mentioned.

Chapter 4
Computational Biology and Bioinformatics

4.1 Introduction

In the past few decades major advances in the fields of molecular biology and genomic technology have led to an explosive growth in the biological information generated by the scientific community. Bioinformatics has evolved as an emerging research direction in response to this deluge of information. It is viewed as the use of computational methods to make biological discoveries, and is almost synonymous with computational biology. The main purpose of bioinformatics or computational biology is to utilize computerized databases to store, organize and index the data and to use specialized computational tools to view and analyze the data [40]. The objective of bioinformatics research is discovering new biological insights and creating a global perspective from which unifying principles in biology can be derived. Sequence analysis, phylogenetic/evolutionary trees, protein classification and analysis of microarray data constitute some typical problems of bioinformatics where mining techniques are required for extracting meaningful patterns. The mining tasks often used for biological data include clustering, classification, prediction and frequent pattern identification [426].

The huge amount of biological data stored in repositories distributed all around the globe is often noisy. Moreover, the same information may be stored in different formats. Therefore, data preprocessing tasks such as cleaning and integration are important in this domain [40, 426]. Clustering and classification of gene expression profiles or microarray data is performed in order to identify the genes that may be responsible for a particular trait [60]. Determining or modeling the evolutionary history of a set of species from genomic DNA or amino acid sequences using phylogenetic trees is widely studied in bioinformatics [143]. Classification of proteins and homology modeling are two important approaches for predicting the structure of proteins, and may be useful in drug design [35, 96]. Motif-based classification of proteins is also another important research direction [221]. A motif is a conserved element of a protein sequence that usually correlates with a particular function. Motifs are identified from a local multiple sequence alignment of proteins corresponding to a region whose function or structure is known. Motif identification from a number of protein sequences is another mining task that is important in bioinformatics.

In this chapter, we first discuss the basic concepts of molecular biology and its components such as DNA, RNA and proteins. Thereafter, different important tasks in bioinformatics and computational biology are discussed with examples. Moreover, an overview of recent approaches under different categories of bioinformatics tasks is provided.

4.2 Basic Concepts of Molecular Biology

Deoxyribonucleic acid (DNA) and *proteins* are biological macromolecules built as long linear chains of chemical components. DNA strands consist of a large sequence of nucleotides, or bases. For example, there are more than three billion bases in human DNA sequences. DNA contains templates for the synthesis of proteins, which are essential molecules for any organism. Moreover, DNA acts as a medium to transmit hereditary information (namely, the building plans of proteins) from generation to generation [50, 380]. Proteins are responsible for structural behavior.

Each *nucleotide*, the unit of DNA, consists of three parts: one of two base molecules (a *purine* or a *pyrimidine*), a sugar (ribose in RNA and deoxyribose DNA), and one or more phosphate groups. The purine nucleotides are *adenine* (A) and *guanine* (G), and the pyrimidines are *cytosine* (C) and *thymine* (T). Nucleotides are sometimes called bases. Since DNA consists of two complementary strands bonded together, these units are often called base pairs. The length of a DNA sequence is often measured in thousands of bases. Nucleotides are usually abbreviated by their first letter, and thus a sequence of nucleotides looks like AGGCTCTAGATT. The nucleotides are linked to each other in the polymer by phosphodiester bonds and this bond is directional. Hence a strand of DNA has a head (called the 5′ end) and a tail (the 3′ end).

DNA forms a double helix, that is, two helical (spiral-shaped) strands of the polypeptide, running in opposite directions, held together by hydrogen bonds. Adenines bond exclusively with the thymines (A-T) and guanines bond exclusively with cytosines (G-C). It may be noted that although the sequence of nucleotide bases in one strand of DNA is completely unrestricted, because of these bonding rules the sequence in the complementary strand can be completely determined. This feature makes it possible to make copies of the information stored in the DNA with high fidelity. It also helps during transcription when DNA is transcribed into complementary strands of ribonucleic acid (RNA), which direct the protein synthesis process. The only difference is that in RNA, uracil (U) takes the place of thymine and bonds to adenine.

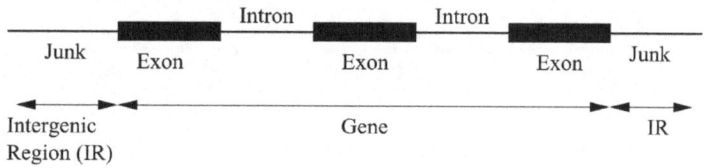

Fig. 4.1 Different parts of DNA

A *gene* is primarily made up of sequences of triplets of the nucleotides (exons). Introns (noncoding sequences) may also be present within genes. Not all portions of DNA sequences are coding. A coding zone indicates that it is a template for a protein. As an example, for the human genome, only 3%–5% of the sequences are coding, i.e., they constitute the genes. The *promoter* is a region before each gene in the DNA that serves as an indication to the cellular mechanism that a gene is ahead. For example, the codon AUG is a protein which codes for methionine and signals the start of a gene. Promoters are key regulatory sequences that are necessary for the initiation of transcription. Transcription is the process in which ribonucleic acid (RNA) is formed from a gene, and through translation, amino acid sequences are formed from RNA. There are sequences of nucleotides within the DNA that are spliced out progressively in the process of transcription and translation. A comprehensive survey of the research done in this field is given in [439]. In brief, the DNA consists of three types of noncoding sequences (see Figure 4.1) as follows [50]:

1. Intergenic regions: Regions between genes that are ignored during the process of transcription.
2. Intragenic regions (or *introns*): Regions within the genes that are spliced out from the transcribed RNA to yield the building blocks of the genes, referred to as *exons*.
3. Pseudogenes: Genes that are transcribed into the RNA and stay there, without being translated, due to the action of a nucleotide sequence.

Proteins are polypeptides, formed within cells as a linear chain of amino acids [380]. Amino acid molecules bond with each other by eliminating water molecules and forming peptides. Twenty different standard *amino acids* (or "residues") are available, which are denoted by 20 different letters of the alphabet. Each of the 20 amino acids is coded by one or more triplets (or *codons*) of the nucleotides making up the DNA. Based on the *genetic code*, the linear string of DNA is translated into a linear string of amino acids, i.e., a protein, via mRNA (messenger RNA) [380]. For example, the DNA sequence GAACTACACACGTGTAAC codes for the amino acid sequence ELHTCN (shown in Figure 4.2) [50].

The process of production of protein from DNA is called the *central dogma* [380] of molecular biology (Figure 4.3). The central dogma of molecular biology states that DNA is transcribed into RNA, which thereafter is translated into pro-

$$\underset{E}{\overset{GAA}{\longleftrightarrow}}\ \underset{L}{\overset{CTA}{\longleftrightarrow}}\ \underset{H}{\overset{CAC}{\longleftrightarrow}}\ \underset{T}{\overset{ACG}{\longleftrightarrow}}\ \underset{C}{\overset{TGT}{\longleftrightarrow}}\ \underset{N}{\overset{AAC}{\longleftrightarrow}}$$

Fig. 4.2 Coding of amino acid sequence from DNA sequence

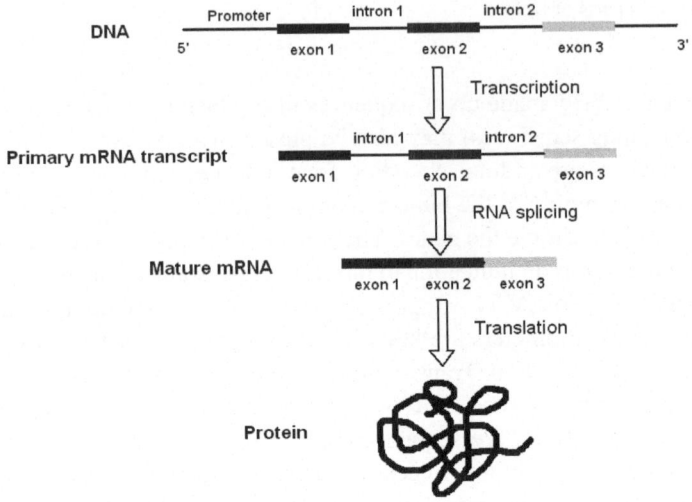

Fig. 4.3 Central dogma of molecular biology

tein. In eukaryotic systems, exons form a part of the final coding sequences (CDSs), whereas introns are transcribed but then edited out by the cellular machinery before the mRNA assumes its final form.

4.3 Important Tasks in Computational Biology

The different tasks in computational biology and bioinformatics can be broadly classified into four different levels, viz., sequence-level, structure-level, expression-based and system-level. In this section, we discuss some important tasks under each category.

4.3.1 Sequence-Level Tasks

Sequence level tasks deal with the biological sequences such as DNA, RNA and proteins. The following is an overview of some sequence level tasks performed in bioinformatics.

4.3.1.1 Genome Sequencing and Fragment Assembly

One important sequence-level task is sequencing the genome, which is the process of determining the precise order of the nucleotide bases, in particular DNA molecules [176, 363]. A substantial amount of manual labor is required for manual sequencing, and it is slow. Therefore it is necessary to develop computational techniques for faster and reliable DNA sequencing.

It is known that it is not possible to read the entire sequence of bases of a DNA molecule or the entire sequence of amino acid residues of a protein molecule, but rather only factors of small length. For example, currently strands of DNA longer than approximately 500 base pairs cannot be sequenced accurately [50, 337]. For sequencing larger strands of DNA, they are first broken into smaller pieces. In the shotgun sequencing method, the DNA is first replicated many times, and then individual strands of the double helix are divided randomly into small fragments. The reconstruction of the original DNA or protein sequence is complicated by other constraints, such as read-errors or unknown orientations of the factors. The problem of reconstruction of the entire DNA or protein sequence from the factors is known as the *fragment assembly problem* (FAP) [192]. FAP is basically a permutation problem, similar in spirit to the traveling salesperson problem (TSP), but with some important differences (circular tours, noise, and special relationships between entities) [380]. FAP plays an important role in genome sequencing. The fragment assembly problem has been tackled by many methods, such as ant colony system [306, 434], neural network [22], and Genetic Algorithms [342, 343].

4.3.1.2 Sequence Alignment

Sequence alignment, another important bioinformatics task, refers to the problem of arranging the sequences of DNA, RNA, or protein to identify sections of similarity that may be a consequence of functional, structural, or evolutionary relationships among the sequences [192, 380]. In an optimal alignment there are the most correspondences and the least differences among the sequences to be aligned. The optimal alignment has the highest score but it may or may not be biologically meaningful. The sequences of nucleotide or amino acid residues to be aligned are typically arranged as rows within a matrix. After that gaps are inserted between the residues in such a manner that identical or similar characters are aligned in corresponding columns.

In multiple sequence alignment [270, 459, 460] a set of sequences are arranged in a manner in which homologous positions are placed in a common column. Among the different conventions for the scoring of a multiple alignment, one approach simply adds up the scores of all the induced pairwise alignments contained in a multiple alignment [50, 337]. For a linear gap penalty, this amounts to scoring each column of the alignment by the sum-of-pairs scores in this column. It is usual not to distinguish between global, local, and other forms of alignment in a multiple alignment, although they could be biologically meaningful. A full set of optimal pairwise alignments in a given set of sequences will usually overdetermine the multiple alignment. For assembling a multiple alignment from pairwise alignments, "closing loops" must be avoided. This means one can put pairwise alignments together as long as no new pairwise alignment is included in a set of sequences which is already part of the multiple alignment.

4.3.1.3 Gene Finding

Another aspect of bioinformatics in sequence analysis is the automatic search for genes and regulatory sequences within a genome. Not all of the nucleotides within a genome are genes [244, 256, 275]. For higher organisms, it has been found that large parts of the DNA do not serve any obvious purpose. These parts of the DNA are called junk DNA. However, junk DNA may contain unrecognized functional elements. An important problem in bioinformatics is the automatic identification of genes from the large DNA sequences. In *extrinsic* (or evidence-based) gene-finding systems, one searches the target genome for sequences that are similar to those with extrinsic evidence. The extrinsic evidence is usually obtained from the known sequence of a messenger RNA (mRNA) or protein product. On the other hand, in *ab initio* gene finding, one searches the genomic DNA sequence alone systematically to find certain telltale signs of protein-coding genes.

A cell mechanism recognizes the beginning of a gene or gene cluster using a promoter and is required for the initiation of the transcription process. The promoter is a region preceding each gene in the DNA, and it serves as an indication of the start of a gene. For example, the codon AUG (which codes for methionine) also signals the start of a gene. Identification of regulatory sites in DNA fragments has become particularly popular with the increasing number of completely sequenced genomes and the mass application of DNA chips. Experimental analyses have spotted no more than 10% of the potential promoter regions, considering that there are at least 30,000 promoters in the human genome, one for each gene [50, 337].

4.3.1.4 Phylogenetic Tree Construction

All species on earth undergo a slow transformation over years, which is called evolution. The study of evolutionary relationships is referred to as phylogenetic analysis. A phylogenetic or evolutionary tree represents the evolutionary relationships among various biological species or other entities that are believed to have a common ancestor [2, 26, 430, 468]. In a phylogenetic tree, each node is called a taxonomic unit. Internal nodes cannot be observed directly and are generally called hypothetical taxonomic units (HTUs). Each node with descendants represents the most recent common ancestor of the descendants. In some trees, lengths of edges correspond to time estimates. The techniques of computational phylogenetics are used to build phylogenetic trees among a nontrivial number of input sequences.

Given the character state matrix or distance matrix for n taxa (objects), the problem of phylogenetic tree reconstruction is to search for the particular permutation of taxa which optimizes the criterion (distance). The problem is equivalent to the problem of traveling salesperson (TSP) [337, 380]. One imaginary city can be associated with each taxon. The distance between two cities can be defined as the data obtained from the data matrix for the corresponding pair of taxa.

4.3.1.5 Protein Sequence Classification

An important problem at the sequence level is the classification of protein sequences. The problem can formally be stated as follows [50, 213, 271, 289, 337]: Given an unlabeled protein sequence S and a known superfamiliy F, determine with certain degree of accuracy whether the protein S belongs to the superfamily F or not. Here, a superfamily refers to a set of proteins that are similar in structure and function. If one has an unclassified protein in hand, the first thing to do is to classify the protein into a known superfamily. This helps us have some idea about the structure and function of the newly discovered protein. The most important real-life application of such a knowledge is possibly in the discovery of new drugs. For example, assume that someone has found sequence S from some disease D and with a classification method it is inferred that S belongs to F. Then, to design a drug for the disease corresponding to D, one may try a combination of existing drugs for F.

Different approaches exist for protein sequence classification [424]. The BLAST (http://blast.ncbi.nlm.nih.gov/) and FASTA (http://fasta.bioch.virginia.edu/) programs have been the major tools to help in analyzing the protein sequence data and interpreting the results in a biologically meaningful manner. BLAST returns a list of *high-scoring segment pairs* from the query sequence and the sequences in the databases. This is done after performing a sequence alignment among them. Faster and more sensitive homology searches based on BLAST have also been devised [287].

Other methods based on Hidden Markov Models (HMMs), neural networks, and multiple sequence alignment techniques are also in vogue. In one of the more recent works [424] Wang et. al. have tried to capture the global and local similarities of protein sequences in inputs to a Bayesian Neural Network (BNN) classifier. The 2-gram encoding scheme, which extracts and counts the occurrences of two consecutive amino acids in a protein sequence, is proposed. They have also compared their technique with BLAST, SAM (http://compbio.soe.ucsc.edu/HMM-apps/) and other iterative methods. In [310], a classification system is presented to create a numerical vector representation of a protein sequence and then classify the sequence into a number of given families. A generalized radial basis function (GRBF) network-based fuzzy rule generation system is proposed in [423] for protein sequence classification.

4.3.2 Structure-Level Tasks

In this level, the main tasks involve three-dimensional structure prediction from one-dimensional sequences, protein folding, structure based protein classification, molecule design and docking etc.

4.3.2.1 DNA Structure Prediction

DNA structure plays an important role in a variety of biological processes [50, 337]. Different dinucleotide and trinucleotide scales have been described to capture various aspects of DNA structure, including base stacking energy, propeller twist angle, protein deformability, bendability, and position preference [33]. The three-dimensional DNA structure and its organization into chromatin fibres is essential for its functions, and is applied in protein binding sites, gene regulation, triplet repeat expansion diseases, and so on. DNA structure depends on the exact sequence of nucleotides and largely on interactions between neighboring base pairs. Different sequences can have different intrinsic structures. The periodic occurrence of bent DNA in phase with the helical pitch causes the DNA to assume a macroscopically curved structure. Compared to rigid and unbent DNA, flexible or intrinsically curved DNA is energetically more favorable for wrapping around histones.

4.3.2.2 RNA Structure Prediction

The single-stranded RNA molecules often need a particular tertiary structure due to their functional form. Secondary structural elements like hydrogen bonds within the molecule act as the scaffold for this structure. This leads to various specialized areas of secondary structure, such as hairpin loops, bulges and internal loops. Finding the three-dimensional structure of the RNA molecule given just the nucleic acid sequence is an important problem. The most popular methods used for solving this problem are the free energy minimization techniques. The lowest free energy structure is determined using dynamic programming algorithms based on a set of nearest neighbor parameters that predict conformational stability [138, 294, 295, 337]. The highest accuracy achieved in free energy minimization is 73% for predicting known base pairs for a diverse set of sequences having a length of 700 nucleotides, with structures determined by comparative analysis [295]. The possibilities of using Genetic Algorithms for the prediction of RNA secondary structure have been studied in [54, 190, 329, 436].

4.3.2.3 Protein Structure Prediction and Protein Folding

Another crucial application of bioinformatics is protein structure prediction. The amino acid sequence of a protein (the primary structure) can be easily determined from the sequence on the gene that codes for it. In many cases, with certain exceptions, the primary structure uniquely determines a structure in its native environment. To understand the function of a protein, knowledge of this structure is very important. The structural information is usually classified as one of secondary, tertiary and quaternary structures (Figure 4.4).

The secondary structure of a protein is the spatial arrangement of the atoms constituting the main protein backbone. Linus Pauling was the first to develop a hypothesis for different potential protein secondary structures [345]. He developed the α-helix structure and later the β-sheet structure for different proteins. An α-helix is a spiral arrangement of the protein backbone in the form of a helix with hydrogen bonding between side chains. The β-sheets consist of parallel or anti-parallel strands of amino acids linked to adjacent strands by hydrogen bonding. Collagen is an example of a protein with β-sheets serving as its secondary structure. The super-secondary structure (or motif) is the local folding pattern built up from particular secondary structures. For example, the EF-hand motif consists of an α-helix, followed by a turn, followed by another α-helix. A tertiary structure is formed by packing secondary structural elements linked by loops and turns into one or several compact globular units called *protein domains*, i.e., by the folding of the entire protein chain. A final protein may contain several protein subunits (more than one amino acid chain) arranged in a quaternary structure [50, 337]. Although there has been immense interest in the problem of protein structure prediction, a general solution to the structure prediction problem has not come out yet.

The process of protein folding refers to a physical process by which a polypeptide folds into its functional three-dimensional structure from a random coil [12]. When first translated from mRNA to a linear sequence of amino acid residues, each protein has a structure of unfolded polypeptide or random coil. Thereafter, amino

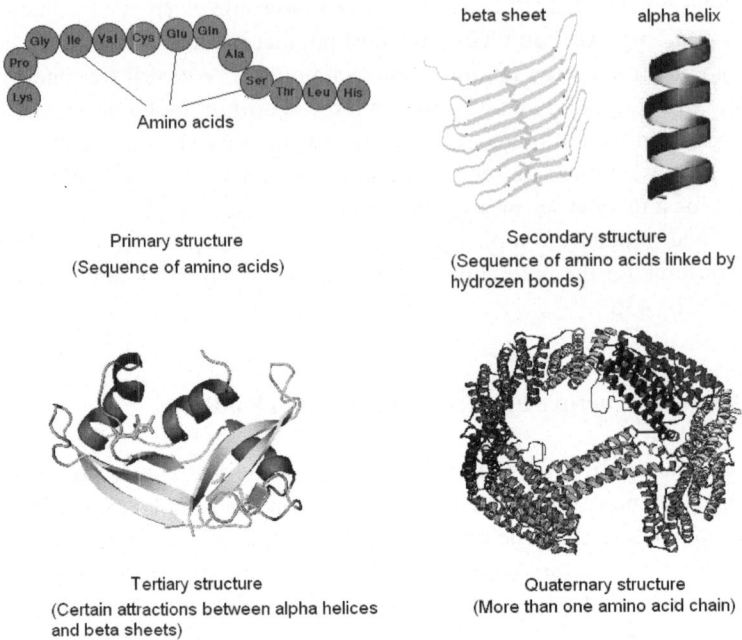

Fig. 4.4 Different levels of protein structures

acids interact with each other through hydrophobic interaction, hydrogen bonding, electrostatic, and other Van der Waals-type interactions to produce a well-defined three-dimensional structure, the folded protein. The final three-dimensional structure is also known as the native state. The amino acids present in the polypeptide chain are responsible for determining the final three-dimensional structure. The intended three-dimensional shape is mandatory for most of the proteins for functioning [149]. Failure in folding into the correct shape sometimes yields inactive proteins with different properties, including toxic prions.

A popular way of predicting protein structure is to use the notion of homology. In genomics, homology is used to predict the function of a gene. If two genes have homologous sequences, then their functions are also likely to be similar. In the same way, homology can be used to determine which parts of a protein are important in structure formation and interaction with other proteins. This information is used to predict the structure of a protein when the structure of a homologous protein is known. The process is called homology modeling. This technique is very reliable. Many other efforts are under way to predict the structure of a protein given its primary sequence [50, 337]. A typical computation of protein folding would require computing all the spatial coordinates of atoms in a protein molecule, starting with an initial configuration and working up to a final minimum-energy folding configuration [380]. Sequence similarity methods can predict the secondary and tertiary structures based on the homology with known proteins. Secondary structure prediction methods include Chou-Fasman [101], neural network [354, 367], nearest neighbor [374, 375] and Garnier-Osguthorpe-Robson (GOR) [170]. Tertiary structure prediction methods are based on energy minimization, molecular dynamics, and stochastic searches of conformational space. GAs and other evolutionary algorithms have also been used in protein structure prediction problems [56, 84, 110, 231, 248, 259, 332, 357, 378].

4.3.2.4 Structure-Based Protein Classification

Although the classification of protein structures is a different task, it has a high overlap with the classification of protein sequences. For sequence classification, the methods of sequence comparison, usually via pairwise alignments of amino acid sequences, are heavily used. Similarly, pairwise structural alignments are the most used techniques for structural classification of proteins. It is very important to devise some methods for automatic classification of protein structures as different aspects of a protein structure are relevant to different biological problems and the large-scale protein structure databases are growing rapidly.

4.3.2.5 Molecular Design and Docking

When two molecules are in close proximity, it can be energetically favorable for them to bind together tightly. The molecular docking problem is the prediction of the energy and the physical configuration of binding between two molecules. A typical application is in drug design, in which one might dock a small molecule that is a described drug to an enzyme one wishes to target [50, 337]. For example, the HIV protease is an enzyme in the AIDS virus that is essential for its replication. The chemical action of the protease takes place at a localized active site on its surface. HIV protease inhibitor drugs are small molecules that bind to the active site in the HIV protease and stay there, so that the normal functioning of the enzyme is prevented. Docking software allows us to evaluate a drug design by predicting whether it will be successful in binding tightly to the active site in the enzyme. Based on the success of the docking, and the resulting docked configuration, designers can refine the drug molecule [274].

Molecular design and docking is a difficult optimization problem, requiring efficient sampling across the entire range of positional, orientational, and conformational possibilities [440]. The major problem in molecular binding is that the search space is very large and the computational cost increases tremendously with the growth of the degrees of freedom. A docking algorithm must deal with two distinct issues: a sampling of the conformational degrees of freedom of molecules involved in the complex, and an objective function (OF) to assess its quality.

4.3.3 Expression-Based Tasks

These tasks deal with computational analysis of expression levels of different biomolecules. Different tasks in this category include measuring the expression of different biomolecules, clustering of gene expression data, classification of gene expression data, and gene ordering.

Gene expression is the process by which the encoded information in a gene is converted into a protein or RNA. Expressed genes include those that are transcribed into mRNA and then translated into protein. Even the genes that are transcribed into RNA but not translated into protein (e.g., transfer and ribosomal RNAs) are considered to be expressed. It should be noted that not all genes are expressed. Gene expression involves the study of the expression level of genes in cells under different experimental conditions or in different kinds of cells. Conventionally, it is expected that gene products that have similar expression profiles are also functionally similar.

The classical approach to genomic research is based on local study and collection of data on single genes. The advent of microarray technology has now made it possible to have a global and simultaneous view of the expression levels for many thousands of genes over different time points during some biological processes. Advances in microarray technology in recent years have major impact in medical diagnosis, characterizing various gene functions, understanding different molecular biological processes, and so on [13, 100, 102, 151, 184]. Microarray technology allows expression levels of thousands of genes to be measured at the same time.

A microarray [85] is a small chip onto which a large number of DNA molecules (probes) are attached in fixed grids. The chip is made of chemically coated glass, nylon, membrane or silicon. Each grid cell of a microarray chip corresponds to a DNA sequence. There are mainly two types of microarrays, viz., two-channel microarrays and single-channel microarrays [142]. In two-channel microarrays (also called two-color microarrays), two mRNA samples are reverse-transcribed into cDNAs (targets) labeled using different fluorescent dyes (red fluorescent dye Cy5 and green fluorescent dye Cy3). Due to the complementary nature of the base pairs, the cDNA binds to specific oligonucleotides on the array. In the subsequent stage, the dye is excited by a laser so that the amount of cDNA can be quantified by measuring the fluorescence intensities. The log ratio of the two intensities is used as the gene expression profile.

$$gene\ expression\ level = \log_2 \frac{Intensity(Cy5)}{Intensity(Cy3)}. \tag{4.1}$$

Although absolute levels of gene expression may be determined using the two-channel microarrays, the system is more useful for the determination of relative differences in gene expression within a sample and between samples.

Single-channel microarrays (also called one-color microarrays) are prepared to estimate the absolute levels of gene expression, thus requiring two separate single-dye hybridizations for the comparison of the two sets of conditions. As only a single dye is used, the data represent absolute values of gene expression. An advantage of single-channel microarrays is that data are more easily compared to arrays from different experiments. However, in the single-channel system, one needs twice as many microarrays to compare the samples within an experiment.

A microarray gene expression dataset consisting of n genes and m time points or experimental conditions is usually expressed as a real-valued $n \times m$ matrix $M = [g_{ij}]$, $i = 1, 2, \ldots, n$, $j = 1, 2, \ldots, m$. Here each element g_{ij} represents the expression level of the ith gene at the jth time point/experimental condition:

$$M = \begin{bmatrix} g_{11} & g_{12} & \cdots & g_{1m} \\ g_{21} & g_{22} & \cdots & g_{2m} \\ \cdots & \cdots & \cdots & \cdots \\ g_{n1} & g_{n2} & \cdots & g_{nm} \end{bmatrix}.$$

The raw gene expression data consists of noise, some variations arising from biological experiments and missing values. Hence before further analysis, the data is preprocessed. Two widely used preprocessing techniques are missing value estimation and standardization. Standardization is a statistical tool for transforming data into a format that can be used for meaningful analysis [381]. Normalization is a useful standardization process by which each row of the matrix M is standardized to have mean 0 and variance 1. This is done by subtracting the mean of each row from each element of that row and dividing it by the standard deviation of the row.

4.3.3.1 Classification in Microarray Gene Expression Data

Supervised classification is usually used to classify the tissue samples into two classes, viz., normal (benign) and cancerous (malignant), or into their subclasses, considering the genes as classification features. For successful diagnosis and treatment of cancer, a reliable classification of tumors is needed. Classical methods for classifying human malignancies depends on various features such as morphological, clinical, and molecular. Despite recent advances, still there are uncertainties in diagnoses. Moreover, the existing classes may be heterogeneous and may comprise diseases that are molecularly distinct. DNA microarrays, which monitor gene expression profiles on a genomic scale, may be utilized to characterize the molecular variations among tumors. This leads to a better and more reliable classification of tumors. It helps to identify the marker genes that are mostly responsible for distinguishing different tumor classes. There are several classification approaches studied by bioinformatics researchers, such as Artificial Neural Networks [109, 227, 370], Support Vector Machines [167, 193], and hybrid methods [11, 425, 428].

4.3.3.2 Clustering and Biclustering in Microarray Gene Expression Data

Clustering is commonly used in microarray experiments to identify groups of genes that share similar expression. Genes that are similarly expressed are often coregulated and involved in the same cellular processes. Therefore, clustering suggests functional relationships between groups of genes. It may also help in identifying promoter sequence elements that are shared among genes. In addition, clustering can be used to analyze the effects of specific changes in experimental conditions, and may reveal the full cellular responses triggered by those conditions. In another study, clustering techniques were used on gene expression data for distinguishing tumor and normal colon tissue probed by oligonucleotide arrays. Here, clustering is done on the experimental conditions or samples, and genes are used as the features. Cluster validation is essential, from both the biological and statistical viewpoints.

Some early works dealt with visual analysis of gene expression patterns to group the genes into functionally relevant classes [13, 100, 102]. However, as these methods were very subjective, standard clustering methods, such as K-means [211], fuzzy C-means [132], hierarchical methods [151], Self-Organizing Maps (SOM) [396], graph-theoretic approaches [207], simulated annealing-based approaches [14, 286] and Genetic Algorithm- (GA-)based clustering methods [288, 313] have been utilized for clustering gene expression data. A detailed review of microarray data clustering algorithms can be found in [241]

As clustering is used to group a set of genes over a complete set of experimental conditions or vice versa, clustering algorithms fail to find a subset of genes that are coregulated and coexpressed in a subset of experimental conditions, which is more practical from a biological point of view. To capture such local patterns, biclustering [318] is used. Biclustering algorithms can find a subset of genes that are similarly expressed across a subset of experimental conditions and hence better reflect the biological reality.

In recent years, several studies have been made by researchers for biclustering in microarray data. Biclustering was first introduced in [206] by the name of direct clustering. The goal was to discover a set of submatrices with zero variance, i.e., with constant values. One of the earlier works on biclustering in the context of microarray data can be found in [94], where the *mean squared residue (MSR)* measure was used to compute the coherence among a group of genes. The algorithm developed by Cheng and Church in [94] was based on a greedy search technique guided by a heuristic. In [173], a coupled two-way clustering (CTWC) method has been proposed that uses hierarchical clustering in both dimensions and combines them to obtain the biclusters. In [272] the authors have used plaid models, where each element of the data matrix is viewed as a superposition of layers (bicluster) and the data matrix is described as a linear function of layers. The concept of layers is utilized to compute the values of the data matrix. An improved version of Cheng and Church's algorithm called Flexible Overlapped biclustering (FLOC), which deals with the missing values, is proposed in [446]. A Random Walk-based biclustering (RWB) method based on a greedy technique and enriched with a local search strategy to escape poor local optima is proposed in [23]. In [69], a Genetic Algorithm- (GA-) based biclustering algorithm has been presented that uses mean squared residue as a fitness function to be minimized. A bipartite graph-based model called Statistical-Algorithmic Method for Bicluster Analysis (SAMBA) has been proposed for biclustering in [398]. In [450], biclustering based on projected clusters is presented. In this work a measure called relevance index that uses local and global variances along dimensions is utilized. A co-clustering technique (COCLUS) based on minimizing the mean squared residue is proposed in [99] to discover non-overlapping biclusters. In [76], a simulated annealing-based technique is presented that is also based on minimizing the MSR.

4.3.3.3 Gene Selection

It has been observed that all the genes in a microarray experiment are not equally important for classification or clustering. Indeed some of them may be redundant or irrelevant. Therefore, it is important to select the set of informative and relevant genes on which the actual classification or clustering algorithms may be applied. Hence, discriminative gene selection is an important task in microarray data analysis. Gene selection can be either supervised using the class information of the samples while selecting the genes [80, 145, 193, 247, 267, 276, 362], or it can be unsupervised not using any class information [160, 251, 279].

4.3.3.4 Gene Ordering

Gene ordering refers to the problem of arranging the genes in a microarray dataset based on their similarity in expression patterns. A good solution to the gene ordering problem (i.e., finding an optimal order of DNA microarray data) will have similar genes grouped together in clusters. A notion of distance must thus be defined in order to measure similarity among genes. A simple measure is the Euclidean distance (other options are possible using Pearson correlation, absolute correlation, Spearman rank correlation, etc.). One can thus construct a matrix of inter-gene distances. Using this matrix one can calculate the total distance between adjacent genes and find that permutation of genes for which the total distance is minimized. The problem can be mapped into the TSP problem as well [337, 380].

4.3.4 System-Level Tasks

The system-level tasks include the dynamics of intra- and intercellular processes that determine cell function, gene regulatory networks and metabolic pathways. Related tasks in this category are the study of survival prediction, cancer prediction, drug response, drug discovery, drug administration, and schedule optimization. These tasks together have formed a new area of bioinformatics research known as *systems biology*.

4.3.4.1 Gene Regulatory Networks and Metabolic Pathway Identification

Inferring a gene regulatory network from gene expression data obtained by a DNA microarray is considered one of the most challenging problems in the field of bioinfomatics [8, 50, 337]. An important and interesting question in biology, regarding the variation of gene expression levels, is how genes are regulated. Since almost all cells in a particular organism have an identical genome, differences in gene expression, and not the genome content, are responsible for cell differentiation during the life of the organism.

For gene regulation, an important role is played by a type of proteins called transcription factors [380] . The transcription factors bind to specific parts of the DNA, called transcription factor binding sites (i.e., specific, relatively short combinations of A, T, C and G), which are located in promoter regions. Specific promoters are associated with particular genes and are generally not too far from their respective genes, although some regulatory effects can be located as far as 30,000 bases away, which makes the definition of the promoter difficult.

Transcription factors control gene expression by binding to the gene's promoter and either activating (switching on) the gene or repressing it (switching it off). Transcription factors are gene products themselves, and therefore, in turn, can be controlled by other transcription factors. Transcription factors can control many genes, and many genes are controlled by combinations of more than one transcription factor. Feedback loops are possible, leading to formation of gene regulation networks. Microarrays and computational methods play a crucial role in attempts to reverse engineer gene networks from various observations.

4.3.4.2 Survival Prediction

Microarray gene expression data have been widely used for survival analysis of cancer patients [416]. However, groups of predictive genes provided by clustering based on differential expressions do not share many overlapping genes. Thus, for independent data, they exhibit less successful predictive power [150]. Therefore, these prediction methods do not provide very reliable and robust ways for survival analysis. In general, gene expression-based prediction models do not consider the existing biological knowledge in the form of biological pathways consisting of interactions between genes and proteins. This pathway information is believed to be more informative than the gene expression information [179].

There have been several recent developments regarding the use of biological pathway and regulatory information along with microarray expression data in predicting the survival of patients. In [92,228], an approach based on the protein-protein interaction network is proposed. This approach effectively discriminates metastatic and non-metastatic breast tumors [103]. A gene pathway-based analysis is used in [139, 179, 360]. In these works, biological pathways are identified by scoring the coherence in the expression changes among their member genes using microarray gene expression data. These methods allow a more biology-driven analysis of microarray data. In [461], a systems biology-based approach is developed by using either combined gene sets and the protein interaction network or the protein network alone to identify common prognostic genes based on microarray gene expression data. The use of biological pathways, regulatory and interaction information along with microarray gene expression data is getting more and more popular in survival prediction analysis.

4.3.4.3 Drug Discovery

Traditional drug discovery processes are not only complex and expensive, they are usually time consuming also. For example, the discovery of a new drug and its commercial application can take up to 5–7 years. Use of computational biology approaches can reduce both the cost and time taken for drug discovery [70,98]. There are various approaches of systems biology for drug discovery, viz., metabolic pathway elucidation [323], disease pathway identification [70], identification of target proteins [31], and identification of bioactive molecules [178]. A comprehensive discussion on this topic is available in [70]. As computational techniques are becoming more sophisticated and automated, their ability to handle huge cell data streams is increasing day by day. Therefore computational systems biology shows a great promise for the development and discovery of new medicines and for having a great impact in biomedical research.

4.4 Summary

The field of bioinformatics has become an emergent area of research and has already generated a vast literature from the sequence level through the systems level. The increasing availability of annotated genomic sequences has resulted in the introduction of computational genomics and proteomics, of large-scale analysis of complete genomes, and of the proteins that they encode for relating specific genes to diseases. The application of computational tools in biology helps us understand different biological processes well and assists experimental analysis by utilizing recent data mining technologies.

In this chapter we have first described the fundamental concepts of molecular biology, such as the structures of DNA, RNA, genes, and proteins and the process of central dogma by which informative parts of DNA are converted into protein. Thereafter, important bioinformatics tasks have been categorized into four different level, viz., sequence-level, structure-level, expression-based and system-level. Under sequence-level tasks, there are several subtasks such as fragment assembly, sequence alignment, gene finding, phylogenetic tree construction and protein sequence classification. The structure-level tasks include 3D structure prediction of DNA, RNA and proteins, structure-based protein classification and molecular design and docking. The expression-based tasks mainly deal with microarray gene expression data, and there are several tools for analyzing gene expression patterns like classification, clustering, gene selection and gene ordering. System-level tasks deal with the dynamics of cellular processes that determine cell function, metabolic pathways and gene regulatory networks. In this chapter, we have reviewed and described these important tasks of bioinformatics.

Chapter 5
Multiobjective Genetic Algorithm-Based Fuzzy Clustering

5.1 Introduction

In this chapter, the problem of clustering is posed as a multiobjective optimization problem where some measures of fuzzy cluster validity (goodness) are optimized. Conventional Genetic Algorithms, a popular search and optimization tool, are usually used to perform the clustering, taking some cluster validity measure as the fitness function. However, there is no single validity measure that works equally well for different kinds of datasets. It may be noted that it is also extremely difficult to combine the different measures into one, since the modality for such combination may not be possible to ascertain, while in some cases the measures themselves may be incompatible. Thus it is natural to simultaneously optimize a number of such measures that can capture different characteristics of the data. Hence it is wise to use some multiobjective optimization (MOO) algorithm to tackle the problem of partitioning where a number of cluster validity indices can be simultaneously optimized.

There are a few recent approaches that use multiobjective optimization for crisp clustering. One of the popular approaches for multiobjective crisp clustering is MOCK (MultiObjective Clustering with automatic K determination) [200]. MOCK uses PESA-II [113] as the underlying multiobjective optimization tool. In MOCK, a locus-based adjacency representation for the chromosomes is used to encode a clustering solution. Two cluster validity measures, namely deviation and connectivity, are optimized simultaneously. The next section discusses this MOCK method in detail. MOCK is a popular method for multiobjective evolutionary clustering; however, it deals with crisp clustering only.

In this book, we mainly deal with fuzzy clustering. Therefore, we have included a detailed description of a recently proposed multiobjective fuzzy clustering technique [41,316] thereafter. Real-coded Multiobjective Genetic Algorithms (MOGAs) are used in this regard in order to determine the appropriate cluster centers and the corresponding partition matrix. NSGA-II [127, 130] is used as the underlying optimization strategy. The Xie-Beni (XB) index [442] and the fuzzy C-Means (FCM) [62] measure (J_m) are used as the objective functions [41, 316]. Note that any other and any number of objective functions could be used instead of the two mentioned above.

Experiments have been done for a number of artificial and real-life datasets. Thereafter clustering results are reported for remote sensing data available as both a set of labeled feature vectors as well as images. Comparison with different well-known clustering techniques is provided. Moreover, Indian remote sensing (IRS) satellite images of parts of Kolkata and Mumbai, two important cities of India, and a Système Probatoire d'Observation de la Terre (SPOT) satellite image of a part of Kolkata have been segmented using the multiobjective fuzzy clustering method. Finally the multiobjective clustering has been applied for clustering microarray gene expression data and the results are demonstrated.

5.2 MOCK Clustering Algorithm

In this section, the popular MOCK (MultiObjective Clustering with automatic K determination) technique [200] is described in detail. This algorithm uses PESA-II as the underlying optimization tool and optimizes two cluster validity measures, namely cluster deviation and connectivity, simultaneously. For encoding, it uses a locus-based adjacency representation for the chromosomes.

5.2.1 Chromosome Representation

In the locus-based adjacency graph representation, each chromosome consists of n genes (n is the number of data points) and each gene can have integer values in $\{1, \ldots, n\}$. If the gene i is assigned a value j, that represents a link between the data points i and j, and in the resulting clustering solution, these two points will belong to the same cluster. Therefore, for decoding a chromosome, it is required we identify all the connected components of the graph. The data points in the same connected component are then assigned to the same cluster. Hence this representation encodes the clustering as well as the number of clusters (number of connected components).

5.2.2 Initial Population

To fill up the initial population, two clustering algorithms, viz., the minimum spanning tree-(MST-)based algorithm and the K-means algorithm, are used. In the MST-based approach, first the complete MST is created using Prim's algorithm [112] and then different numbers of *interesting* links are deleted to create many clustering solutions with different numbers of clusters. A link between two data items is called *interesting* if neither of them belongs to the other's set of L nearest neighbors, L being a user input. Half of the initial population is filled up using MST-based clustering solutions. The remaining half is filled up from the clustering solutions generated using K-means clustering (only a few iterations) with different numbers of clusters. Finally these K-means solutions are converted into chromosome genotypes by replacing the links of the MST that cross cluster boundaries with a link to a randomly chosen neighbor in the nearest neighbor list.

5.2.3 Objective Functions

Two cluster validity measures that have been optimized simultaneously are *overall deviation* and *connectivity*. The overall deviation of a clustering solution is computed as the overall summed distances between data points and their corresponding cluster center (Equation 5.1) [200]:

$$Dev(C) = \sum_{C_k \in C} \sum_{x_i \in C_k} D(z_k, x_i), \tag{5.1}$$

where C represents the set of clusters, C_k represents the kth cluster, x_i is the ith data point, z_k is the cluster center of the kth cluster, and $D(.,.)$ is the distance metric used. Overall deviation has to be minimized in order to obtain compact clusters. This objective is practically same as the objective of K-means clustering.

The other objective, *connectivity*, represents the connectedness of the clusters and is defined as [200]

$$Conn(C) = \sum_{i=1}^{n} \left(\sum_{j=1}^{L} x_{i,nn_{ij}} \right), \tag{5.2}$$

where $x_{r,s} = \frac{1}{j}$ if $\nexists C_k : r \in C_k \wedge s \in C_k$, and $x_{r,s} = 0$ otherwise. Here nn_{ij} denotes the jth nearest neighbor of the data point i and L is a user parameter which determines the number of neighbors that contribute to the connectivity measure. It needs to compute the complete nearest neighbor list for each data point in the initialization phase in order to make the algorithm faster during execution. The connectivity measure is also to be minimized in order to get highly connected clusters. Hence the goal is to minimize both the objectives simultaneously.

5.2.4 Crossover and Mutation

As the crossover operation, MOCK uses the conventional uniform crossover, where a random binary mask is generated and the genes in the corresponding '1' positions of the mask are swapped between the two parents for generating two offspring chromosomes.

A special mutation operator, called neighborhood-biased mutation [200], is used to make the convergence faster despite the size of the large search space (n^n for a dataset of size n), as the chromosome length is same as the number of data points. In this mutation, each point i is linked to its L nearest neighbors $\{nn_{i1}, nn_{i2}, \ldots, nn_{iL}\}$, and thus the effective search space is reduced to L^n. The mutation probability is also determined adaptively. It can be thought that 'longer links' in the encoding are expected to be less favorable. For example, a link $i \to j$ with $j = nn_{il}$ (nn_{il} denotes the lth nearest neighbor of the ith data point) should be preferred over a link $i \to j'$ where $j' = nn_{ik}$ and $l < k$. This is utilized as the bias to the mutation probability p_m of individual links $i \to j$, which is defined as [200]:

$$p_m = \frac{1}{n} + \left(\frac{l}{n}\right)^2, \tag{5.3}$$

where $j = nn_{il}$ and n is the number of points in the dataset. This helps in discarding unfavorable links quickly.

5.2.5 Selecting Final Solution

By the nature of the multiobjective optimization, the final solution set consists of a number of non-dominated solutions. For selecting the final solution an approach motivated by the GAP statistic [406] has been used. The non-dominated front consists of clustering solutions corresponding to different trade-offs between the overall deviation and connectivity and contains different numbers of clusters. For selecting a particular solution, an automated technique is developed that is based on the shape of the Pareto front. More specifically, this technique aims to find the 'knees' in the Pareto front. These knees correspond to promising solutions of the clustering problem. This is done by comparing the generated Pareto front with control fronts generated by applying MOCK on random control data. The solution that corresponds to the maximum distance between the generated Pareto front and the control fronts is selected as the final solution.

MOCK has the advantage that it can find clusters with different shapes and automatically determine the number of clusters. However, there have been a few criticisms about MOCK. Firstly, the chromosome length of MOCK is same as the number of data points. Thus for large datasets, the effective search space becomes large and the genetic operations like crossover take more time. This makes the convergence slower. However, to make it faster the special mutation operator (neighborhood-biased mutation) is used. It is dependent on a user-specified parameter L, which might be sensitive in many cases. Furthermore, the initialization phase may take considerable time and space for large datasets, because it is required to pre-compute the list of nearest neighbors for each data point. Moreover, the selection of the final solution also takes some time in order to compute the control fronts and distances of the Pareto front solutions from the corresponding control front solutions. Finally, MOCK is designed for crisp clustering only. It has been found that fuzzy clustering can handle noisy data and overlapping clusters in a better way than crisp clustering. In the next section, a recent multiobjective fuzzy clustering technique is described in detail.

5.3 Multiobjective Fuzzy Clustering Technique

This section describes the use of NSGA-II [130] for evolving a set of near-Pareto-optimal non-degenerate fuzzy partition matrices. The Xie-Beni index [442] and the J_m [62] measure are considered as the objective functions that must be minimized simultaneously. The technique is described below in detail.

5.3.1 Solution Representation

Here the chromosomes are made up of real numbers which represent the coordinates of the cluster centers. If a chromosome encodes the centers of K clusters in d-dimensional space then its length l will be $d \times K$. For example, in four-dimensional space, the chromosome
$$<1.3\ 11.4\ 53.8\ 2.6\ 10.1\ 21.4\ 0.4\ 5.3\ 35.6\ 0.0\ 10.3\ 17.6>$$
encodes three cluster centers, (1.3, 11.4, 53.8, 2.6), (10.1, 21.4, 0.4, 5.3) and (35.6, 0.0, 10.3, 17.6). Each cluster center is considered to be indivisible.

5.3.2 Initial Population

In the initial population, each string i encodes the centers of some K number of clusters. The K centers encoded in a chromosome of the initial population are randomly selected distinct points from the input dataset.

5.3.3 Computing the Objectives

Here the Xie-Beni (XB) index [442] and the J_m measure [62] (described in Section 1.2.6.2) are taken as the two objectives that need to be simultaneously optimized. For computing the measures, the centers encoded in a chromosome are first extracted. Let these be denoted as $Z = \{z_1, z_2, \ldots, z_K\}$. The membership values u_{ik}, $i = 1, 2, \ldots, K$ and $k = 1, 2, \ldots, n$ are computed as follows [62]:

$$u_{ik} = \frac{1}{\sum_{j=1}^{K} \left(\frac{D(z_i, x_k)}{D(z_j, x_k)}\right)^{\frac{2}{m-1}}}, \quad \text{for } 1 \leq i \leq K; \ 1 \leq k \leq n, \quad (5.4)$$

where $D(z_i, x_k)$ denotes the distance of point x_k from the center of ith cluster. m is the weighting coefficient. (Note that while computing u_{ik} using Equation 5.4, if $D(z_j, x_k)$ is equal to 0 for some j, then u_{ik} is set to 0 for all $i = 1, \ldots, K$, $i \neq j$, while u_{jk} is set equal to 1.) Subsequently, the centers encoded in a chromosome are updated using the following equation [62]:

$$z_i = \frac{\sum_{k=1}^{n} (u_{ik})^m x_k}{\sum_{k=1}^{n} (u_{ik})^m}, \quad 1 \leq i \leq K, \quad (5.5)$$

and the cluster membership values are recomputed. Also the chromosome is updated using the newly computed centers. Thereafter, using the new centers and the membership matrix the XB and J_m indices are computed. As discussed in Chaper 1, both the XB and the J_m indices are to be minimized for achieving better clustering. Hence, both the objective functions are minimized here to obtain compact and well-separated clusters.

As multiobjective clustering deals with simultaneous optimization of more than one clustering objective, its performance depends highly on the choice of these objectives. A careful choice of objectives can produce remarkable results, whereas arbitrary or unintelligent objective selection can unexpectedly lead to bad situations. The selection of objectives should be such that they can balance each other critically and are possibly contradictory in nature. Contradiction in the objective functions is beneficial since it leads to a global optimum solution. It also ensures that no single clustering objective is optimized, leaving other probable significant objectives unnoticed.

In this work, the XB and J_m validity indices have been chosen as the two objectives to be optimized. From Equation 1.12 it can be noted that J_m calculates the global cluster variance, i.e., it considers the within-cluster variance summed up over all the clusters. Lower values of J_m indicates better compactness of the clusters. On the other hand, the XB index (Equation 1.24) is a combination of global (numerator) and particular (denominator) situations. The numerator is similar to J_m, but the denominator has a factor that gives the separation between two minimum distant clusters. Hence, this factor only considers the worst case, i.e., which two clusters are closest to each other, and forgets about the other partitions. Here, a larger value of the denominator (a lower value of the whole index) signifies a better solution. Hence it is evident that both the J_m and the XB indices should be minimized in order to get good solutions. Although the numerator of the XB index (σ in Equation 1.24) is similar to J_m index, the denominator contains an additional term (sep) representing the separation between the two nearest clusters. Therefore, the XB index is minimized by minimizing σ (or J_2) and maximizing sep. These two terms may not attain their best values for the same partitioning when the data has complex and overlapping clusters. Hence, considering J_m and XB (or in effect σ and sep) will provide a set of alternate partitionings of the data.

Figure 5.1 shows, for the purpose of illustration, the final Pareto-optimal front (composed of non-dominated solutions) of one of the runs of the multiobjective fuzzy clustering algorithm for the LANDSAT dataset to demonstrate the contradictory nature of the J_m and the XB indices. The final front indicates that the solution which provides the minimum J_m value is not the same to that which provides the minimum XB value and vice versa.

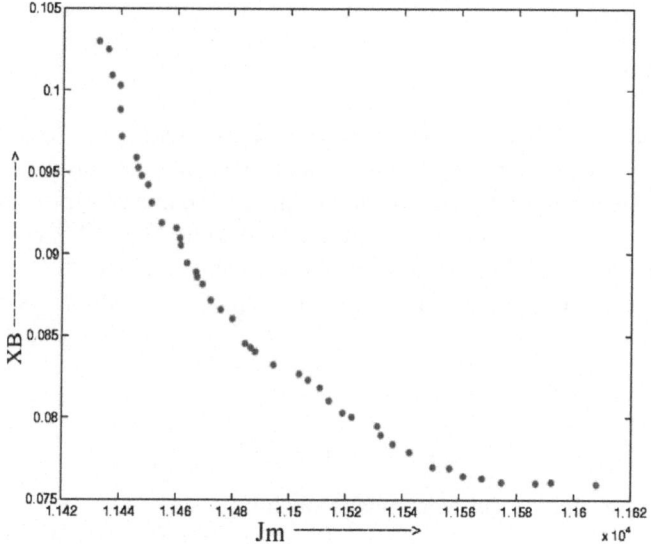

Fig. 5.1 Non-dominating Pareto front for the LANDSAT dataset

5.3.4 Genetic Operators

The genetic operations used here are selection, crossover, mutation and elitism. These are described below.

5.3.4.1 Selection

The selection process selects chromosomes from the mating pool directed by the *survival of the fittest* concept of natural genetic systems. As the selection operator, the crowded binary tournament selection method is used, as in NSGA-II [130].

5.3.4.2 Crossover

Crossover is a probabilistic process that exchanges information between two parent chromosomes for generating two child chromosomes. In this work, single-point crossover with a fixed crossover probability of p_c is used. For chromosomes of length l, a random integer, called the crossover point, is generated in the range [1, l-1]. The portions of the chromosomes lying to the right of the crossover point are exchanged to produce two offspring. The crossover operator ensures that the crossover point always fall between two cluster centers, i.e., each cluster center is considered indivisible.

5.3.4.3 Mutation

Mutation is a sudden change in a chromosome. This is required to keep diversity in the population and allow getting out of the stagnant position. In this work the following mutation operator is adopted depending on the fixed mutation probability p_m. In this method, a random center of the chromosome to be mutated is chosen. Thereafter a random number δ in the range [0, 1] is generated with uniform distribution. If the value of the center in the dth dimension is z_d, after mutation it becomes

$$(1 \pm 2.\delta).z_d, \quad \text{when} \quad z_d \neq 0,$$

and

$$(\pm 2.\delta), \quad \text{when} \quad z_d = 0.$$

The '+' and '−' signs occur with equal probability.

5.3.4.4 Elitism

Elitism is used to keep track of the better chromosomes obtained so far. As discussed in Chapter 2, in NSGA-II, elitism is implemented by combining the parent and child populations and then performing a non-dominated sorting followed by the crowded comparison operator to carry the P fittest solutions to the next generation. Here P denotes the population size.

5.3.5 Termination Criteria

In this work the processes of fitness computation, selection, crossover, and mutation are executed for a maximum number of generations. The final generation provides a set of non-dominated solutions (Rank 1 chromosomes), none of which can be further improved with regard to any one objective without degrading another. The final clustering solution is selected using a third cluster validity index (index $\mathcal{I}$ in this case [299]).

5.3.6 Selecting a Solution from the Non-dominated Set

In the final generation, the multiobjective clustering method produces a near-Pareto-optimal non-dominated set of solutions. Hence, it is necessary to choose a particular solution from the set of non-dominated solutions. Here, a popular cluster validity index $\mathcal{I}$ [299] (described in Section 1.2.6.2) is used for this purpose. As discussed in Chapter 1, a larger value of the $\mathcal{I}$ index implies a better solution.

To obtain the final solution, for each non-dominated chromosome, first the centers encoded in it are extracted and then the corresponding membership matrix is computed, as per Equation 5.4. Thereafter, the centers are recomputed as per Equation 5.5. Again, the membership values are recalculated. Finally, the $\mathcal{I}$ index value corresponding to that solution is computed. The solution providing the highest $\mathcal{I}$ index value is chosen to be the final clustering solution.

5.4 Experimental Results

Two artificial datasets (Data 1, Data 2) and two real-life datasets (Iris, Cancer) are considered for the experiments. These are described below.

5.4.1 Artificial Datasets

5.4.1.1 Data 1

This is an overlapping two-dimensional dataset where the number of clusters is five. It has 250 points. The value of K is chosen to be 5. The dataset is shown in Figure 5.2.

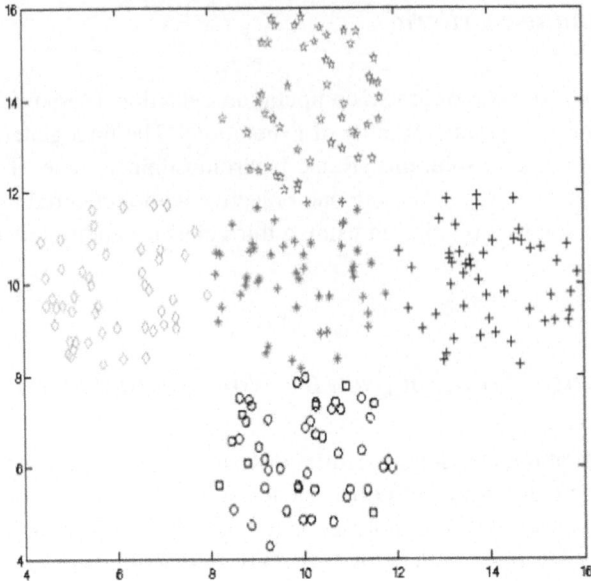

Fig. 5.2 Data 1: True clustering

5.4.1.2 Data 2

This is an overlapping two-dimensional triangular distribution of data points having nine classes where all the classes are assumed to have equal *apriori* probabilities ($= \frac{1}{9}$). It has 900 data points. The $X - Y$ ranges for the nine classes are as follows:

Class 1: [-3.3, -0.7] × [0.7, 3.3],
Class 2: [-1.3, 1.3] × [0.7, 3.3],
Class 3: [0.7, 3.3] × [0.7, 3.3],
Class 4: [-3.3, -0.7] × [-1.3, 1.3],
Class 5: [-1.3, 1.3] × [-1.3, 1.3],
Class 6: [0.7, 3.3] × [-1.3, 1.3],
Class 7: [-3.3, -0.7] × [- 3.3, -0.7],
Class 8: [-1.3, 1.3] × [- 3.3, -0.7],
Class 9: [0.7, 3.3] × [- 3.3, -0.7].

Thus the domain of the triangular distribution for each class and for each axis is 2.6. Consequently, the height will be $\frac{1}{1.3}$ (since $\frac{1}{2} \times 2.6 \times height = 1$). This dataset is shown in Figure 5.3.

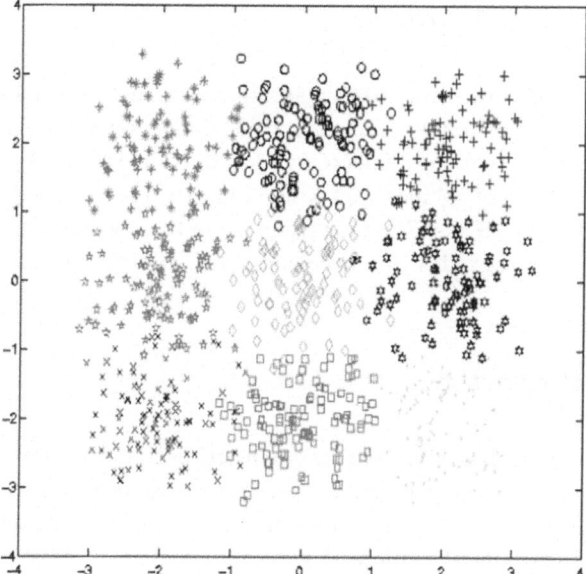

Fig. 5.3 Data 2: True clustering

5.4.2 Real-Life Datasets

5.4.2.1 Iris

This data consists of 150 patterns divided into three classes of Iris flowers, namely, Setosa, Virginia and Versicolor. The data is in four-dimensional space (sepal length, sepal width, petal length and petal width).

5.4.2.2 Cancer

It has 683 patterns in nine features (clump thickness, cell size uniformity, cell shape uniformity, marginal adhesion, single epithelial cell size, bare nuclei, bland chromatin, normal nucleoli and mitoses), and two classes malignant and benign. The two classes are known to be linearly inseparable.

The real-life datasets mentioned above were obtained from the UCI Machine Learning Repository (http://archive.ics.uci.edu/ml/datasets.html).

5.4.3 Performance Metrics

The clustering results are evaluated using both external and internal validation. For external evaluation, Minkowski score (described in Section 1.2.6.1) and percentage of correctly clustered pairs of points (*CP*) (described in Section 1.2.6.1) are used. For internal evaluation of solutions, a validity index $\mathcal{I}$ [299] is used. One may note that for computing the *MS* or *CP*, knowledge about the true partitioning of the data is necessary. In contrast, index $\mathcal{I}$ can be computed without any such information.

5.4.4 Input Parameters

The parameters of the multiobjective fuzzy clustering algorithm are as follows: population size = 50, number of generations = 100, crossover probability = 0.8, mutation probability = $\frac{1}{length\ of\ chromosome}$. Results reported in the tables are average values obtained over ten runs of the algorithms. The single objective technique has parameters exactly the same as the multiobjective one, and FCM is executed for a maximum of 100 iterations, with m, the fuzzy exponent, equal to 2.0.

5.4.5 Performance

Tables 5.1, 5.2, 5.3 and 5.4 show the comparative results obtained for the Data 1, Data 2, Iris and Cancer datasets respectively. Results comparing the performance of the multiobjective fuzzy clustering technique with that of the single objective Genetic Algorithm-based clustering having XB index as the objective, FCM and the average linkage algorithm have been presented. Tables only show the best solution obtained from the final set of non-dominated solutions averaged over 10 runs.

It is seen from the tables that the multiobjective fuzzy clustering method consistently outperforms the single objective version, FCM, as well as the average linkage algorithm in terms of the *MS* and *CP* scores. Also the $\mathcal{I}$ index values for the multiobjective method are better than those of the other algorithms. It is also evident from the tables that individually neither the XB index (which the single objective approach optimizes) nor the J_m index (which the FCM optimizes) is sufficient for proper clustering. For example, for Data 1 (Table 5.1), the multiobjective fuzzy clustering method achieves an XB value of 0.0854 while the single objective version provides a better value of 0.0737. However, when compared, in terms of the *MS* score and the *CP* value, to the actual clustering information, the former is found to perform better. Similarly for J_m values, as FCM minimizes the J_m metric, it produces a J_m value 330.7508, whereas the multiobjective fuzzy clustering technique gives a value 331.4524. However, *MS* and *CP* scores are better for the multiobjective technique compared to FCM. Note that the internal validity measure $\mathscr{I}$ index is larger for multiobjective clustering (19.8953) than that for FCM, single objective GA and average linkage algorithm. Similar results are obtained for the other datasets. This underlines the importance of utilizing multiple criteria and optimizing them simultaneously rather than using only a single optimizing objective. Figure 5.4 shows the final Pareto-optimal fronts (composed of non-dominated solutions) produced by the multiobjective fuzzy clustering algorithm for the different datasets to demonstrate the contradictory nature of the J_m and XB indices. The final front indicates that the solution which provides the minimum J_m value is not the same to that which provides minimum the XB value and vice versa.

5.5 Application to Remote Sensing Imagery

Here we have considered two types of remote sensing data. Numeric image data and IRS satellite images. The multiobjective fuzzy clustering algorithm is applied on these data and the results obtained are discussed.

5.5.1 Results for Numeric Remote Sensing Data

Two numeric satellite image data obtained from Système Probatoire d'Observation de la Terre (SPOT) (a part of Kolkata) and LANDSAT (a part of the Chhotanagpur area of Bihar) are considered here for experimenting. Numeric image data means that some pixels from several known land cover types are extracted from an image. The land cover type serves as the class label, and the intensity values (in multiple bands) serve as the different feature values. Therefore, these datasets can be displayed visually as presented in Figures 5.5 and 5.6. Moreover, in contrast to the normal image data, in the numeric SPOT and LANDSAT data, no spatial information is available as the pixel locations are not retained. As the distance measure, Euclidean measure is used here. The datasets are first described below.

Table 5.1 Results of different algorithms for Data 1

Method	MS	$\%CP$	XB	J_m	$\mathscr{I}$
Multiobjective clustering	0.3305	97.8121	0.0854	331.4524	19.8953
Single objective clustering	0.3519	97.5197	0.0737	-	19.6704
FCM	0.4404	96.1060	-	330.7508	18.9412
Average linkage	0.4654	95.6627	0.1118	335.7685	17.8401

Table 5.2 Results of different algorithms for Data 2

Method	MS	%CP	XB	J_m	$\mathscr{I}$
Multiobjective clustering	0.4874	97.3363	0.0724	254.3789	5.2836
Single objective clustering	0.5563	96.5280	0.0722	-	5.0143
FCM	0.5314	96.8335	-	253.4357	4.9881
Average linkage	0.6516	94.8573	0.1031	254.2831	4.7662

Table 5.3 Results of different algorithms for the Iris dataset

Method	MS	%CP	XB	J_m	$\mathscr{I}$
Multiobjective clustering	0.5112	91.2394	0.0890	73.6361	27.8762
Single objective clustering	0.5309	90.5503	0.0811	-	27.1622
FCM	0.5912	87.9732	-	60.5057	24.0332
Average linkage	0.5583	89.2260	0.1138	61.1608	25.6591

Table 5.4 Results of different algorithms for the Cancer dataset

Method	MS	%CP	XB	J_m	$\mathscr{I}$
Multiobjective clustering	0.3863	91.8567	0.1093	14924.1120	151.7352
Single objective clustering	0.4224	90.2620	0.1021	-	149.3001
FCM	0.3925	91.5888	-	14916.6839	149.7819
Average linkage	0.4445	89.2161	0.1354	14919.2846	143.0022

5.5.1.1 SPOT

This is a three-dimensional dataset (corresponding to green, red, and near-infrared bands) that consists of 932 samples partitioned into seven distinct classes of turbid water (TW), pond water (PW), concrete (Concr), vegetation (Veg), habitation (Hab), open space (OS), and roads (including bridges) (B/R). More details of this dataset is available in [49]. Figure 5.5 shows the scatter plot of the dataset, from where it can be seen that the clusters are highly complex and overlapping in nature.

5.5.1.2 LANDSAT

This dataset is obtained by a multispectral scanner (MSS) used in LANDSAT-V for recording remote sensing images [49]. The image is taken in four bands: green, red, near-infrared and infrared. These four bands are taken as four features of the dataset. Since the features are highly correlated, the feature space is reduced to two dimensions by using principal component analysis. The dataset contains 795 samples with five classes, viz., Manda Granite, Romapahari Granite, Vegetation, Black Phillite and Alluvium. Figure 5.6 shows such a two-dimensional feature space representing satellite imagery data of various rocks, soil and vegetation.

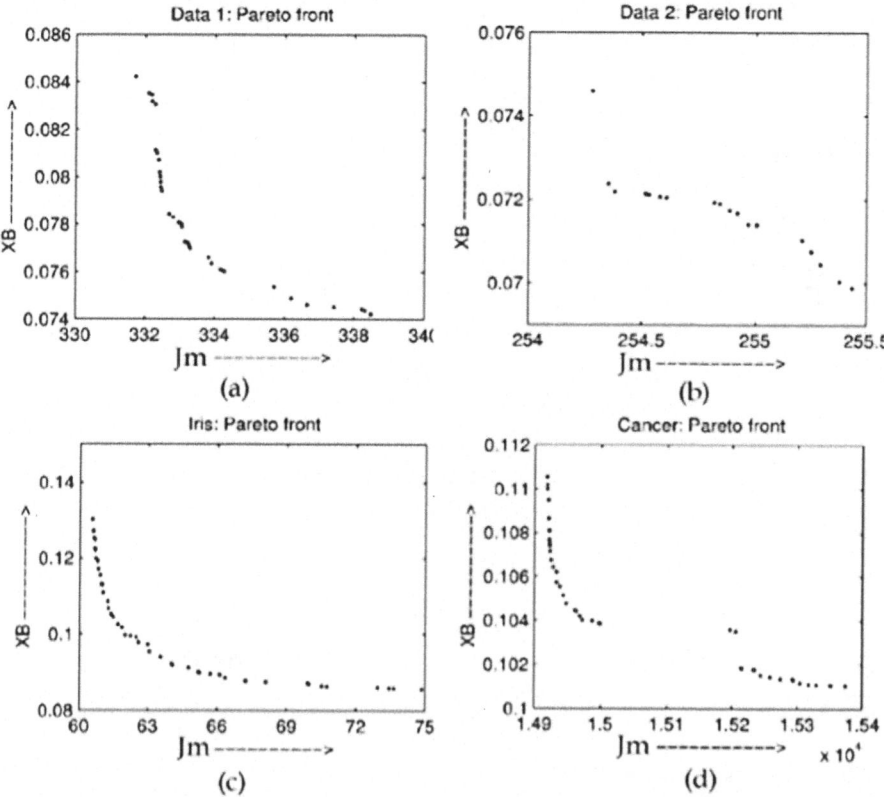

Fig. 5.4 Non-dominating Pareto fronts obtained by multiobjective fuzzy clustering for different datasets

5.5.1.3 Performance

Tables 5.5 and 5.6 present the performance of the multiobjective fuzzy clustering method for clustering the SPOT and LANDSAT numeric image data (which, as mentioned earlier, correspond to pixel values of some known land cover types) respectively. The solution providing the best value of the $\mathscr{I}$ index is selected from the set of non-dominated solutions. The J_m, XB and $\mathscr{I}$ index values, along with the percentage of correctly classified pairs (%CP), are reported corresponding to this solution. For the purpose of comparison, three other clustering algorithms, viz., a single objective genetic clustering optimizing the XB index only, FCM and average linkage, are considered.

The single objective genetic clustering algorithm minimizes the XB index only. Hence, as expected, it provides the best XB index value (Tables 5.5 and 5.6). How-

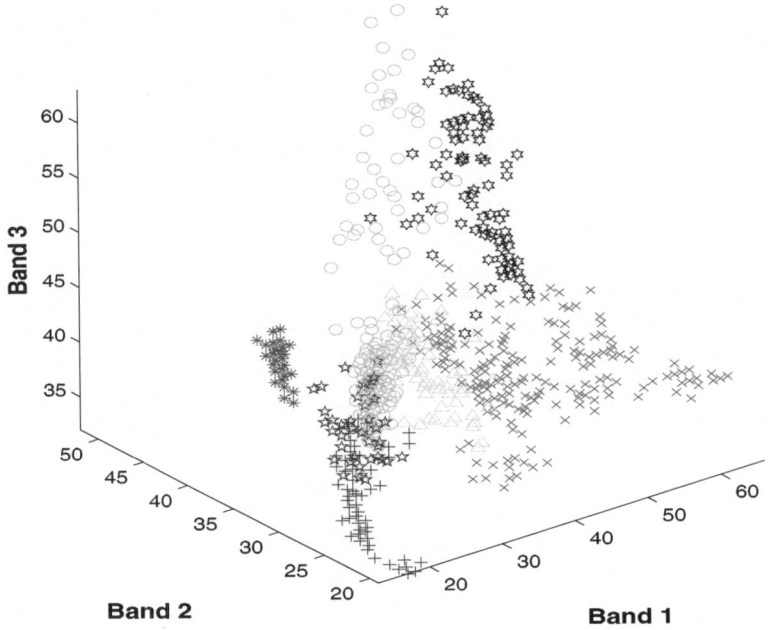

Fig. 5.5 Scatter plot of numeric SPOT data of Kolkata having 932 samples and seven classes

Table 5.5 Results for numeric SPOT data

Method	XB	J_m	$\mathscr{I}$	%CP
Multiobjective clustering	0.0818	11535.7695	689.6587	89.12
Single objective clustering	0.0714	11984.5623	635.3235	87.35
FCM	0.1712	11425.6113	634.3225	87.26
Average linkage	1.3476	14890.8848	517.8401	79.81

Table 5.6 Results for numeric LANDSAT data

Method	XB	J_m	$\mathscr{I}$	%CP
Multiobjective clustering	0.1148	23226.9785	38334.5092	91.18
Single objective clustering	0.0924	23627.5443	37901.9102	88.42
FCM	0.1591	23014.8878	38242.7409	88.94
Average linkage	2.2355	24080.2842	34253.9776	84.33

ever, the J_m value reported for this algorithm (as attained by the solution with the best XB index) is quite poor. Again, since FCM optimizes the J_m index, it provides the best value for this index. The corresponding XB index value (as attained by the solution with the best J_m value) is once again quite poor. The multiobjective method is found to provide values of the XB and J_m indices that are only slightly poorer than

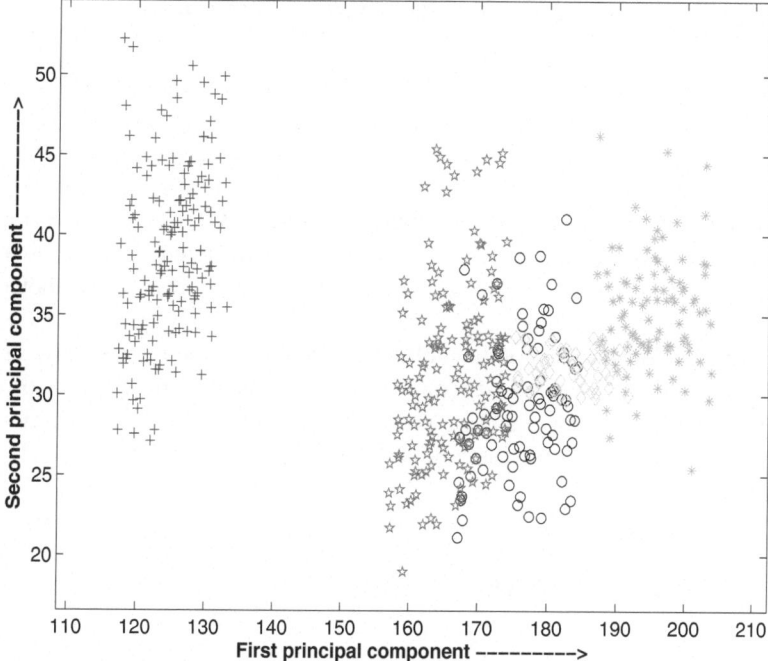

Fig. 5.6 Scatter plot of numeric LANDSAT data having 795 samples and five classes

the best values. Interestingly, in terms of the $\mathscr{I}$ index, which none of the algorithms optimizes directly, the multiobjective method significantly outperforms the other methods. Also, the multiobjective scheme provides better classification accuracy in terms of the %CP scores. This signifies that in terms of the algorithm-independent measure of clustering goodness (or, validity), i.e., the $\mathscr{I}$ index, just optimizing the XB index (asthe single objective version does) or the J_m index (as FCM does) is not good enough. A trade-off solution, provided by the multiobjective scheme, that balances both the objectives, appears to be better. The average linkage method provides the poorest scores, which could be due to the highly overlapping nature of the clusters as evident from Figures 5.5 and 5.6.

This section highlights the importance of utilizing multiple criteria and optimizing them simultaneously rather than using only a single optimizing objective. The results were demonstrated quantitatively for some numeric data extracted from satellite images. The following section provides performance results that are obtained by applying the above-mentioned algorithms on full satellite images of parts of Kolkata and Mumbai.

5.5.2 Application to IRS Satellite Image Segmentation

This section describes three image datasets used for the experiments, and the results obtained by application of the multiobjective GA-based fuzzy clustering algorithm on them. Two Indian Remote Sensing (IRS) satellite images of parts of the cities of Kolkata and Mumbai are considered. The third image is also a satellite image of a part of Kolkata acquired from the French satellite SPOT. Each image is of size 512×512 (pixels), i.e., the size of the dataset to be clustered in all the images is 262,144. These are fairly large datasets. The segmentation results are shown here both visually and numerically.

To show the effectiveness of the multiobjective technique quantitatively, the cluster goodness index $\mathscr{I}$ has been examined. Besides this, the effectiveness of multiobjective genetic clustering can also be verified visually from the clustered images presented here by comparing them with the available ground knowledge about the land cover areas.

5.5.2.1 IRS Image of Kolkata

Fig. 5.7 IRS image of Kolkata in the NIR band with histogram equalization

The data used here was acquired from the Indian Remote Sensing satellite (IRS-1A) [1] using the LISS-II sensor, which has a resolution of 36.25m $\times$ 36.25m. The image is contained in four spectral bands, namely, a blue band of wavelength 0.45 −

0.52 μm, a green band of wavelength $0.52 - 0.59$ μm, a red band of wavelength $0.62 - 0.68$ μm, and a near-infrared band of wavelength $0.77 - 0.86$ μm.

Figure 5.7 shows the Kolkata image in the near-infrared band. Some characteristic regions in the image are the river Hooghly cutting across the middle of the image, several fisheries observed towards the lower right portion, and a township, Salt Lake, to the upper left-hand side of the fisheries. This township is bounded on the top by a canal. Two parallel lines observed towards the upper right-hand side of the image correspond to the airstrips of the Dumdum airport. Other than these there are several water bodies, roads, etc. in the image. From our ground knowledge, it is known the image has four clusters [300], and these four clusters correspond to the classes turbid water (TW), pond water (PW), concrete (Concr) and open space (OS).

Figures 5.8 and 5.9 show the Kolkata image partitioned using the multiobjective fuzzy clustering and the FCM algorithm respectively. From Figure 5.8, it appears that the water class has been differentiated into turbid water (the river Hooghly) and pond water (canal, fisheries etc.) because they differ in their spectral properties. Here, the class turbid water contains sea water, river water, etc., where the soil content is more than that of pond water. The Salt Lake township has come out partially as a combination of concrete and open space, which appears to be correct, since this particular region is known to have several open spaces. The canal bounding Salt Lake from the upper portion has also been correctly classified as PW. Also, the airstrips of Dumdum airport are classified rightly as belonging to the class concrete. The presence of some small areas of PW beside the airstrips is correct again as these correspond to the several ponds around the region. The predominance of concrete on both sides of the river, particularly towards the bottom of the image, is also correct. This region corresponds to the central part of the city of Kolkata.

From Figure 5.9, it can be noted that the river Hooghly and the city region have been incorrectly classified as belonging to the same class. Another flaw apparent is that the whole Salt Lake city has been put into one class. It is evident that although some portions such as the canals, parts of the airstrips, and the fisheries are correctly identified, a significant amount of confusion lies in the FCM clustering result.

The superiority of the multiobjective fuzzy clustering scheme can be verified from the $\mathscr{I}$ index values that are reported in Table 5.7. The $\mathscr{I}$ index values for the multiobjective GA and FCM algorithms are tabulated along with single objective GA, which tries to minimize the XB validity index. The value of $\mathscr{I}$ reported for multiobjective GA is for the clustering solution that provides the maximum $\mathscr{I}$ index value, from the final non-dominating set of solutions. From the table, it is found that these values are 96.2788, 31.1697 and 81.5934 respectively. As a higher value of the $\mathscr{I}$ index indicates a better clustering result, it follows that the multiobjective fuzzy clustering method outperforms both its single objective counterpart and the FCM algorithm.

Fig. 5.8 Clustered IRS image of Kolkata using multiobjective GA

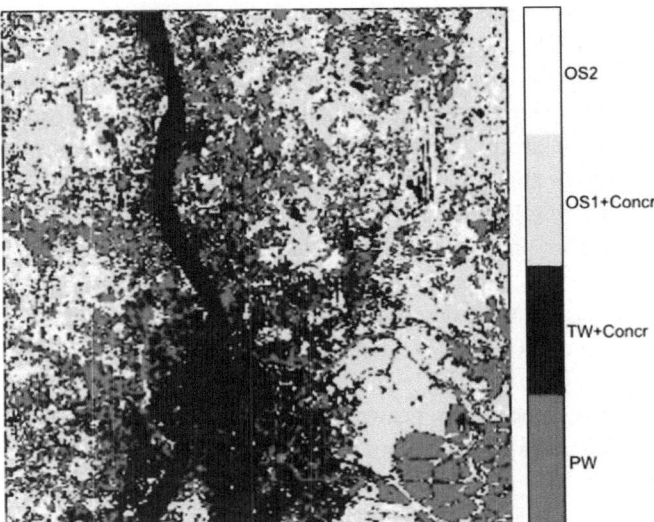

Fig. 5.9 Clustered IRS image of Kolkata using FCM

5.5.2.2 IRS Image of Mumbai

As with the Kolkata image, the IRS image of Mumbai was also obtained using the
LISS-II sensor. It is available in four bands, viz., blue, green, red and near-infrared.
Figure 5.10 shows the IRS image of a part of Mumbai in the near-infrared band.
As can be seen, the elongated city area is surrounded on three sides by the Arabian
Sea. Towards the bottom right of the image, there are several islands, including the

well-known Elephanta Island. The dockyard is situated on the southeastern part of Mumbai, which can be seen as a set of three finger-like structures. This image has been classified into seven clusters, namely concrete (Concr), open spaces (OS1 and OS2), vegetation (Veg), habitation (Hab), and turbid water (TW1 and TW2) [300].

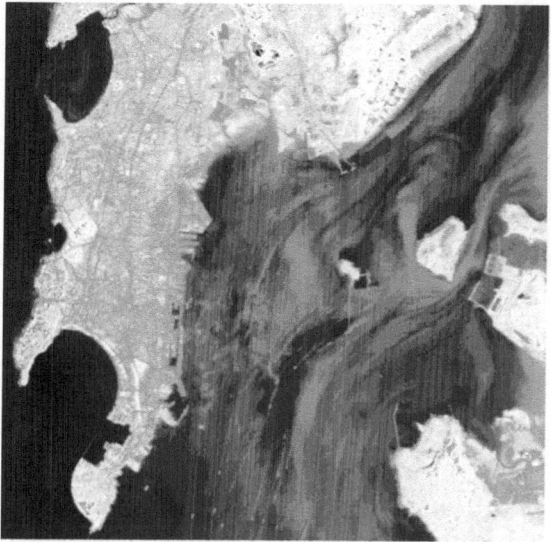

Fig. 5.10 IRS image of Mumbai in the NIR band with histogram equalization

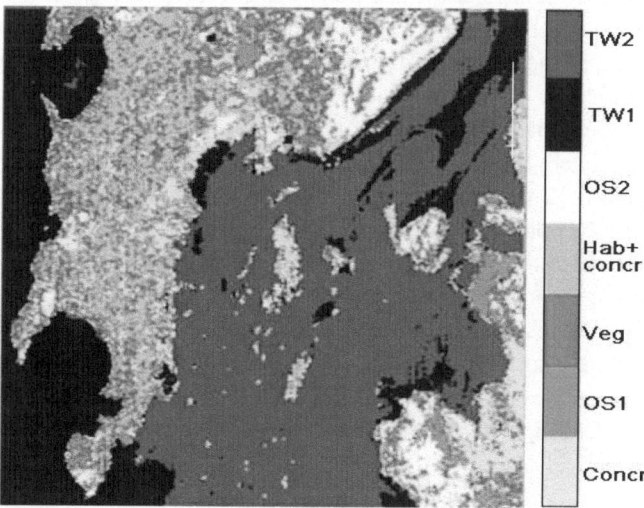

Fig. 5.11 Clustered IRS image of Mumbai using multiobjective GA

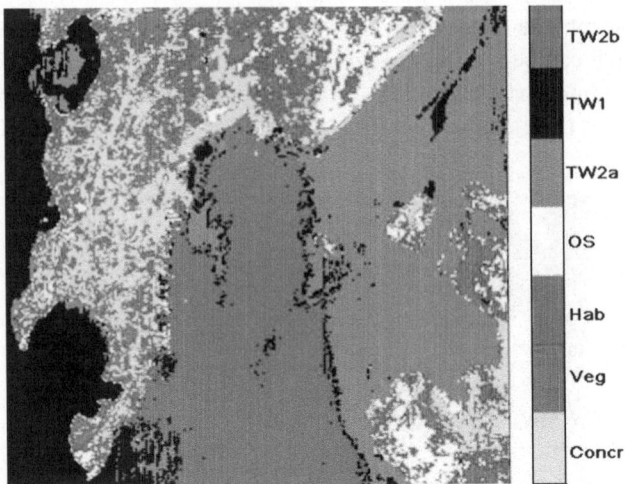

Fig. 5.12 Clustered IRS image of Mumbai using FCM

The result of the application of the multiobjective fuzzy clustering technique on the Mumbai image is presented in Figure 5.11. According to the available ground knowledge, the different clusters are labeled as concrete (Concr), open spaces (OS1 and OS2), vegetation (Veg), habitation (Hab) and turbid water (TW1 and TW2). Here, the class habitation refers to the regions which have concrete structures and buildings, but relatively lower density than the class concrete. Thus these two classes share common properties. From the result it can be seen that the large water body of the Arabian Sea has been categorized into two classes named TW1 and TW2. It is evident from Figure 5.11, that the sea water has two distinct regions with different spectral properties. Hence a clustering result providing two partitions for this region is expected. The islands, dockyard, and several road structures have mostly been correctly identified in the image. Within the islands, as expected, there is a high proportion of open space and vegetation. The southern part of the city, which is heavily industrialized, has been classified as primarily belonging to habitation and concrete. Figure 5.12 demonstrates the Mumbai image clustered using the FCM technique. It can be noted from the figure that the water of the Arabian Sea has been partitioned into three regions, rather than two, as earlier. The other regions appear to be classified more or less correctly for this data.

The $\mathscr{I}$ index values reported in Table 5.7 for the Mumbai image also speak for the multiobjective fuzzy clustering technique. Here the multiobjective GA produces 183.7727, whereas the single objective GA with XB index as objective to be minimized and the FCM algorithm are able to provide $\mathscr{I}$ index values of 180.4512 and 178.0322, respectively.

5.5.2.3 SPOT Image of Kolkata

This image has been acquired from the French satellite SPOT. This image has three bands in the multispectral mode, viz., a green band of wavelength $0.50 - 0.59$ μm, a red band of wavelength $0.61 - 0.68$ μm, and a near-infrared band of wavelength $0.79 - 0.89$ μm.

Figure 5.13 shows the SPOT image of Kolkata in the near-infrared band with histogram equalization. Some important land cover regions are seen from the image and they can be mostly identified with the knowledge about the area. The prominent black stretch through the image is the river Hooghly. Around the center of the image, a bridge (second Hooghly bridge) can be noticed across the river. The bridge was being constructed when the picture was taken. Below the river, on the left side, two distinguished black patches are water bodies. The one on the right is the Khidirpore dockyard, and the one on the left is the Garden Reach Lake. Just on the right of those water bodies, a very thin black line is noticed, starting from the river and going towards the bottom edge of the picture. This is a canal called the Talis nala. The triangular patch on right side of Talis nala is the race course. There is a thin line starting from the top right-hand corner of the picture and stretching towards the middle of it. This is the Beleghata canal, with a road beside it. Several roads are found at the right side of the picture, mainly near the top and middle portions. Another bridge, called Rabindra Setu, cuts the river near the top of the image. The image has seven classes: turbid water (TW), pond water (PW), concrete (Concr), vegetation (Veg), habitation (Hab), open space (OS), and roads (including bridges) (B/R) [41].

Fig. 5.13 SPOT image of Kolkata in NIR band with histogram equalization

Figure 5.14 and 5.15 show how the multiobjective technique and FCM algorithm perform on the SPOT image of Kolkata respectively. As seen from Figure 5.14, the multiobjective fuzzy clustering method has correctly identified most of the classes. However, it seems from the figure that there is some confusion between the classes Concr and B/R and between Concr and Hab. This is expected as these classes have a large amount of overlapping in practice. On the other hand, FCM algorithm performs poorly (Figure 5.15) and it produces a great amount of confusion among the classes TW, PW, Veg and Concr. For example, the Khidirpore dockyard has come out as a combination of PW and Concr. Also, FCM cannot distinguish between the two water classes (TW and PW). The Talis nala is also not very prominent in this case. Note that these areas have been clearly identified by the multiobjective fuzzy clustering method (Figure 5.14). Another interesting observation is that the multiobjective fuzzy clustering algorithm rightly detects the Rabindra Setu as a thin line (combination of B/R and Concr), whereas FCM fails to recognize it properly.

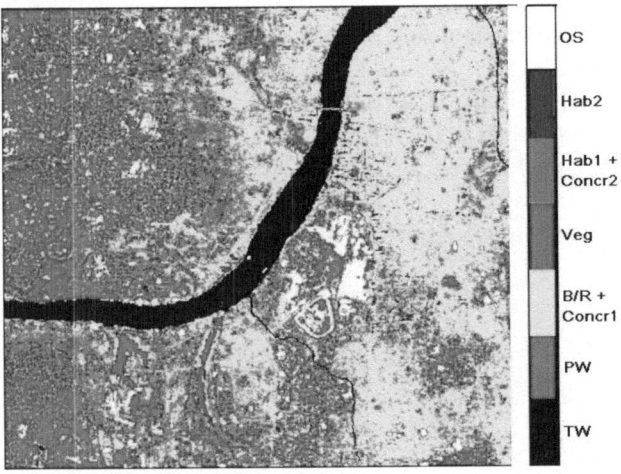

Fig. 5.14 Clustered SPOT image of Kolkata using multiobjective GA

For the purpose of demonstration, the segmentation results of the race course part are magnified in Figure 5.16 for different clustering algorithms. Though all of them can locate the race course, only the multiobjective method is able to identify a lighter outline within it properly. This line corresponds to the tracks of the race course and belongs to the class open space (OS).

Table 5.7 shows the $\mathscr{I}$ index values for this image. It indicates the superiority of the multiobjective technique, which produces a far better value of the $\mathscr{I}$ index (97.6453) than do the single objective clustering optimizing XB (88.9346) and the FCM algorithm (82.2081). Therefore the multiobjective fuzzy genetic clustering technique clearly outperforms its competitors for this image.

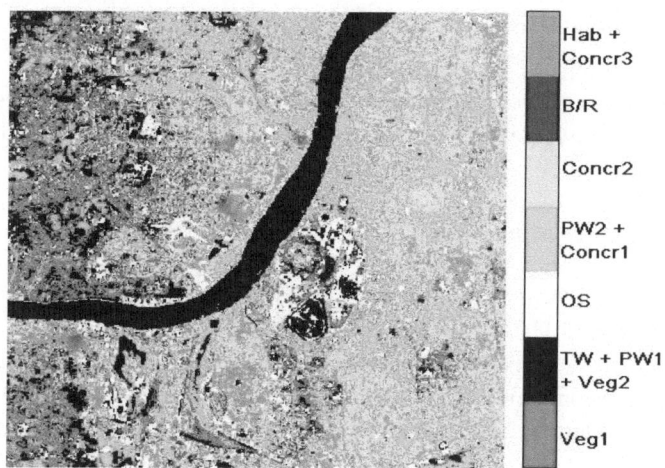

Hab +
Concr3

B/R

Concr2

PW2 +
Concr1

OS

TW + PW1
+ Veg2

Veg1

Fig. 5.15 Clustered SPOT image of Kolkata using FCM

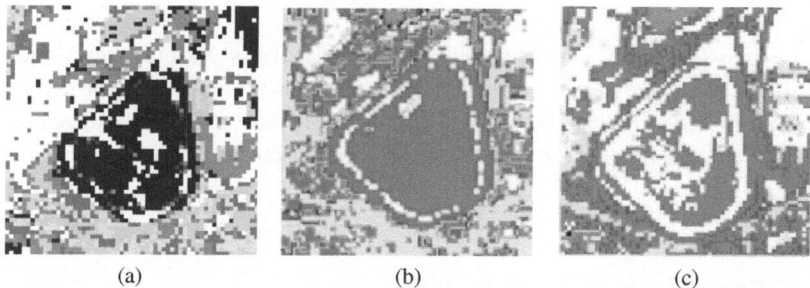

(a)   (b)   (c)

Fig. 5.16 Segmented SPOT image of Kolkata (zooming the race course) using (a) FCM, (b) single objective clustering, (c) multiobjective clustering

Table 5.7 $\mathscr{I}$ index values for Kolkata and Mumbai images for different algorithms

Method	Kolkata (IRS)	Mumbai (IRS)	Kolkata (SPOT)
Multiobjective clustering	96.2788	183.7727	97.6453
Single objective clustering	81.5934	180.4512	88.9346
FCM	31.1697	178.0322	82.2081

5.6 Application to Microarray Gene Expression Data

In this section, the multiobjective GA-based fuzzy clustering algorithm has been applied to cluster some real-life microarray gene expression datasets. The datasets and experimental results are described below.

5.6.1 Microarray Datasets and Preprocessing

Two publicly available gene expression datasets: *yeast sporulation data* and *human fibroblasts serum data* are used for experiments. These datasets are described below.

5.6.1.1 Yeast Sporulation Data

Microarray data on the transcriptional program of sporulation in budding yeast collected and analyzed [102] has been considered here. The sporulation dataset is publicly available at the Web site http://cmgm.stanford.edu/pbrown/sporulation. A DNA microarray containing 97% of the known and predicted genes involved, 6,118 in total, is used. During the sporulation process, the mRNA levels were obtained at seven time points, 0, 0.5, 2, 5, 7, 9 and 11.5 hours. The ratio of each gene's mRNA level (expression) to its mRNA level in vegetative cells before transfer to the sporulation medium is measured. Consequently, the ratio data are then log-transformed. From the 6,118 genes, those whose expression levels did not change significantly during harvesting have been ignored from further analysis. This is determined with a threshold level of 1.6 for the root mean squares of the log2-transformed ratios. The resulting set consists of 474 genes. The resulting data is then normalized so that each row has mean 0 and variance 1. The number of clusters chosen is seven, as in [102].

5.6.1.2 Human Fibroblasts Serum Data

This dataset [236] contains the expression levels of 8,613 human genes. The dataset is obtained as follows: First, human fibroblasts were deprived of serum for 48 hours and then stimulated by addition of serum. After the stimulation, expression levels of the genes were computed over 12 time points and an additional data point was obtained from a separate unsynchronized sample. Hence the dataset has 13 dimensions. A subset of 517 genes whose expression levels changed substantially across the time points have been chosen [151]). The data is then log2-transformed and subsequently, the expression vectors are normalized to have mean 0 and variance 1. The serum dataset is clustered into ten clusters, as in [236]. This dataset can be downloaded from http://www.sciencemag.org/feature/data/984559.shl.

5.6.2 Performance Metrics

Performances of the clustering algorithms are examined both quantitatively and visually. A cluster validity index called Silhouette index [369] (described in Section 1.2.6.2) has been used for quantitative analysis. A larger value of the Silhouette index $s(C)$ (near 1) indicates that most of the genes are correctly clustered, and this in turn reflects a better clustering solution. The Silhouette index can be used with any kind of distance metric. For visual demonstration, two cluster visualization tools, Eisen plot and cluster profile plot, are utilized. These are described below.

5.6.2.1 Eisen Plot

The Eisen plot (see Figure 5.17 for example) was introduced by M. B. Eisen [151]. The matrix M representing the expression data is plotted as a table where the element g_{ij} (i.e., the expression value of the ith gene at the jth time point) is denoted by coloring the corresponding cell of the table with a color similar to the original color of its spot on the microarray. Red and green represent high and low expression values respectively. Black denotes a zero expression value, which indicates the absence of a differential expression. The rows are ordered to keep similar genes together. In our representation the cluster boundaries are identified by white-colored blank rows.

5.6.2.2 Cluster Profile Plot

The cluster profile plot (see Figure 5.18 for example) shows for each cluster the normalized gene expression values (light green) of the genes of that cluster with respect to the time points. Along with this, the average expression values of the genes of the cluster over different time points are shown as a black line, together with the standard deviation within the cluster at each time point. The cluster profile plot demonstrates how the plots of the expression values of the genes across the time points differ for each cluster. Also, the genes within a cluster are similar in terms of the expression profile.

5.6.3 Input Parameters

The parameters for the GA-based clustering techniques (for both single objective and multiobjective) are fixed as follows: number of generations = 100, population size = 50, crossover probability = 0.8, mutation probability = $\frac{1}{length\ of\ chromosome}$. The number of iterations for K-means, K-medoids and fuzzy C-means algorithms is taken as 200 unless they converge before that. All the algorithms are executed several times and only the best solutions obtained in terms of the Silhouette index are considered here.

5.6.4 Results on Microarray Datasets

This section provides the results of experiments on the above-described microarray datasets both quantitatively and visually.

5.6.4.1 Quantitative Assessments

The performance of the multiobjective clustering technique has been compared with that of several well-known clustering algorithms. The algorithms are differentiated into three fundamental categories, viz., partitional, hierarchical and GA-based algorithms. K-means, K-medoids and FCM are in the first category. The three hierarchical agglomerative linkage clustering algorithms (single, average and complete linkage) are in the second category. The GA-based algorithms may be single objective (with objective functions either J_m or XB) or multiobjective (which optimizes both J_m and XB simultaneously). As the multiobjective algorithm generates a set of Pareto-optimal solutions, the solution producing the best Silhouette index value is chosen and reported here. The Silhouette index values for all the algorithms for the two datasets are reported in Table 5.8.

Table 5.8 Silhouette index values of clustering solutions obtained by different algorithms for yeast sporulation and human fibroblasts serum datasets

Clustering type	Algorithm	datasets	
		Sporulation	Serum
Partitional	K-means	0.5733	0.3245
	K-medoids	0.5936	0.2609
	Fuzzy C-means	0.5879	0.3304
Hierarchical	Single linkage	-0.4913	-0.3278
	Average linkage	0.5007	0.2977
	Complete linkage	0.4388	0.2776
GA based	Single objective GA (J_m)	0.5886	0.3434
	Single objective GA (XB)	0.5837	0.3532
	Multiobjective GA	0.6465	0.4135

As can be seen from the table, for both the datasets, the multiobjective clustering algorithm produces the best Silhouette index values. For yeast sporulation data, the K-medoids algorithm is the best of the partitional methods, producing a Silhouette index value of 0.5936. For hierarchical methods, average linkage has the largest Silhouette index (0.5007). In the case of GA-based methods, the multiobjective technique gives a Silhouette value of 0.6465, whereas the single objective methods with objectives J_m and XB provide Silhouette values of 0.5886 and 0.5837 respectively. For the fibroblasts serum data, the multiobjective GA-based clustering technique produces a Silhouette value of 0.4135, which is better than those of all other methods. For this dataset, FCM is the best partitional method (Silhouette value 0.3304). The average linkage algorithm again provides the best Silhouette value (0.2977) among the hierarchical methods. Hence it is evident from the table that the multiobjective algorithm outperforms all other clustering methods.

Tables 5.9 and 5.10 show some interesting observations for the yeast sporulation data and the serum data respectively. For example, Table 5.9 reports the J_m and XB index values for the GA-based clustering techniques for the Sporulation data. It can be observed that the single objective GA that minimizes the J_m index produces the best final J_m value of 2.2839, whereas the multiobjective algorithm obtains a J_m value of 2.3667. Similarly, the XB index produced by the single objective algorithm (minimizing XB) is 0.0148, but the multiobjective technique provides a slightly larger XB index value of 0.0164. However, the Silhouette index value (which no algorithm optimizes directly) for the multiobjective method (0.6465) is better than those produced by the single objective method optimizing J_m (0.5886) and the single objective algorithm optimizing XB (0.5837). Note that similar results are obtained for the serum dataset. These results imply that neither J_m nor XB alone is sufficient for generating good clustering solutions. Both the indices should be optimized simultaneously in order to obtain superior clustering results.

Table 5.9 Silhouette index and objective function values of different GA-based algorithms for yeast sporulation data

Clustering Algorithm	J_m	XB	Sil
Single objective GA (J_m)	2.2839	0.0178	0.5886
Single objective GA (XB)	2.3945	0.0148	0.5837
Multiobjective GA	2.3667	0.0164	0.6465

Table 5.10 Silhouette index and objective function values of different GA-based algorithms for human fibroblasts serum data

Clustering Algorithm	J_m	XB	Sil
Single objective GA (J_m)	7.8041	0.0516	0.3434
Single objective GA (XB)	8.0438	0.0498	0.3532
Multiobjective GA	7.9921	0.0500	0.4135

5.6.4.2 Visualization of Results

In this section the clustering solutions provided by different clustering algorithms are visualized using the Eisen plot and cluster profile plots. The best clustering results obtained by the algorithms in each category of clustering methods (partitional, hierarchical and GA-based methods) for both the datasets are visualized. Figure 5.17

shows the Eisen plots of clustering solutions obtained for the yeast sporulation dataset for the K-medoids, average linkage and multiobjective GA-based clustering algorithms respectively.

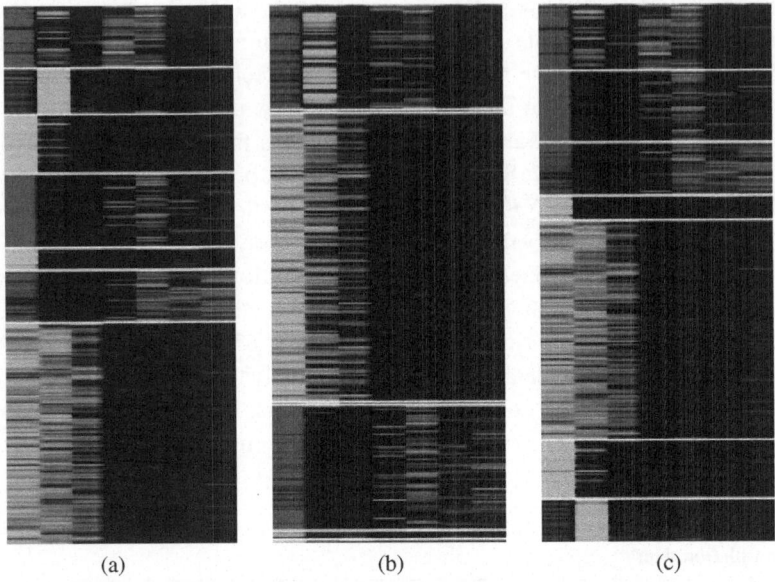

(a)   (b)   (c)

Fig. 5.17 Clustering results obtained on yeast sporulation data using (a) K-medoids algorithm, (b) average linkage algorithm, (c) multiobjective GA optimizing J_m and XB

It is evident from Figure 5.17 that the average linkage algorithm performs more poorly than the other two algorithms. In fact this algorithm produces two singleton clusters (clusters 2 and 4) and a cluster with only two genes (cluster 7). Also the expression pattern within a cluster is not similar across different genes. The K-medoids algorithm produces clusters that have corresponding clusters in the multiobjective technique. For example, cluster 7 of the K-medoids algorithm corresponds to cluster 5 of the multiobjective method. However, the expression patterns are more prominent for the multiobjective algorithm and with no doubt, it provides the best solution.

For the purpose of illustration, the cluster profile plots for the clusters produced by the multiobjective algorithm for the yeast sporulation data are shown in Figure 5.18. The cluster profile plots demonstrate how the gene expression profiles differ for different clusters. Also it can be noted that expression patterns are similar within any particular cluster.

The Eisen plots for the clustering solutions obtained on the fibroblasts serum data for fuzzy C-Means, average linkage and multiobjective GA-based clustering methods are shown in Figure 5.19. It can be observed from the figure that the performance of the average linkage algorithm is not up to the mark. Here also it produces a sin-

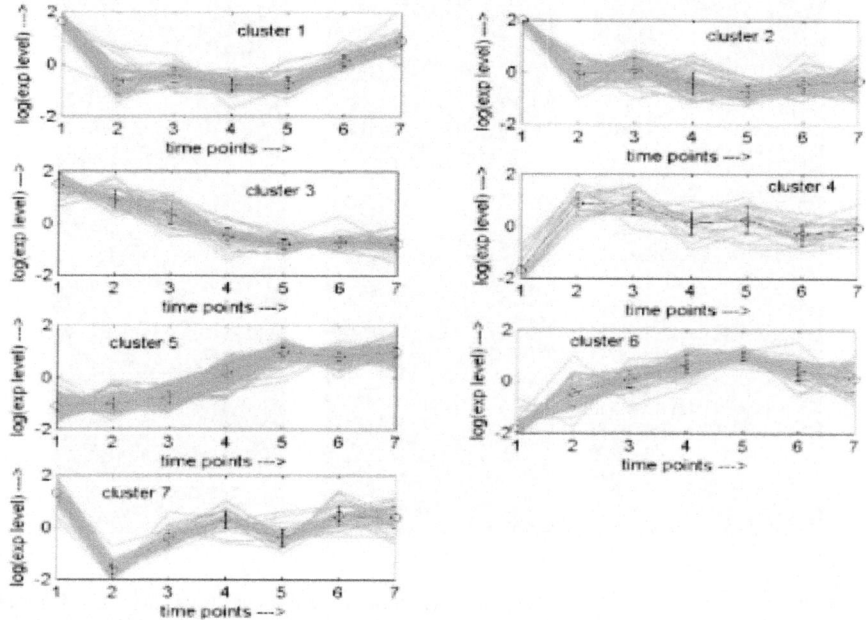

Fig. 5.18 Cluster profile plots for clustering solution obtained by the multiobjective GA-based clustering algorithm for yeast sporulation data

gleton cluster (cluster 6). Besides this, the overall expression pattern is not uniform throughout the same clusters. The performances of the fuzzy C-Means and multiobjective techniques are good as different clusters have different expression patterns and within a cluster, the expression pattern is uniform. However, there remains some confusion among several classes. This is probably due to the fact that the same gene may have various functions. The cluster profile plots for the multiobjective clustering results found on the fibroblasts serum data is illustrated in Figure 5.20. These plots demonstrate the superior performance of the multiobjective GA clustering algorithm.

Considering all the experimental results, it can be concluded that the said multiobjective clustering algorithm outperforms all the partitional and hierarchical algorithms, as well as its single objective counterpart. It also produces clusters consisting of similar genes whose expression patterns are similar to each other.

5.7 Summary

In this chapter, the problem of clustering has been posed as a multiobjective optimization problem. First a popular multiobjective crisp clustering tech-

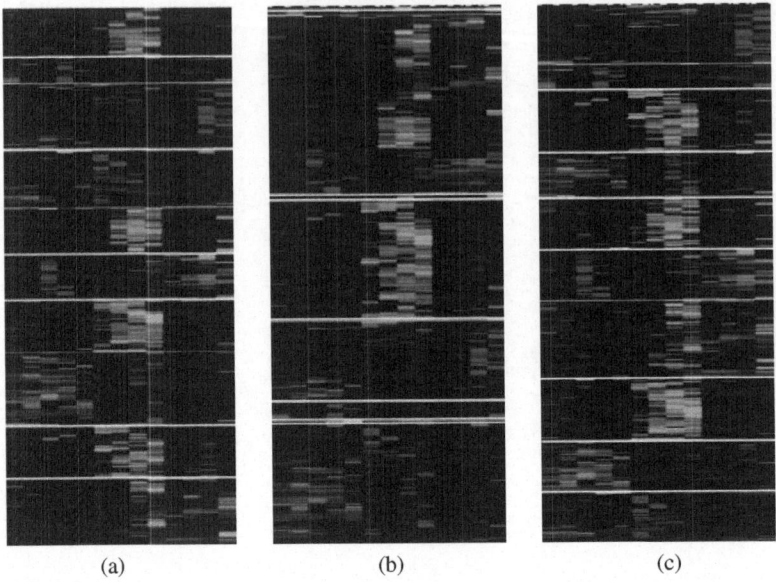

(a)   (b)   (c)

Fig. 5.19 Clustering results obtained on human fibroblasts serum data using (a) fuzzy C-means algorithm, (b) average linkage algorithm, (c) multiobjective GA optimizing J_m and XB

nique called MOCK has been discussed. MOCK uses PESA-II as the underlying optimization tool and optimizes two cluster validation measures called cluster deviation and connectivity. As the crisp clustering method is not well equipped for handling noisy data such as remote sensing and microarray data with overlapping clusters, we next discuss a recently proposed multiobjective GA-based fuzzy clustering technique. This algorithm simultaneously optimizes two cluster validity indices, namely, J_m and XB. Real-coded chromosomes representing coordinates of the cluster centers are used. The algorithm is designed on the framework of NSGA-II.

Experiments have been done for a number of artificial and real-life datasets. Thereafter, clustering results are reported for remote sensing data available as both a set of labeled feature vectors as well as images. A comparison with different well-known clustering techniques is provided. Moreover, Indian Remote Sensing (IRS) satellite images of parts of the cities of Kolkata and Mumbai and a SPOT satellite image of a part of the city of Kolkata have been segmented using the multiobjective fuzzy clustering method. Finally, the multiobjective fuzzy clustering has been applied for clustering two publicly available real-life benchmark microarray gene expression datasets and the results are demonstrated quantitatively and visually. Experimental results indicate the superiority of the multiobjective fuzzy clustering scheme over other well-known clustering methods.

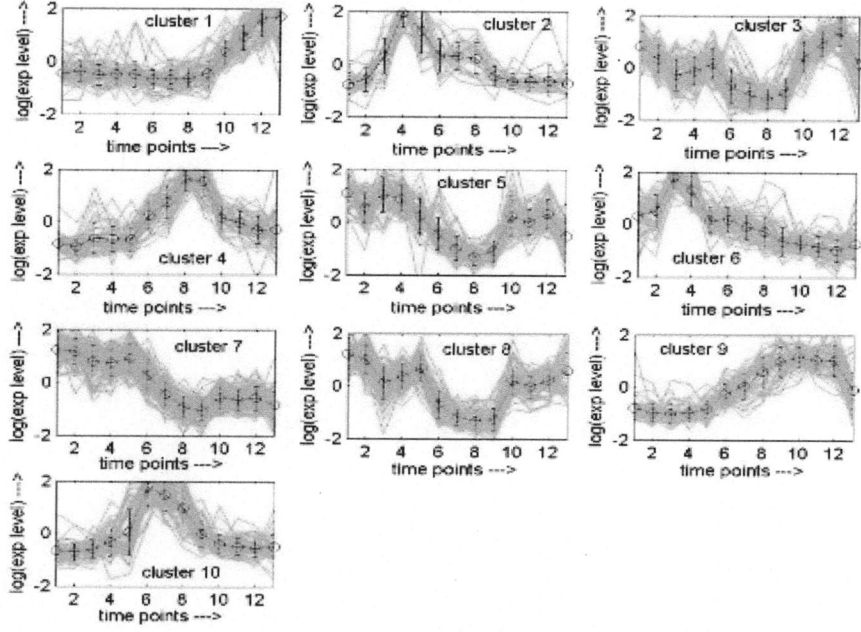

Fig. 5.20 Cluster profile plots for clustering solution obtained by the multiobjective GA-based clustering algorithm for human fibroblasts serum data

Chapter 6
Combining Pareto-Optimal Clusters Using Supervised Learning

6.1 Introduction

The multiobjective fuzzy genetic clustering technique described in the previous chapter simultaneously optimizes the Xie-Beni (XB) index [442] and the fuzzy C-means (FCM) [62] measure (J_m). In multiobjective optimization (MOO), a search is performed over a number of, often conflicting, objective functions. Instead of yielding a single best solution, MOO yields the final solution set containing a number of non-dominated Pareto-optimal solutions. A characteristic of the multiobjective fuzzy clustering approach is that it often produces a large number of Pareto-optimal solutions, from which selecting a particular solution is difficult. The existing methods use the characteristics of the Pareto-optimal surface or some external measure for this purpose. This problem has been addressed in several recent research works [73, 116, 126, 297, 394], where search is focussed on identifying the solutions situated at the 'knee' regions of the non-dominated front. In [199, 200, 202], the authors proposed a post-processing approach as described in the MOCK algorithm in the previous chapter, where the most complete Pareto front approximation set is obtained first, and then it is reduced to a single solution. The method is motivated by the GAP statistic [406] and makes use of several domain-specific considerations. In the multiobjective fuzzy clustering approach described in the previous chapter, the final solution was picked up based on some cluster validity index. Thus it is sensitive to the choice of the validity measure. Moreover, these approaches pick up one solution from the Pareto-optimal set as the final solution, although evidently all the solutions in this set have some information that is inherently good for the problem at hand.

In this chapter, we describe a recently proposed method to obtain the final solution while considering all the Pareto-optimal solutions by utilizing the input data as a guiding factor [302, 314]. The approach is to integrate the multiobjective clustering technique with support vector machine- (SVM-) [417] based classifier to obtain the final solution from the Pareto-optimal set. The procedure involves utilizing the points which are given a high membership degree to a particular class by a majority of the non-dominated solutions. These points are taken as the training points to train the SVM classifier. The remaining points are then classified by the trained SVM classifier to yield the class labels for these points.

The performance of the Multiobjective GA (MOGA) clustering followed by SVM classification (MOGA-SVM) [302, 314] is demonstrated on two numeric image datasets. MOGA-SVM is also applied on two IRS satellite images, viz., those of Kolkata and Mumbai. Moreover, the MOGA-SVM clustering has been applied for clustering the genes of microarray gene expression data. Statistical and biological significance tests are also conducted. The superiority of the MOGA-SVM technique, as compared to other well-known clustering algorithms, is demonstrated both quantitatively and qualitatively.

6.2 Combining Multiobjective Fuzzy Clustering With SVM Classifier

This section describes the scheme for combining the NSGA-II-based multiobjective fuzzy clustering algorithm described in Chapter 5 with the SVM classifier. The extended version of the two-class SVM that deals with the multi-class classification problem by designing a number of one-against-all two-class SVMs [115, 219] is used here. For example, a K-class problem is handled with K two-class SVMs, each of which is used to separate a class of points from all the remaining points.

In the combined approach, named MOGA-SVM, each non-dominated solution is given equal importance and a fuzzy majority voting technique is applied. This is motivated by the fact that due to the presence of training points, supervised classification usually performs better than unsupervised classification or clustering. This advantage has been exploited while selecting some training points using fuzzy majority voting on the non-dominated solutions produced by multiobjective fuzzy clustering. The fuzzy majority voting technique chooses a set of points which are assigned to some clusters with high membership degree by most of the non-dominated solutions. Hence these points can be thought to be clustered properly and thus can be used as the training points of the classifier. The remaining low-confidence points are thereafter classified using the trained classifier [302, 314].

6.2.1 Consistency Among Non-dominated Solutions

First it is ensured that the non-dominated solutions are consistent with each other, i.e., cluster 1 of the first solution should correspond to the cluster 1 of all other solutions, and so on. This is done as follows: First the clustering label vectors are computed from the unique non-dominated solutions produced by the multiobjective technique. This is done by assigning each of the data points to the cluster to which it has the highest membership. Let $\mathscr{X} = \{l_1, l_2, \ldots, l_n\}$ (n is the number of points) be the label vector of the first non-dominated solution, where each $l_i \in \{1, 2, \ldots, K\}$ is the cluster label of the point x_i. At first, $\mathscr{X}$ is relabeled so that the first point is labeled 1 and the subsequent points are labeled accordingly. To relabel $\mathscr{X}$, first a vector $\mathscr{L}$ of length K is formed that stores the unique class labels from $\mathscr{X}$ in the order of their first appearance in $\mathscr{X}$. The vector $\mathscr{L}$ is computed as follows:

$k = 1, \mathscr{L}_k = l_1, lab = \{\mathscr{L}_1\}.$

for $i = 2, \ldots, n$

    if $l_i \notin lab$ **then**

        $k = k + 1.$

        $\mathscr{L}_k = l_i.$

        $lab = lab \cup \{l_i\}.$

    end if

end for

Then a mapping $\mathscr{M} : \mathscr{L} \rightarrow \{1, \ldots, K\}$ is defined as follows:

$$\forall i = 1, \ldots, K : \mathscr{M}[\mathscr{L}_i] = i. \tag{6.1}$$

Next a temporary vector $\mathscr{T}$ of length n is obtained by applying the above mapping on $\mathscr{X}$ as follows:

$$\forall i = 1, 2, \ldots, n : \mathscr{T}_i = \mathscr{M}[l_i]. \tag{6.2}$$

After that, $\mathscr{X}$ is replaced with $\mathscr{T}$. This way $\mathscr{X}$ is relabeled. For example, let initially $\mathscr{X} = \{3\ 3\ 1\ 1\ 1\ 4\ 4\ 2\}$. After relabeling, it would be $\{1\ 1\ 2\ 2\ 2\ 3\ 3\ 4\}$.

Now the label vector of each of the other non-dominated solutions is modified by comparing it with the label vector of the first solution as follows: Let $\mathbb{N}$ be the set of non-dominated solutions (label vectors) produced by the multiobjective clustering technique and $\mathscr{X}$ be the relabeled cluster label vector of the first solution. Suppose $\mathscr{Y} \in \mathbb{N} - \mathscr{X}$ (i.e., $\mathscr{Y}$ is a label vector in $\mathbb{N}$ other than $\mathscr{X}$) is another label vector which is to be relabeled in accordance with $\mathscr{X}$. This is done as follows: first for each unique class label l in $\mathscr{X}$, all the points $\mathscr{P}_l$ that are assigned the class label l in $\mathscr{X}$ are found. Thereafter, observing the class labels of these points from $\mathscr{Y}$, the class label b from $\mathscr{Y}$ that is assigned for the maximum number of points in $\mathscr{P}_l$ is obtained. Then a mapping $\mathscr{M}ap_b$ is defined as $\mathscr{M}ap_b : b \rightarrow l$. This process is repeated for each unique class label $l \in \{1, \ldots, K\}$ in $\mathscr{X}$. After getting all the mappings $\mathscr{M}ap_b$ for all unique class labels $b \in \{1, \ldots, K\}$ in $\mathscr{Y}$, these are applied on $\mathscr{Y}$ to relabel $\mathscr{Y}$ in accordance with $\mathscr{X}$. All the non-dominated solutions $\mathscr{Y} \in \mathbb{N} - \mathscr{X}$ are relabeled in accordance with $\mathscr{X}$ as discussed above.

Note that the mapping $\mathcal{M}ap$ should be one-to-one to ensure that after relabeling, $\mathcal{Y}$ contains all the $\mathcal{K}$ class labels. This constraint may be violated while finding b, specially in cases of ties. This situation is handled as follows: If a one-to-one mapping cannot be obtained, all possible relabelings of $\mathcal{Y}$ are tried to be matched with $\mathcal{X}$. Thus, from the $K!$ number of relabelings of $\mathcal{Y}$, the relabeling that best matches with $\mathcal{X}$ is kept.

Consider the following example. Let $\mathcal{X}$ be $\{1\ 1\ 2\ 2\ 2\ 3\ 3\ 4\}$ and two other label vectors be $\mathcal{Y} = \{2\ 2\ 4\ 4\ 4\ 1\ 1\ 3\}$ and $\mathcal{Z} = \{4\ 2\ 3\ 3\ 3\ 2\ 2\ 1\}$. If $\mathcal{Y}$ and $\mathcal{Z}$ are relabeled to make them consistent with $\mathcal{X}$, then relabeled $\mathcal{Y}$ becomes $\{1\ 1\ 2\ 2\ 2\ 3\ 3\ 4\}$ and relabeled $\mathcal{Z}$ becomes $\{1\ 3\ 2\ 2\ 2\ 3\ 3\ 4\}$.

After relabeling all the non-dominated solutions, cluster centers corresponding to all non-dominated solutions are recomputed by taking the means of the points in their respective clusters. Thereafter, the fuzzy membership matrices are also recomputed as per the membership computation rule described in Equation 5.4.

6.2.2 MOGA-SVM Clustering Technique

Here, the steps of MOGA-SVM procedure is discussed in detail.

Procedure MOGA-SVM

1. Apply multiobjective fuzzy clustering on the given dataset to obtain a set $\mathbb{N} = \{s_1, s_2, \ldots, s_N\}$, $N \leq P$ (P is the population size), of non-dominated solution strings consisting of cluster centers.
2. Using Equation 5.4, compute the fuzzy membership matrix $U^{(i)}$ for each of the non-dominated solutions s_i, $1 \leq i \leq N$.
3. Reorganize the membership matrices to make them consistent with each other as per the technique described in Section 6.2.1, i.e., cluster j in the first solution should correspond to cluster j in all the other solutions.
4. Mark as training points, the points whose maximum membership degree (to cluster j, $j \in \{1, 2, \ldots, K\}$) is greater than or equal to a membership threshold α ($0 < \alpha \leq 1$) for at least βN solutions. Here β ($0 < \beta \leq 1$) is the threshold of the fuzzy majority voting. These points are labeled with class j.
5. Train the multi-class SVM classifier (i.e., K one-against-all two-class SVM classifiers, K being the number of clusters) using the selected training points.
6. Predict the class labels for the remaining points (test points) using the trained SVM classifier.
7. Combine the label vectors corresponding to training and testing points to obtain the final clustering for the complete dataset.

The sizes of the training and testing sets depend on the two threshold parameters α and β. Here α is the membership threshold, i.e., it is the maximum membership degree above which a point can be considered as a training point. Hence if α is increased, the size of the training set will decrease, but the confidence of the training points will increase. On the other hand, if α is decreased, the size of the training set will increase but the confidence of the training points will decrease. The parameter β determines the minimum number of non-dominated solutions that agree with each other in the fuzzy voting context. If β is increased, the size of the training set will decrease but it indicates that more non-dominated solutions agree with each other. In contrast, if β is decreased, the size of the training set increases but it indicates fewer of non-dominated solutions have agreement among them. Hence both the parameters α and β need to be tuned in such a way so that a trade-off is achieved between the size and confidence of the training set of the SVM. To achieve this, after several experiments, both the parameters are set to a value of 0.5.

6.3 Application to Remote Sensing Imagery

In this section we have discussed the performance of MOGA-SVM clustering on both numeric satellite image data and IRS satellite images.

6.3.1 Results for Numeric Remote Sensing Data

Two numeric satellite image datasets obtained from SPOT (a part of the city of Kolkata, India) and LANDSAT (a part of the Chhotanagpur area in Bihar, India) are considered [41] (described in Section 5.5.1). For computing the distance between two points, the Euclidean distance measure has been used.

6.3.1.1 Performance Metrics

The two performance metrics used are $\%CP$ and the adjusted Rand index (ARI) score (described in Section 1.2.6.1). Both are external validity indices and used to compare a clustering solution with the true clustering.

6.3.1.2 Input Parameters

The parameter values of MOGA are as follows: Number of generations $G = 100$, population size $P = 50$, crossover probability $p_c = 0.8$, mutation probability $p_m = 0.01$ and the fuzzy exponent $m = 2$. The values of both α and β are 0.5. FCM is run for a maximum of 200 iterations. The SVM with an RBF kernel is used. The SVM parameter C and the RBF kernel parameter γ are fixed at 100 and 0.1, respectively. The parameters are chosen experimentally.

6.3.1.3 Results

Table 6.1 reports the performance of the MOGA-SVM method for clustering the SPOT and LANDSAT data. For both *%CP* and *ARI*, a larger value indicates better clustering. For comparison, multiobjective clustering (MOGA) [41], a single objective genetic clustering optimizing the *XB* index only (SGA) [300] and FCM [62] are considered. Moreover, besides SVM, two other classification techniques, viz., a three-layer feed-forward Artificial Neural Network (ANN) classifier [16], which uses hyperbolic tangent function for the hidden layer and the Softmax function for output layer, and the well-known k-nearest neighbor (k-nn) classifier are combined with MOGA (named MOGA-ANN and MOGA-knn, respectively). The average *%CP* and *ARI* scores obtained from ten consecutive runs of the algorithms are reported. The standard deviations of the performance scores are also given. Interestingly, for both the image datasets, use of a classifier consistently improves the results of multiobjective clustering. The SVM classifier performs the best. This illustrates the usefulness of combining the SVM classifier with MOGA. In the next section, MOGA-SVM is used for clustering the two IRS remote sensing images.

Table 6.1 Average *%CP* and *ARI* scores (with standard deviations) in ten runs of different algorithms for SPOT and LANDSAT datasets

Method	SPOT		LANDSAT	
	%CP	*ARI*	*%CP*	*ARI*
MOGA-SVM	$90.33 \pm .12$	$0.77 \pm .01$	$89.82 \pm .08$	$0.75 \pm .01$
MOGA-ANN	$89.12 \pm .15$	$0.75 \pm .01$	$89.01 \pm .12$	$0.73 \pm .01$
MOGA-knn	$88.32 \pm .43$	$0.72 \pm .02$	$88.34 \pm .07$	$0.72 \pm .01$
MOGA	$87.94 \pm .35$	$0.68 \pm .01$	$88.21 \pm .05$	$0.71 \pm .03$
SGA	$86.24 \pm .28$	$0.60 \pm .03$	$87.67 \pm .07$	$0.68 \pm .04$
FCM	$86.78 \pm .22$	$0.61 \pm .02$	$87.39 \pm .11$	$0.67 \pm .07$

6.3.2 Application to IRS Satellite Image Segmentation

The use of SVM in satellite image classification is available in the literature. In [429], the classification of multi-sensor data based on the fusion of SVMs is proposed. In [174], a semi-supervised multi-temporal classification technique based on constrained multiobjective GA is presented that updates the ground truth information through an automatic estimation process. In [57], multiobjective PSO is used for model selection of SVMs for semi-supervised regression that exploits the unlabeled samples from satellite images. A study on the classification of hyperspectral satellite images by SVM is done in [307]. In [72], a technique for running a semi-supervised SVM for change detection in satellite imagery is presented. In [81], kernel methods for the integration of heterogeneous sources of information for multi-temporal classification of satellite images is proposed. A context-sensitive clustering based on a graph-cut-initialized Expectation Maximization (EM) algorithm for satellite image classification is proposed in [411]. In [90] and [154], urban satellite image classification has been done by the fuzzy possibilistic classifier and a fusion of multiple classifiers, respectively. A transductive SVM-based semi-supervised classifier for remote sensing imagery is proposed in [75]. Unlike these supervised and semi-supervised approaches, the MOGA-SVM method is a completely unsupervised multiobjective classification technique that utilizes the strength of the SVM classifier and is applied for satellite image segmentation.

Two IRS satellite images of parts of the cities of Kolkata and Mumbai as described in Chapter 5, are used for demonstrating unsupervised pixel classification. The number of clusters for these two images are taken as 4 and 7, respectively, as per the available domain knowledge and the previous results [300]. The results obtained by MOGA-SVM clustering have been reported and compared with that of MOGA, single objective GA and FCM clustering results. For validation, two cluster validity indices $\mathscr{I}$ [299] and K-index [266] (described in Chapter 1.2.6.2) are examined.

6.3.2.1 Results for IRS Image of Kolkata

As can be seen from Table 6.2, the maximum and minimum of the $\mathscr{I}$ and K-index scores respectively, produced by MOGA-SVM are better than those of the other algorithms. This indicates that MOGA-SVM outperforms all other algorithms in terms of $\mathscr{I}$ and K-index scores.

Table 6.2 $\mathscr{I}$ and K-index values for Mumbai and Kolkata images

Method	Mumbai		Kolkata	
	$\mathscr{I}-index$	$K-index$	$\mathscr{I}-index$	$K-index$
MOGA-SVM	198.85	6.182e4	102.45	2.886e4
MOGA	183.77	8.326e4	96.28	4.463e4
SGA	180.45	8.963e4	81.59	5.683e4
FCM	178.03	9.744e4	31.17	8.968e4

Figures 6.1, 6.2 and 6.3 show the Kolkata image clustered using MOGA-SVM, MOGA and FCM, respectively. From Figure 6.1, it appears that the water class has been differentiated into turbid water (the river Hooghly) and pond water (canals, fisheries, etc.) as they differ in their spectral properties. Salt Lake township has come

out partially as combination of concrete and open space, which appears to be correct, since this region is known to have several open spaces. The canal bounding Salt Lake has also been correctly classified as PW. Moreover, the airstrips of Dumdum airport are classified rightly as belonging to the class concrete. The presence of some small PW areas beside the airstrips is also correct as these correspond to the several ponds around the region. The concrete regions on both sides of the river, particularly towards the bottom of the image, is also correct. This region refers to the central part of Kolkata.

From Figure 6.2, which shows the clustered Kolkata image using MOGA only, a similar cluster structure is noticed as that in Figure 6.1. However, careful observation reveals that MOGA seems to have some confusion regarding the classes concrete and PW (towards the bottom of Figure 6.2, which represents the city area).

It appears from Figure 6.3 that the river Hooghly and the city region have been incorrectly classified as belonging to the same class by FCM. Also, the whole of Salt Lake city has wrongly been put into one class. Hence the FCM result involves a significant amount of confusion.

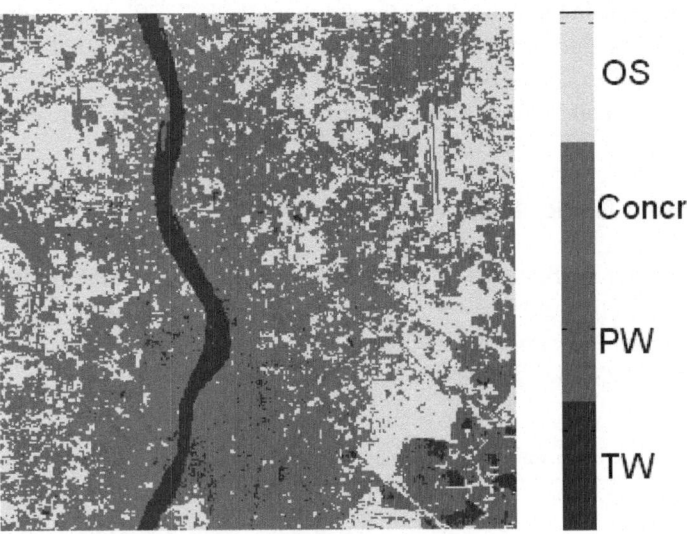

Fig. 6.1 Clustered IRS image of Kolkata using MOGA-SVM

6.3.2.2 Results for IRS Image of Mumbai

As is evident from Table 6.2, for this image, MOGA-SVM provides the highest value of the $\mathscr{I}$ index and the lowest value of K-index. This indicates that MOGA-SVM outperforms the other algorithms for this image.

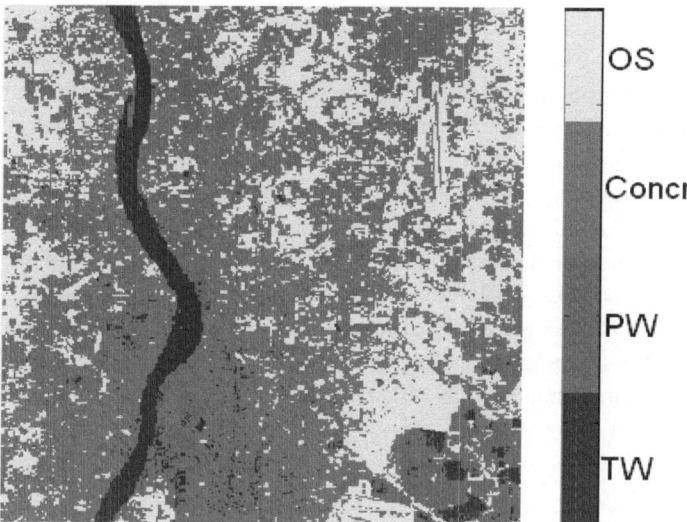

Fig. 6.2 Clustered IRS image of Kolkata using MOGA

Fig. 6.3 Clustered IRS image of Kolkata using FCM

The result of the application of MOGA-SVM on the Mumbai image is presented in Figure 6.4. According to the available ground knowledge about the area and by visual inspection, the different clusters are labeled as concrete (Concr), open spaces (OS1 and OS2), vegetation (Veg), habitation (Hab) and turbid water (TW1 and TW2). The classes concrete and habitation have some common properties as habitation means low density of concrete. As is evident from Figure 6.4, the Ara-

bian Sea is differentiated into two classes (named TW1 and TW2) due to different
spectral properties. The islands, dockyard, and several road structures have mostly
been correctly identified. A high proportion of open space and vegetation is noticed
within the islands. The southern part of the city, which is heavily industrialized, has
been classified as primarily belonging to habitation and concrete.

Figure 6.5 shows the clustered Mumbai image using MOGA clustering. Visually
it provides a similar clustering structure as that provided by MOGA-SVM. However,
the dockyard is not so clear as compared to that of the MOGA-SVM method and
there are more overlaps between the two turbid water classes.

Figure 6.6 illustrates the Mumbai image clustered using the FCM technique. It
appears from the figure that the water of the Arabian Sea has been partitioned into
three regions, rather than two, as obtained earlier. The other regions appear to be
classified more or less correctly.

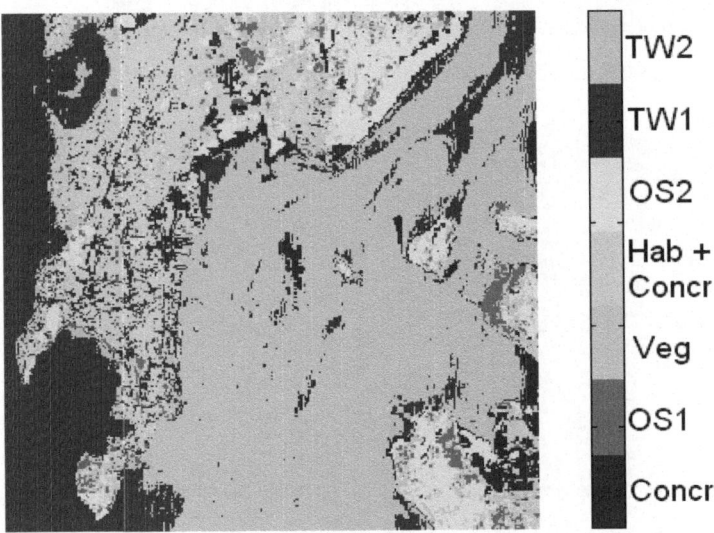

Fig. 6.4 Clustered IRS image of Mumbai using MOGA-SVM

6.4 Application to Microarray Gene Expression Data

The performance of the MOGA-SVM clustering has been evaluated on five publicly available real-life gene expression datasets, viz., Yeast Sporulation, Yeast Cell
Cycle, Arabidopsis Thaliana, Human Fibroblasts Serum and Rat CNS, for all the
algorithms. The correlation-based distance measure is used. First, the effect of the
parameter β (majority voting threshold) on the performance of MOGA-SVM clus-

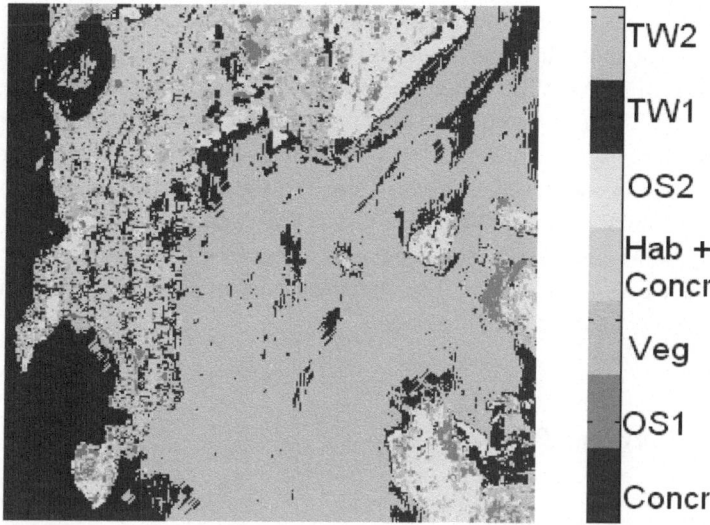

Fig. 6.5 Clustered IRS image of Mumbai using MOGA

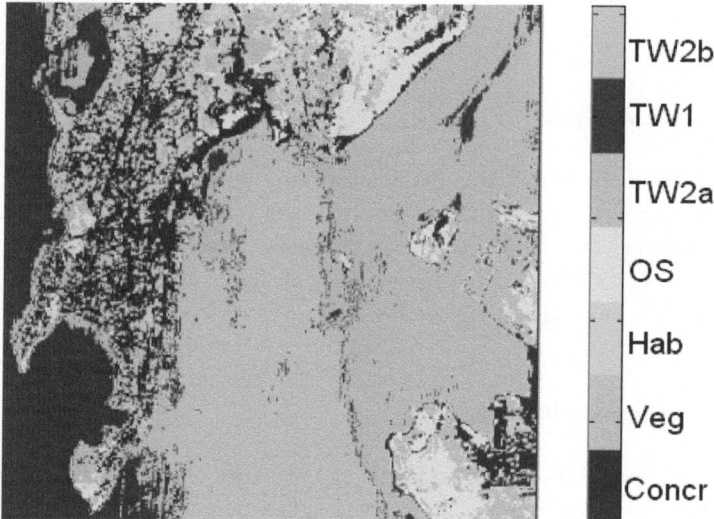

Fig. 6.6 Clustered IRS image of Mumbai using FCM

tering has been examined. Thereafter, the use of different kernel functions is examined and their performances are compared. The performance of the MOGA-SVM technique has also been compared with that of fuzzy MOGA clustering (without SVM) [41, 316], FCM [62], single objective genetic clustering scheme that minimizes the XB validity measure (SGA) [300], average linkage method [408], Self Organizing Map (SOM) [396] and Chinese Restaurant Clustering (CRC) [355].

Moreover, a crisp version of MOGA-SVM clustering (MOGA$_{crisp}$-SVM) is considered for comparison in order to establish the utility of incorporating fuzziness. Unlike fuzzy MOGA-SVM, which uses the FCM-based chromosome update, in MOGA$_{crisp}$-SVM, chromosomes are updated using the K-means-like center update process and the crisp versions of J_m and XB indices are optimized simultaneously. To obtain the final clustering solution from the set of non-dominated solutions, a procedure similar to fuzzy MOGA-SVM is followed. Note that in the case of MOGA$_{crisp}$-SVM, as membership degrees are either 0 or 1, the membership threshold parameter α is not required. The statistical and biological significance of the clustering results has also been evaluated.

6.4.1 Datasets and Preprocessing

6.4.1.1 Yeast Sporulation

This dataset [102] consists of 474 genes over seven time points. It has been described in Section 5.6.1.1.

6.4.1.2 Yeast Cell Cycle

The yeast cell cycle dataset was extracted from a dataset that shows the fluctuation of expression levels of approximately 6,000 genes over two cell cycles (17 time points). Out of these 6,000 genes, 384 genes have been selected to be cell-cycle regulated [100]. This dataset is publicly available at the following Web site: http://faculty.washington.edu/kayee/cluster.

6.4.1.3 Arabidopsis Thaliana

This dataset consists of expression levels of 138 genes of Arabidopsis Thaliana. It contains expression levels of the genes over eight time points, viz., 15 min, 30 min, 60 min, 90 min, 3 hours, 6 hours, 9 hours, and 24 hours [366]. It is available at http://homes.esat.kuleuven.be/ sistawww/bioi/thijs/Work/ClusterData.txt.

6.4.1.4 Human Fibroblasts Serum

This dataset [236] consists of 517 genes over 13 time points. It has been described in Section 5.6.1.2.

6.4.1.5 Rat CNS

The Rat Central Nervous System (CNS) dataset has been obtained by reverse transcription-coupled PCR to examine the expression levels of a set of 112 genes during rat central nervous system development over nine time points [433]. This dataset is available at http://faculty.washington.edu/kayee/cluster.

All the datasets are normalized so that each row has mean 0 and variance 1.

6.4.2 Effect of Majority Voting Threshold β

In this section, how the parameter β (majority voting threshold) affects the performance of the MOGA-SVM clustering technique is analyzed. The algorithm has been executed for a range of β values from 0.1 to 0.9 with a step size of 0.05 for all the datasets. The results reported in this section are for the Radial Basis Function (RBF) [115, 417] kernel. Experiments with other kernel functions are also found to reveal similar behavior. For each value of β, the average value of the Silhouette index ($s(C)$) scores over 20 runs has been considered. The parameter α (membership threshold) has been kept constant at 0.5. The variation of average $s(C)$ scores for different values of β is demonstrated in Figure 6.7 for the five datasets.

It is evident from Figure 6.7 that for all the datasets, MOGA-SVM behaves similarly in terms of variation of average $s(C)$ over the range of β values. The general trend is that first the average of $s(C)$ scores improves with increasing β value, then remains almost constant in the range of around 0.4 to 0.6, and then deteriorates with further increase in the β value. This behavior is quite expected, as for a small value of β, the training set will contain a lot of low-confidence points, which causes the class boundaries to be defined incorrectly for SVM. On the other hand, when the β value is very high, the training set is small and contains only a few high-confidence points. Thus the hyperplanes between the classes cannot be properly defined. In some range of β (around 0.4 to 0.6), a trade-off is obtained between the size of the training set and its confidence level. Hence, in this range, MOGA-SVM provides the best $s(C)$ index scores. With this observation, in all the experiments hereafter, the β value has been kept constant at 0.5.

6.4.3 Performance of MOGA-SVM for Different Kernels

Four kernel functions, viz., linear, polynomial, sigmoidal and RBF are considered here. In this section, a study has been made on how the different kernel functions perform for the five datasets. Table 6.3 reports the Silhouette index score $s(C)$ (averaged over 20 runs) produced by MOGA-SVM with the four different kernel

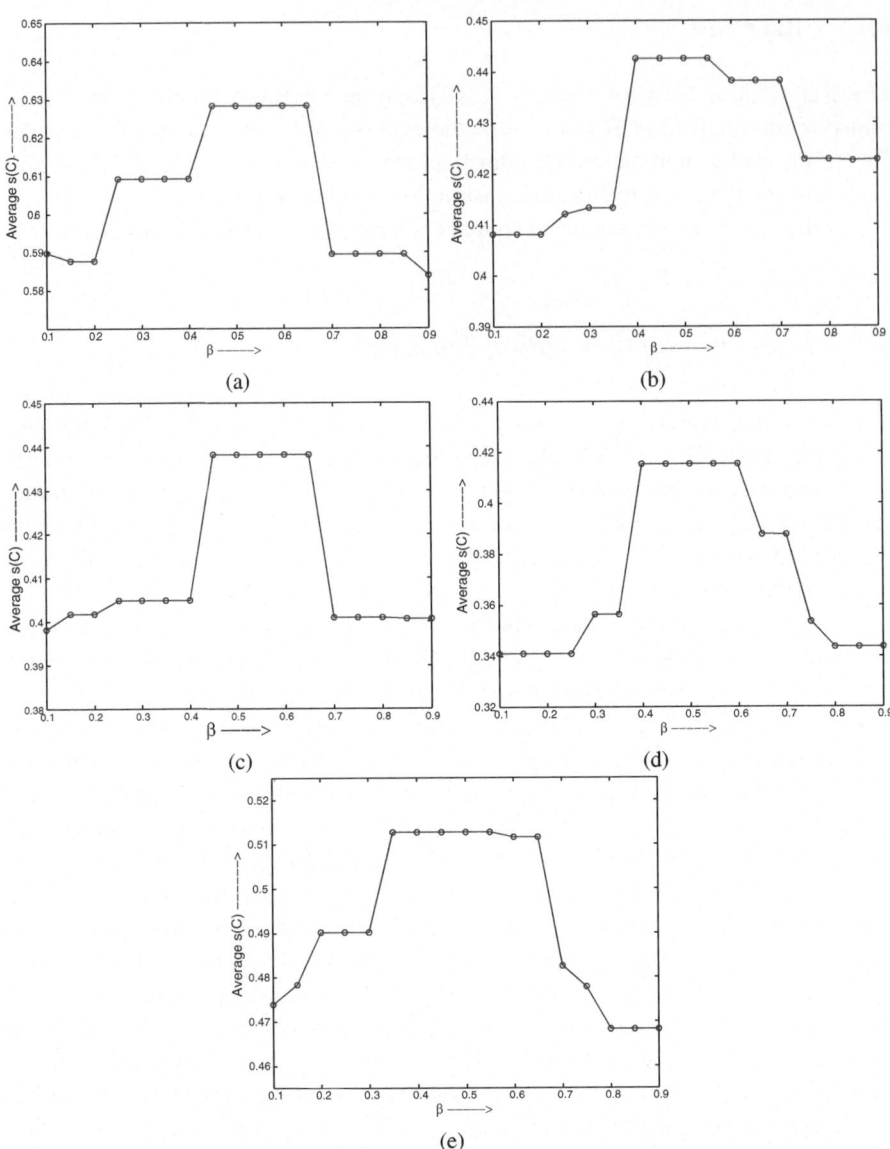

Fig. 6.7 Variation of average $s(C)$ index values produced by MOGA-SVM (RBF) clustering over different β values ranging from 0.1 to 0.9 for the datasets (a) Sporulation, (b) Cell cycle, (c) Arabidopsis, (d) Serum, (e) Rat CNS

functions for the five datasets. The average $s(C)$ score provided by MOGA (without SVM) over 20 runs is also reported for each dataset. Moreover, the number of clusters K (corresponding to the solution providing the best Silhouette index score) found for the different datasets has been shown.

As is evident from the table, irrespective of the kernel function considered, use of SVM provides a better $s(C)$ score than MOGA (without SVM). This is expected since the MOGA-SVM techniques provide equal importance to all the non-dominated solutions. Thus, through fuzzy voting, the core group of genes for each cluster is identified and the class labels of the remaining genes are predicted by the SVM. It can also be observed from the table that the Silhouette index produced by the RBF kernel is better than that produced by the other kernels. This is because RBF kernels are known to perform well in the case of spherically shaped clusters, which is very common in case of gene expression datasets. Henceforth, MOGA-SVM will indicate MOGA-SVM with RBF kernel only.

Table 6.3 Average Silhouette index scores over 20 runs of MOGA-SVM with different kernel functions for the five gene expression datasets along with the average Silhouette index score of the MOGA (without SVM)

Algorithm	Spor	Cell	Thaliana	Serum	Rat
	$K = 6$	$K = 5$	$K = 4$	$K = 6$	$K = 6$
MOGA-SVM (linear)	0.5852	0.4398	0.4092	0.4017	0.4966
MOGA-SVM (polynomial)	0.5877	0.4127	0.4202	0.4112	0.5082
MOGA-SVM (sigmoidal)	0.5982	0.4402	0.4122	0.4112	0.5106
MOGA-SVM (RBF)	0.6283	0.4426	0.4312	0.4154	0.5127
MOGA (without SVM)	0.5794	0.4392	0.4011	0.3947	0.4872

6.4.4 Comparative Results

Table 6.4 reports the average $s(C)$ index values provided by MOGA-SVM (RBF), MOGA (without SVM), MOGA$_{crisp}$-SVM (RBF), FCM, SGA, average linkage, SOM and CRC clustering over 20 runs of the algorithms for the five real-life datasets considered here. Also, the number of clusters K obtained corresponding to the maximum $s(C)$ index score for each algorithm is reported. The values reported in the tables show that for all the datasets, MOGA-SVM provides the best $s(C)$ index score. MOGA$_{crisp}$-SVM (RBF) also provides reasonably good $s(C)$ index scores, but is outperformed by MOGA-SVM for all the datasets. This indicates the utility of incorporating fuzziness in MOGA clustering. Interestingly, while incorporation of SVM-based training improves the performance of MOGA clustering, the latter also provides, in most cases, better $s(C)$ values than SGA and the other non-genetic approaches. Only for Yeast Sporulation and Arabidopsis Thaliana datasets are the results for MOGA (without SVM) slightly inferior to those of SOM and CRC, respectively. However, the performance of the MOGA-SVM is the best for all the datasets.

MOGA has determined six, five, four, six and six clusters for the Sporulation, Cell cycle, Arabidopsis, Serum and Rat CNS datasets, respectively. This conforms to the findings in the literature [102, 384, 433, 443]. Hence it is evident from the table that while MOGA (without SVM) and $MOGA_{crisp}$-SVM (RBF) are generally superior to the other methods, MOGA-SVM is the best among all the competing methods for all the datasets considered here.

Table 6.4 Average Silhouette index scores over 20 runs of different algorithms for the five gene expression datasets

Algorithm	Sporulation		Cell cycle		Thaliana		Serum		Rat CNS	
	K	$s(C)$	K	$s(C)$	K	$s(C)$	K	$s(C)$	K	$s(C)$
MOGA-SVM	6	0.6283	5	0.4426	4	0.4312	6	0.4154	6	0.5127
MOGA	6	0.5794	5	0.4392	4	0.4011	6	0.3947	6	0.4872
$MOGA_{crisp}$-SVM	6	0.5971	5	0.4271	4	0.4187	6	0.3908	6	0.4917
FCM	7	0.4755	6	0.3872	4	0.3642	8	0.2995	5	0.4050
SGA	6	0.5703	5	0.4221	4	0.3831	6	0.3443	6	0.4486
Average linkage	6	0.5007	4	0.4388	5	0.3151	4	0.3562	6	0.4122
SOM	6	0.5845	6	0.3682	5	0.2133	6	0.3235	5	0.4430
CRC	8	0.5622	5	0.4288	4	0.4109	10	0.3174	4	0.4423

To demonstrate visually the result of MOGA-SVM clustering, Figures 6.8-6.12 show the Eisen plot and cluster profile plots provided by MOGA-SVM on the five datasets, respectively. For example, the six clusters of the Yeast Sporulation data are very prominent, as shown in the Eisen plot (Figure 6.8(a)). It is evident from the figure that the expression profiles of the genes of a cluster are similar to each other and they produce similar color patterns. The cluster profile plots (Figure 6.8(b)) also demonstrate how the expression profiles for the different groups of genes differ from each other, while the profiles within a group are reasonably similar. Similar results are obtained for the other datasets.

The MOGA-SVM technique performs better than the other clustering methods mainly because of the following reasons. First of all, this is a multiobjective clustering method. Simultaneous optimization of multiple cluster validity measures helps to cope with different characteristics of the partitioning and leads to higher quality solutions and an improved robustness in terms of the different data properties. Secondly, the strength of supervised learning has been integrated with the multiobjective clustering efficiently. As each of the solutions in the final non-dominated set contains some information about the clustering structure of the dataset, combining them with the help of majority voting followed by supervised classification yields a high-quality clustering solution. Finally, incorporation of fuzziness makes the MOGA-SVM technique better equipped for handling overlapping clusters.

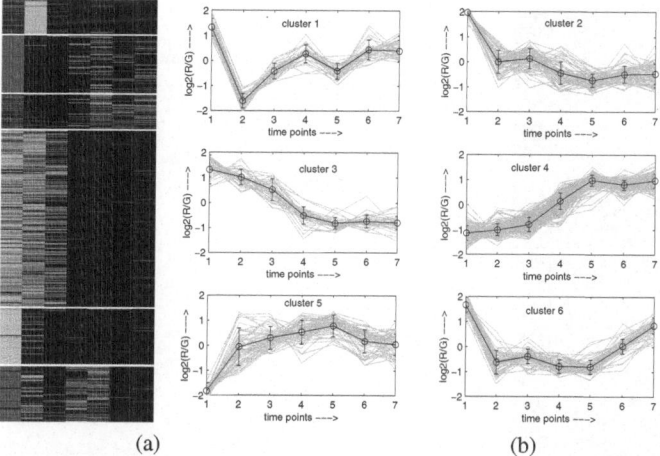

Fig. 6.8 Yeast Sporulation data clustered using the MOGA-SVM clustering method. (a) Eisen plot, (b) cluster profile plots

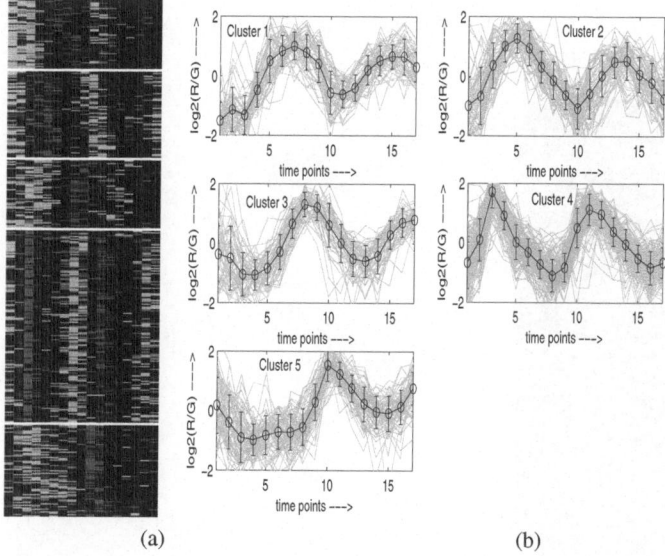

Fig. 6.9 Yeast Cell Cycle data clustered using the MOGA-SVM clustering method. (a) Eisen plot, (b) cluster profile plots

6.4.5 Statistical Significance Test

To establish that MOGA-SVM is significantly superior to the other algorithms, a non-parametric statistical significance test called Wilcoxon's rank sum test for independent samples [215] has been conducted at the 5% significance level. Except for

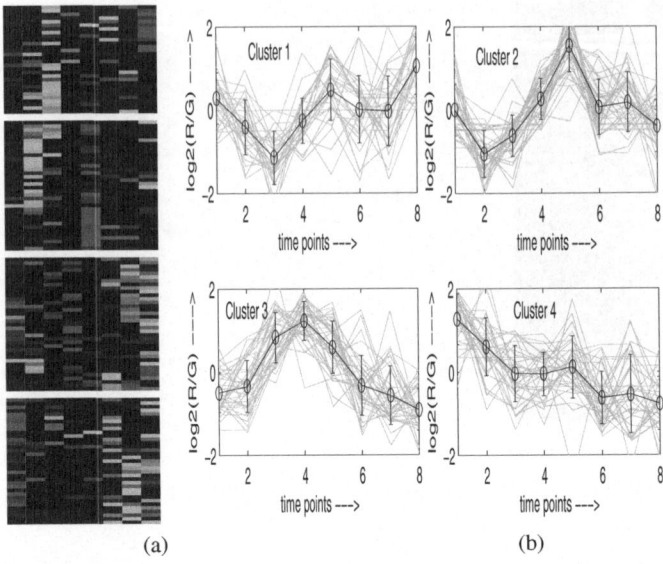

Fig. 6.10 Arabidopsis Thaliana data clustered using the MOGA-SVM clustering method. (a) Eisen plot, (b) cluster profile plots

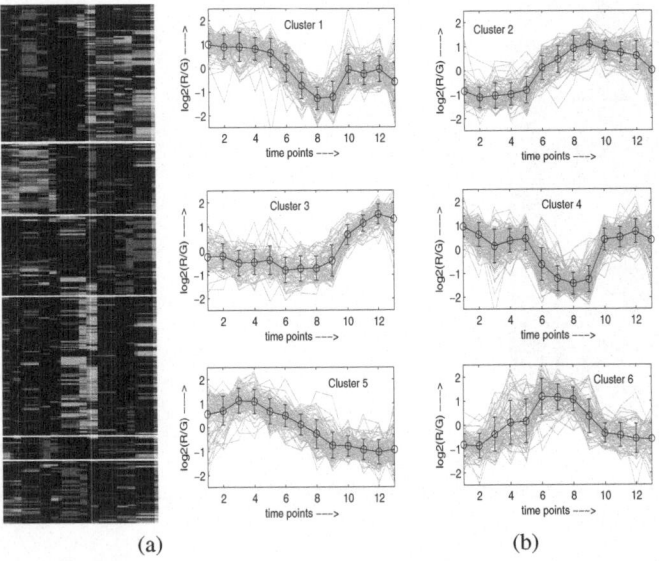

Fig. 6.11 Human Fibroblasts Serum data clustered using the MOGA-SVM clustering method. (a) Eisen plot, (b) cluster profile plots

average linkage, all other methods considered here are probabilistic in nature, i.e., they may produce different clustering results in different runs depending on the initialization. It has been found that in all the runs, MOGA-SVM produces a better $s(C)$

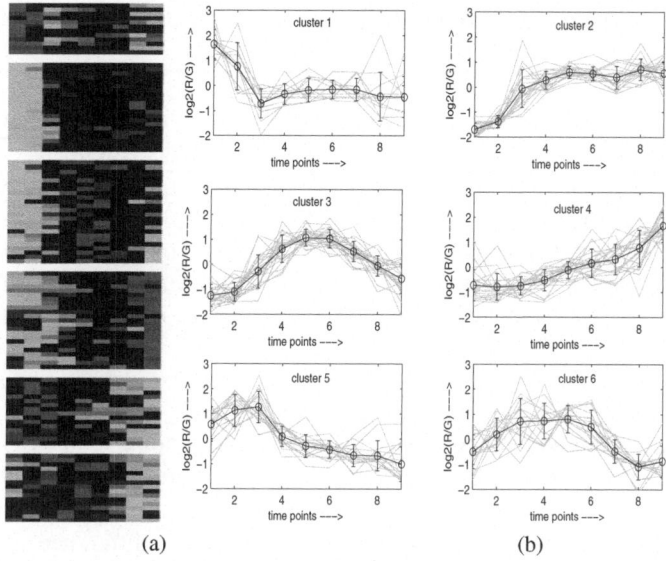

(a)   (b)

Fig. 6.12 Rat CNS data clustered using the MOGA-SVM clustering method. (a) Eisen plot, (b) cluster profile plots

index score than the average linkage algorithm. Therefore, the average linkage algorithm is not considered in the statistical test conducted. Seven groups, corresponding to the seven algorithms MOGA-SVM, MOGA (without SVM), MOGA_{crisp}-SVM (RBF), FCM, SGA, SOM, and CRC, have been created for each dataset. Each group consists of the $s(C)$ index scores produced over 20 runs of the corresponding algorithm. The median values of each group for all the datasets are reported in Table 6.5.

Table 6.5 Median values of Silhouette index scores over 20 consecutive runs of different algorithms

Algorithm	Sporulation	Cell cycle	Arabidopsis	Serum	Rat CNS
MOGA-SVM	0.6288	0.4498	0.4329	0.4148	0.5108
MOGA	0.5766	0.4221	0.4024	0.3844	0.4822
MOGA_{crisp}-SVM	0.6002	0.4301	0.4192	0.3901	0.4961
FCM	0.4686	0.3812	0.3656	0.3152	0.4113
SGA	0.5698	0.4315	0.3837	0.3672	0.4563
Average linkage	0.5007	0.4388	0.3151	0.3562	0.4122
SOM	0.5786	0.3823	0.2334	0.3352	0.4340
CRC	0.5619	0.4271	0.3955	0.3246	0.4561

As is evident from Table 6.5, the median values of $s(C)$ scores for MOGA-SVM are better than those for the other algorithms. To establish that this goodness is statistically significant, Table 6.6 reports the p-values produced by Wilcoxon's rank

Table 6.6 The p-values produced by Wilcoxon's rank sum test comparing MOGA-SVM with other algorithms

Data	p-values (comparing median $s(C)$ scores of MOGA-SVM with other algorithms)					
Sets	MOGA	FCM	MOGA$_{crisp}$-SVM	SGA	SOM	CRC
Spor	2.10E-03	2.17E-05	1.32E-03	2.41E-03	11.5E-03	5.20E-03
Cell	2.21E-03	1.67E-05	2.90E-05	1.30E-04	1.44E-04	1.90E-04
Thaliana	1.62E-03	1.43E-04	1.78E-03	5.80E-05	2.10E-03	1.08E-05
Serum	1.30E-04	1.52E-04	3.34E-04	1.48E-04	1.44E-04	1.39E-04
Rat	1.53E-04	1.08E-05	2.10E-04	1.53E-04	1.43E-04	1.68E-04

sum test for comparison of two groups (a group corresponding to MOGA-SVM and a group corresponding to some other algorithm) at a time. As a null hypothesis, it is assumed that there is no significant difference in the median values of the two groups; the alternative hypothesis is that there is significant difference in the median values of the two groups. All the p-values reported in the table are less than 0.05 (5% significance level). This is strong evidence against the null hypothesis, indicating that the better median value of the performance metric produced by MOGA-SVM is statistically significant and has not occurred by chance.

6.4.6 Biological Significance

The biological relevance of a cluster can be verified based on the statistically significant Gene Ontology (GO) annotation database (http://db.yeastgenome.org/cgi-bin/GO/goTermFinder). This is used to test the functional enrichment of a group of genes in terms of three structured, controlled vocabularies (ontologies), viz., associated biological processes, molecular functions and biological components. The degree of functional enrichment (p-value) is computed using a cumulative hypergeometric distribution. This measures the probability of finding the number of genes involved in a given GO term (i.e., function, process, component) within a cluster. From a given GO category, the probability p of getting k or more genes within a cluster of size n can be defined as [401]

$$p = 1 - \sum_{i=0}^{k-1} \frac{\binom{f}{i}\binom{g-f}{n-i}}{\binom{g}{n}}, \tag{6.3}$$

where f and g denote the total number of genes within a category and within the genome, respectively. Statistical significance is evaluated for the genes in a cluster by computing the p-value for each GO category. This signifies how well the genes in the cluster match with the different GO categories. If the majority of genes in a cluster have the same biological function, then it is unlikely that this takes place by chance, and the p-value of the category will be close to 0.

The biological significance test for the Yeast Sporulation data has been conducted at the 1% significance level. For different algorithms, the number of clusters for which the most significant GO terms have a p-value less than 0.01 (1% significance level) are as follows: MOGA-SVM, 6; MOGA (without SVM), 6; $MOGA_{crisp}$-SVM (RBF), 6; FCM, 4; SGA, 6; average linkage, 4; SOM, 4; and CRC, 6. In Figure 6.13, the boxplots of the p-values of the most significant GO terms of all the clusters having at least one significant GO term as obtained by the different algorithms are shown. The p-values are log-transformed for better readability. It is evident from the figure that the boxplot corresponding to the MOGA-SVM method has lower p-values (i.e., higher $-\log_{10}$(p-value)). This indicates that the clusters identified by MOGA-SVM are more biologically significant and functionally enriched than those by the other algorithms.

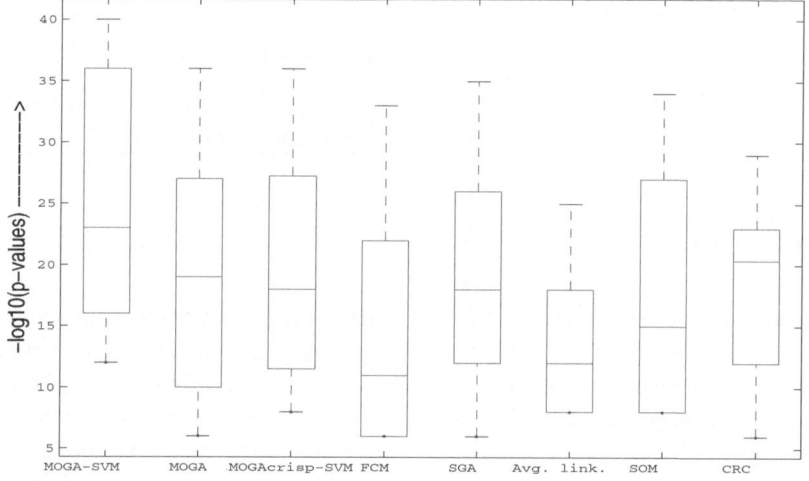

Fig. 6.13 Boxplots of the p-values of the most significant GO terms of all the clusters having at least one significant GO term as obtained by different algorithms for Yeast Sporulation data. The p-values are log-transformed for better readability

As an illustration, Table 6.7 reports the three most significant GO terms (along with the corresponding p-values) shared by the genes of each of the six clusters identified by the MOGA-SVM technique (Figure 6.8). As is evident from the table, all the clusters produced by the MOGA-SVM clustering scheme are significantly enriched with some GO categories, since all the p-values are less than 0.01 (1% significance level). This establishes that the MOGA-SVM clustering scheme is able to produce biologically relevant and functionally enriched clusters.

Table 6.7 The three most significant GO terms and the corresponding p-values for each of the six clusters of Yeast Sporulation data as found by the MOGA-SVM clustering technique

Clusters	Significant GO term	p -value
	ribosome biogenesis and assembly - GO:0042254	1.4E-37
1	intracellular non-membrane-bound organelle - GO:0043232	1.38E-23
	organelle lumen - GO:0043233	9.46E-21
	nucleotide metabolic process - GO:0009117	1.32E-8
2	glucose catabolic process - GO:0006007	2.86E-4
	external encapsulating structure - GO:0030312	3.39E-4
	organic acid metabolic process - GO:0006082	1.86E-14
3	amino acid and derivative metabolic process - GO:0006519	4.35E-4
	external encapsulating structure - GO:0030312	6.70E-4
	spore wall assembly (sensu fungi) - GO:0030476	8.97E-18
4	sporulation - GO:0030435	2.02E-18
	cell division - GO:0051301	7.92E-16
	M phase of meiotic cell cycle - GO:0051327	1.71E-23
5	M phase - GO:0000279	1.28E-20
	meiosis I - GO:0007127	5.10E-22
	cytosolic part - GO:0044445	1.4E-30
6	cytosol - GO:0005829	1.4E-30
	ribosomal large subunit assembly and maintenance - GO:0000027	7.42E-8

6.5 Summary

This chapter describes a recently proposed method for obtaining a final solution from the set of non-dominated solutions produced by an NSGA-II-based real-coded multiobjective fuzzy clustering scheme, that optimizes the Xie-Beni (XB) index and the J_m measure simultaneously. In this regard, a fuzzy majority voting technique followed by SVM classification has been utilized. The fuzzy majority voting technique is based on two user-specified parameters α (membership threshold) and β (majority voting threshold). Note that the MOGA-SVM technique described here keeps the values of α and β constant at 0.5. However, one can think of choosing these parameters adaptively, which can be a way for further improvement of the MOGA-SVM algorithm.

Results on remote sensing imagery (both numeric and image data) and five real-life gene expression datasets have been presented. The use of different kernel methods is investigated and the RBF kernel is found to perform the best. The performance of the MOGA-SVM technique has been compared with that of different clustering methods. The results have been demonstrated both quantitatively and visually. The MOGA-SVM clustering technique consistently outperformed the other algorithms considered here as it integrates multiobjective optimization, fuzzy clustering and supervised learning in an effective manner. For the gene expression data, statistical superiority has been established through statistical significance tests. Moreover, biological significance tests have been conducted to establish that the clusters identified by the MOGA-SVM technique are biologically significant.

Chapter 7
Two-Stage Fuzzy Clustering

7.1 Introduction

It has been observed that, in general, the performance of clustering algorithms degrades if the clusters in a dataset overlap, i.e., there are many points in the dataset that have significant membership in multiple classes (SiMM). It leads to a lot of confusion regarding their cluster assignments. Hence, it may be beneficial if these points are first identified and excluded from consideration while clustering the dataset. In the next stage, they can be assigned to a cluster using some classifier trained by the remaining points. In this work, a Support Vector Machine- (SVM-)based classifier [417] has been utilized for this purpose.

In this chapter, a two-stage fuzzy clustering algorithm has been described that utilizes the concept of points having significant membership in more than one cluster. Performance of the two-stage clustering method is compared to that of conventional FCM, K-means, hierarchical average linkage and the Genetic Algorithm-based method on two artificial datasets. Also it is applied to cluster two Indian Remote Sensing (IRS) satellite images [1] of parts of the cities of Kolkata and Mumbai. The superiority of the clustering technique described here, as compared to the well-known FCM, K-means and single-stage GA-based clustering is demonstrated both visually (by showing the clustered images) and using the cluster validity index $\mathscr{I}$ [299]. Moreover, statistical significance tests have been conducted to establish the significant superiority of the two-stage clustering technique.

The performance of the SiMM-TS clustering method is also compared with that of well-known gene expression data clustering methods, viz., average linkage, SOM and CRC. One artificial and three real-life gene expression datasets are considered for experiments. Statistical and biological significance tests have also been conducted.

7.2 SiMM-TS: Two-Stage Clustering Technique

In this section, the SiMM points-based two-stage (SiMM-TS) clustering algorithm is discussed. First, the technique for identifying the SiMM points has been described. Subsequently, the clustering algorithm based on SiMM point identification has been discussed.

7.2.1 Identification of SiMM Points

The matrix $U(X,K)$ produced by some fuzzy clustering technique (X and K denote the dataset and the number of clusters, respectively) is used to find out the points which have significant multi-class membership (SiMM), i.e., the points which are situated at the overlapping regions of two or more clusters, and hence cannot be assigned to any cluster with a reasonable amount of certainty. Let us assume that a particular point $x_j \in X$ has the highest membership value in cluster C_q, and the next highest membership value in cluster C_r, i.e., $u_{qj} \geq u_{rj} \geq u_{kj}$ where $k = 1, \ldots, K$, $k \neq q$, and $k \neq r$. Suppose the difference in the membership values u_{qj} and u_{rj} is δ, i.e., $\delta = u_{qj} - u_{rj}$. Let $\mathscr{B}$ be the set of points lying on the overlapping regions of two or more clusters (SiMM points) and $Prob(x_j \in \mathscr{B})$ denotes the probability of x_j belonging to $\mathscr{B}$. Evidently, as δ increases, x_j can be assigned more confidently to cluster C_q and hence less confidently to $\mathscr{B}$. Therefore,

$$Prob(x_j \in \mathscr{B}) \ \propto \ \frac{1}{\delta}. \tag{7.1}$$

For each point in the dataset X, the value δ is calculated. Now the data points are sorted in ascending order of their δ values. Hence, in the sorted list, the probability $Prob(x_j \in \mathscr{B})$ decreases as we move towards the tail of the list. A tuning parameter $\mathscr{P}$ is defined on the sorted list such that the first $\mathscr{P}\%$ points from the sorted list are chosen to be the SiMM points. The value of $\mathscr{P}$ should be chosen carefully so that the appropriate set of SiMM points is identified.

7.2.2 SiMM-TS Clustering

The algorithm described here has two different stages.

7.2.2.1 Stage I

In the first stage, the underlying dataset is clustered using either an iterated version of FCM (IFCM) or a variable string length GA-based fuzzy clustering algorithm minimizing the XB index (VGA) [300] to evolve the number of clusters K as well as the fuzzy partition matrix $U(X,K)$. In IFCM, the dataset X is clustered using FCM with different values of K from 2 to $\sqrt{n}$, where n is the number of points in the dataset. The solution producing the best XB index value is considered and the corresponding partition matrix $U(X,K)$ is used for further processing. Using the resulting partition matrix obtained by IFCM or VGA, the SiMM points are identified using the technique discussed in Section 7.2.1.

7.2.2.2 Stage II

In the second stage of SiMM-TS, the SiMM points are excluded from the dataset and the remaining points are re-clustered into K clusters using any of FCM, fixed length single objective GA [298] and MOGA-based clustering algorithms (described in Chapter 5) [41]. The fuzzy clustering result is defuzzified by assigning each of these remaining points to the cluster to which it has the highest membership degree. Next, the SVM classifier is trained by these points. Thereafter, each of the SiMM points that was identified in the first stage, and excluded from consideration in the second, is assigned to a cluster as predicted by the trained SVM classifier. Here, SVM classifier with RBF kernel has been used.

7.3 Illustration of the SiMM-TS Clustering Process

The experimental results of clustering using the SiMM-TS clustering scheme are provided for two artificial datasets, Data 1 and Data 2, described in Chapter 5. The Minkowski score is used to evaluate the performance of the clustering methods.

7.3.1 Input Parameters

The GA-based algorithms have been run for 100 generations with population size 50. The crossover and mutation probabilities were taken as 0.8 and 0.1, respectively. The FCM and K-means algorithms have been executed for 100 iterations unless they converge before that. The fuzziness parameter m is taken as 2.0. For SiMM-TS clustering, the tuning parameter $\mathscr{P}$ is chosen experimentally, i.e., SiMM-TS is run for different values of $\mathscr{P}$ ranging from 5% to 15%, and the best solution has been

chosen. Each algorithm has been run 20 times and the best performance index value
has been reported.

7.3.2 Results

Tables 7.1 and 7.2 show the comparative results in terms of Minkowski scores obtained by the different algorithms for the two datasets, respectively. As can be seen
from the tables, irrespective of the clustering methods used in the two-stage clustering algorithm, the performance improves after the application of the second stage
of clustering. For example, in the case of Data 1, the Minkowski score after the application of VGA in the first stage of SiMM-TS is 0.4398, while this gets improved
to 0.3418 (with MOGA at the second stage) at the end. Similarly, when IFCM is
applied in the first stage, the score is 0.4404, which gets improved to 0.3527 (with
MOGA at the second stage). It also appears that both VGA and IFCM correctly identify the number of clusters in the datasets. The final Minkowski scores are also better than those obtained using the average linkage and K-means clustering methods.
Similar results are found for Data 2. The results demonstrate the utility of adopting
the approach presented in this paper, irrespective of the clustering method used.

Table 7.1 Results for Data 1

Stage I Algorithm	K	Stage I MS	Stage II Algorithm	Final MS	Average linkage (K=5)	K-means (K=5)
VGA	5	0.4398	FCM	0.4037		
			GA	0.3911		
			MOGA	0.3418	0.4360	0.4243
IFCM	5	0.4404	FCM	0.4146		
			GA	0.3661		
			MOGA	0.3527		

Table 7.2 Results for Data 2

Stage I Algorithm	K	Stage I MS	Stage II Algorithm	Final MS	Average linkage (K=9)	K-means (K=9)
VGA	9	0.5295	FCM	0.4804		
			GA	0.4718		
			MOGA	0.4418	0.6516	0.6161
IFCM	9	0.5314	FCM	0.5022		
			GA	0.5022		
			MOGA	0.4639		

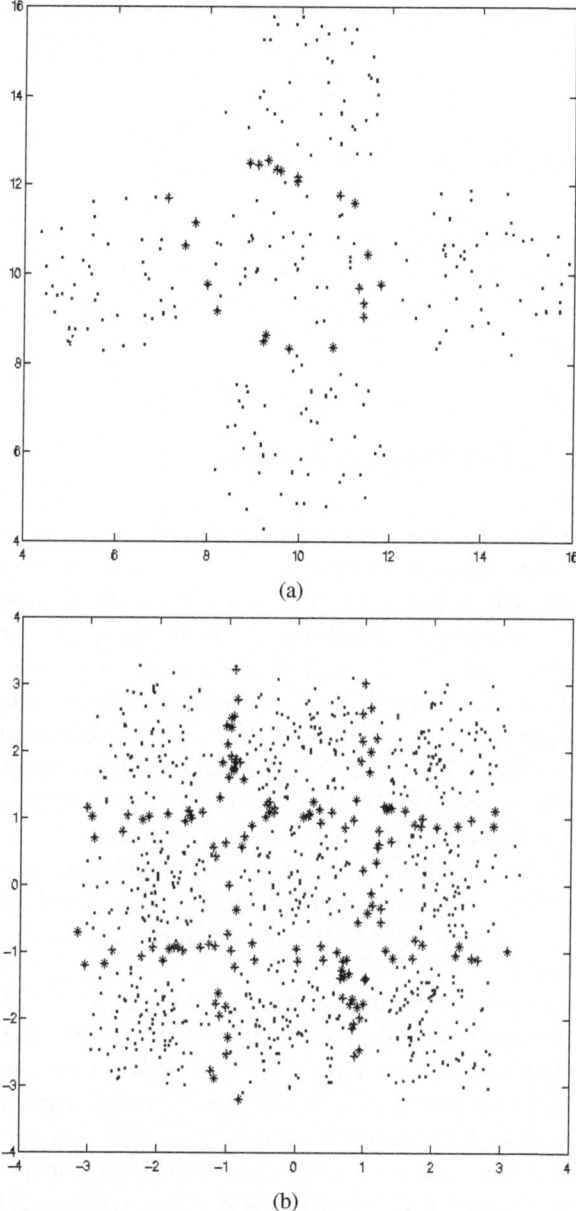

Fig. 7.1 Artificial dataset with the identified SiMM points marked as '*': (a) Data 1, (b) Data 2

Figure 7.1 shows, for the purpose of illustration, the SiMM point identified by one of the runs of the SiMM points identification method for Data 1 and Data 2. As is evident from the figure, these points are situated at the overlapping regions of two or more clusters.

7.4 Application to Remote Sensing Imagery

This section presents the results of application of SiMM-TS clustering on the two IRS images of parts of the cities of Kolkata and Mumbai, as described in Chapter 5. Results are demonstrated both visually and using the cluster validity index $\mathscr{I}$ [299] described in Chapter 1.

7.4.1 Results for IRS Image of Kolkata

Figure 7.2 shows the Kolkata image partitioned using the two-stage fuzzy algorithm (VGA-MOGA combination in the two stages). It appears that the water class has been differentiated into turbid water (the river Hooghly) and pond water (canal, fisheries, etc.). This is expected because they differ in their spectral properties. Here, the class turbid water contains sea water, river water, etc., where the soil content is more than that of pond water. The Salt Lake township has come out as a combination of concrete and open space, which appears to be correct, as the township is known to have several open spaces. The canal bounding Salt Lake from the upper portion has also been correctly classified as PW. The airstrips of Dumdum airport are classified correctly as belonging to the class concrete. The presence of some small areas of PW beside the airstrips is also correct as these correspond to the several ponds. The high proportion of concrete on both sides of the river, particularly towards the lower region of the image, is also correct. This region corresponds to the central part of the city of Kolkata.

Figure 7.3 demonstrates the clustering result of the FCM algorithm on the IRS Kolkata image. It appears from the figure that the river Hooghly and the city region have been incorrectly classified into the same class. It is also evident that the whole Salt Lake city has been wrongly put into one class. It is also apparent that although some portions such as the fisheries, the canals, parts of the airstrips are identified correctly, there is a significant amount of confusion in the FCM clustering result.

The K-means clustering result for the Kolkata image is shown in Figure 7.4. The figure implies that for K-means clustering also, a lot of confusion exists between the classes turbid water and concrete. Also the open spaces and concrete classes have some amount of overlapping. It appears also that the airstrips of Dumdum airport are not very clear.

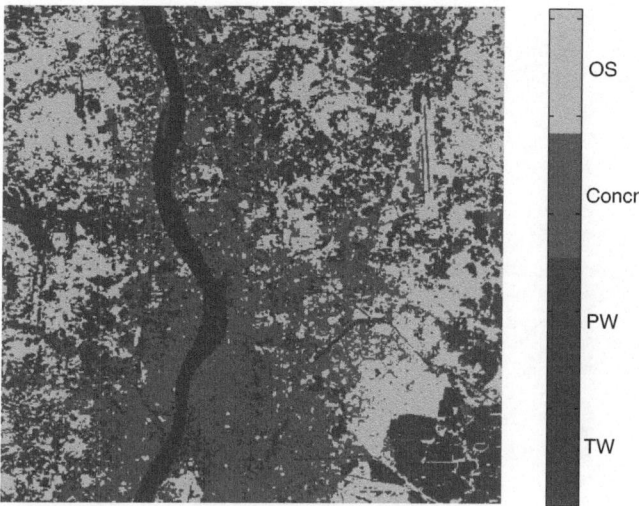

Fig. 7.2 Clustered IRS image of Kolkata using SiMM-TS (VGA-MOGA) clustering

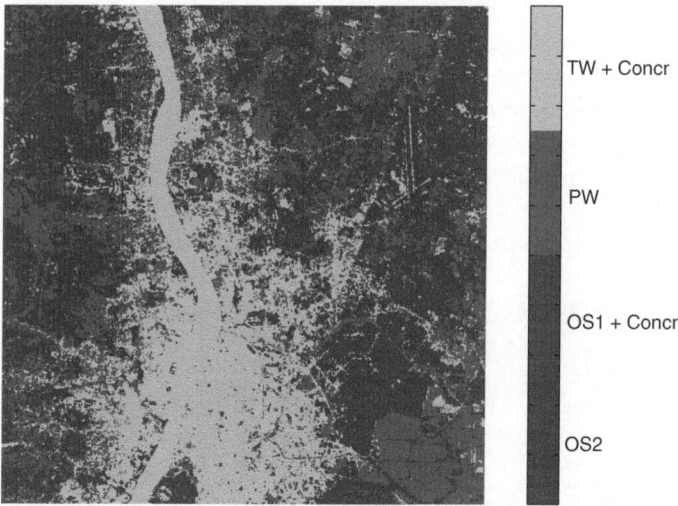

Fig. 7.3 Clustered IRS image of Kolkata using FCM clustering

7.4.2 Results for IRS Image of Mumbai

The result of the application of the SiMM-TS (VGA-MOGA combination) cluster-ing technique on the Mumbai image is shown in Figure 7.5. The southern part of the city, which is heavily industrialized, has been classified as primarily belonging to habitation and concrete. Here, the class habitation represents the regions having concrete structures and buildings, but with relatively lower density than the class

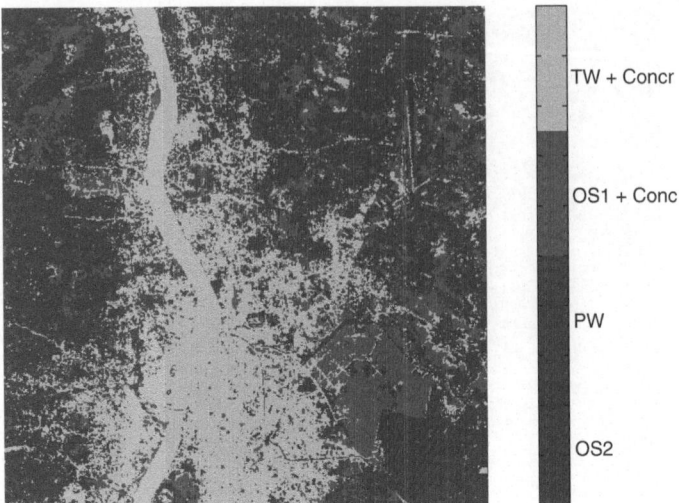

Fig. 7.4 Clustered IRS image of Kolkata using K-means clustering

concrete. Hence these two classes share common properties. From the result, it appears that the large water body of the Arabian Sea is grouped into two classes (TW1 and TW2). It is evident from the figure that the sea water has two distinct regions with different spectral properties. Hence the clustering result providing two partitions for this region is quite expected. Most of the islands, the dockyard, and several road structures have been correctly identified in the image. As expected, there is a high proportion of open space and vegetation within the islands.

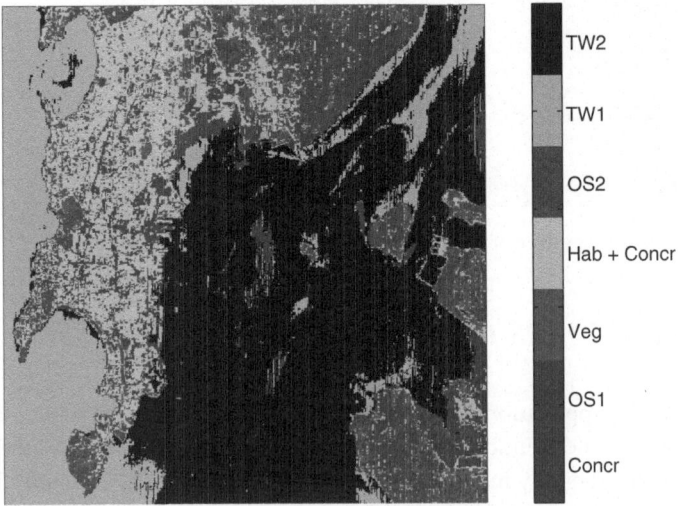

Fig. 7.5 Clustered IRS image of Mumbai using SiMM-TS (VGA-MOGA) clustering

Figure 7.6 demonstrates the Mumbai image clustered using the FCM clustering algorithm. It can be noted from the figure that the water of the Arabian Sea has been wrongly clustered into three regions, rather than two, as obtained earlier. It appears that the other regions in the image have been classified more or less correctly for this data.

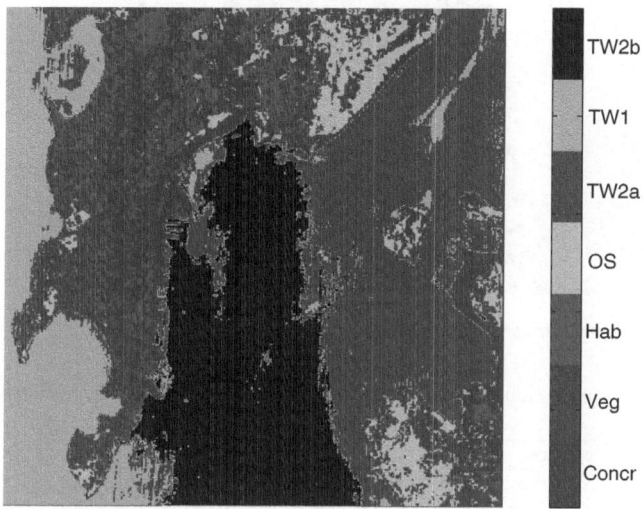

Fig. 7.6 Clustered IRS image of Mumbai using FCM clustering

In Figure 7.7, the Mumbai image clustered using K-means clustering has been shown. It appears from the figure that the sea area is wrongly classified into four different regions. Also, there is overlapping between the classes turbid water and concrete, as well as between open space and vegetation.

7.4.3 Numerical Results

In Table 7.3, the best $\mathscr{I}$ index values over 20 consecutive runs of the SiMM-TS clustering technique (with different combinations of VGA and IFCM in the first stage and FCM, GA and MOGA in the second stage), along with single-stage VGA, FCM and K-means algorithms, are tabulated. It is evident from the table that the two-stage clustering algorithm with the VGA-MOGA combination performs the best, far outperforming the single stage VGA, FCM and K-means algorithms. Hence it follows from the results that the SiMM-TS clustering algorithm outperforms the single-stage VGA, FCM and K-means clustering algorithms.

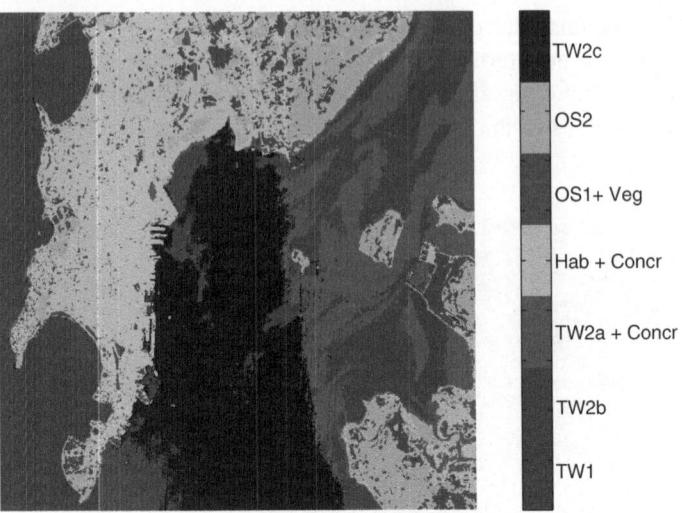

Fig. 7.7 Clustered IRS image of Mumbai using K-means clustering

Table 7.3 Best $\mathscr{I}$ index values for IRS Kolkata and Mumbai images for different algorithms over 20 runs

Method	Kolkata	Mumbai
Two-stage (VGA-FCM) clustering	102.3039	193.9961
Two-stage (VGA-GA) clustering	103.8810	198.1145
Two-stage (VGA-MOGA) clustering	**111.3328**	**209.8299**
Two-stage (IFCM-FCM) clustering	85.5597	188.3759
Two-stage (IFCM-GA) clustering	89.2094	191.8271
Two-stage (IFCM-MOGA) clustering	94.3211	192.7360
VGA clustering	81.5934	180.4512
FCM clustering	31.1697	178.0322
K-means clustering	54.3139	161.3783

7.4.4 Statistical Significance Test

Table 7.4 reports the average values of the $\mathscr{I}$ index scores produced by different algorithms over 20 consecutive runs for the Mumbai and Kolkata images. It is evident from the table that for both the images, the SiMM-TS clustering technique with the VGA-MOGA combination produces the best average $\mathscr{I}$ index scores. To establish that the better average $\mathscr{I}$ index scores provided by VGA-MOGA is statistically significant and does not come by chance, it is necessary to conduct a statistical significance test.

A statistical significance test called t-test [159] has been carried out at the 5% significance level to compare the average $\mathscr{I}$ index scores produced by different algorithms. Nine groups, corresponding to the nine algorithms (VGA-FCM, VGA-GA, VGA-MOGA, IFCM-FCM, IFCM-GA, IFCM-MOGA, VGA, FCM and K-means),

Table 7.4 Average $\mathscr{I}$ index values for IRS Kolkata and Mumbai images for different algorithms over 20 runs

Method	Kolkata	Mumbai
Two-stage (VGA-FCM) clustering	98.3677	190.9300
Two-stage (VGA-GA) clustering	99.1122	194.6330
Two-stage (VGA-MOGA) clustering	**105.2473**	**204.2614**
Two-stage (IFCM-FCM) clustering	82.1449	185.6765
Two-stage (IFCM-GA) clustering	86.0098	187.7561
Two-stage (IFCM-MOGA) clustering	90.9404	189.9294
VGA clustering	80.1971	177.5296
FCM clustering	28.5166	173.8489
K-means clustering	50.8785	157.5376

have been created for each of the two image datasets considered here. Each group consists of $\mathscr{I}$ index scores produced by 20 consecutive runs of the corresponding algorithm. Two groups are compared at a time, one corresponding to the SiMM-TS (VGA-MOGA) scheme and the other corresponding to some other algorithm considered here.

Table 7.5 The t-test results for the IRS Kolkata image

Test No.	Null hypothesis $(H_0 : \mu_1 = \mu_2)$	t-test statistic	p-value	Accept/ Reject
1	$\mu_{VGA-MOGA} = \mu_{VGA-FCM}$	4.7217	1.7016E-004	Reject
2	$\mu_{VGA-MOGA} = \mu_{VGA-GA}$	4.9114	1.1251E-004	Reject
3	$\mu_{VGA-MOGA} = \mu_{IFCM-FCM}$	16.7409	2.0255E-012	Reject
4	$\mu_{VGA-MOGA} = \mu_{IFCM-GA}$	14.2634	2.9843E-011	Reject
5	$\mu_{VGA-MOGA} = \mu_{IFCM-MOGA}$	10.1479	7.1216E-009	Reject
6	$\mu_{VGA-MOGA} = \mu_{VGA}$	20.4079	6.7946E-014	Reject
7	$\mu_{VGA-MOGA} = \mu_{IFCM}$	59.3758	6.8433E-042	Reject
8	$\mu_{VGA-MOGA} = \mu_{K-means}$	36.1834	4.2241E-030	Reject

Table 7.6 The t-test results for the IRS Mumbai image

Test No.	Null hypothesis $(H_0 : \mu_1 = \mu_2)$	t-test statistic	p-value	Accept/ Reject
1	$\mu_{VGA-MOGA} = \mu_{VGA-FCM}$	12.7033	2.0065E-010	Reject
2	$\mu_{VGA-MOGA} = \mu_{VGA-GA}$	9.0797	3.8615E-008	Reject
3	$\mu_{VGA-MOGA} = \mu_{IFCM-FCM}$	19.4688	1.5321E-013	Reject
4	$\mu_{VGA-MOGA} = \mu_{IFCM-GA}$	14.2710	2.9582E-011	Reject
5	$\mu_{VGA-MOGA} = \mu_{IFCM-MOGA}$	14.6817	1.8445E-011	Reject
6	$\mu_{VGA-MOGA} = \mu_{VGA}$	16.0858	3.9810E-012	Reject
7	$\mu_{VGA-MOGA} = \mu_{IFCM}$	26.1787	8.8818E-016	Reject
8	$\mu_{VGA-MOGA} = \mu_{K-means}$	30.2918	5.1039E-029	Reject

Tables 7.5 and 7.6 report the results of the t-test for the IRS Kolkata and Mumbai image, respectively. The null hypothesis (the means of two groups are equal) are shown in the tables. The alternative hypothesis is that the mean of the first group is larger than the mean of the second group. For each test, the degree of freedom is $M + N - 2$, where M and N are the sizes of the two groups considered. Here $M = N = 20$. Hence the degree of freedom is 38. Also, the values of the t-statistic and the probability p-value of accepting the null hypothesis are shown in the tables. It is clear from the tables that the p-values are much less than 0.05 (5% significance level), which is strong evidence for rejecting the null hypothesis. This proves that the better average $\mathscr{I}$ index values produced by the SiMM-TS clustering with the VGA-MOGA scheme is statistically significant and has not come by chance.

7.5 Application to Microarray Gene Expression Data

In this section, the performance of the SiMM-TS (using the VGA-MOGA combination) is compared with that of well-known gene expression data clustering methods, viz., average linkage, SOM and Chinese Restaurant Clustering (CRC), and also with VGA and IFCM applied independently. One artificial and three real-life gene expression datasets are considered for experiments. For evaluating the performance of the clustering algorithms, the adjusted Rand index (ARI) [449] and the Silhouette index ($s(C)$) [369] are used for artificial (where true clustering is known) and real-life (where true clustering is unknown) gene expression datasets, respectively. Also two cluster visualization tools, namely the Eisen plot and cluster profile plot, have been utilized.

7.5.1 Input Parameters

The GA-based clustering techniques (both single objective and multiobjective) are executed for 100 generations with a fixed population size of 50. The crossover and mutation probabilities are chosen to be 0.8 and 0.01 respectively. The number of iterations for the FCM algorithm is taken as 200 unless it converges before that. The fuzzy exponent m is chosen as in [132,249], and the values of m for the datasets AD400_10_10, Sporulation, Serum and Rat CNS are obtained as 1.32, 1.34, 1.25 and 1.21, respectively. The parameter $\mathscr{P}$ for SiMM-TS clustering is chosen by experiment, i.e., the algorithm is run for different values of $\mathscr{P}$ ranging from 5% to 15%, and the best solution in terms of adjusted Rand index or Silhouette index is chosen. All the algorithms have been executed ten times and the average performance scores (the ARI) values for artificial data where the true clustering is known and the $s(C)$ values for real-life data where the true clustering is unknown) are reported.

7.5.2 Datasets

One artificial dataset, AD400_10_10, and three real-life datasets, viz., Yeast Sporulation, Human Fibroblasts Serum and Rat CNS, are considered for experiment.

7.5.2.1 AD400_10_10

This dataset consists of expression levels of 400 genes across ten time points. It is generated as in [448]. The dataset has ten clusters, each containing 40 genes, representing ten different expression patterns.

7.5.2.2 Real-Life Datasets

Three real-life gene expression datasets, viz., Yeast Sporulation, Human Fibroblasts Serum and Rat CNS (described in Sections 5.6.1 and 6.4.1), have been used for experiments.

Each dataset is normalized so that each row has mean 0 and variance 1 (Z normalization) [249].

7.5.3 Results for AD400_10_10 Data

To establish the fact that the SiMM-TS clustering algorithm (VGA followed by MOGA) performs better in terms of partitioning as well as determining the correct number of clusters, it was compared to the other combinations of algorithms in the two stages discussed in Section 7.5 for AD400_10_10. The average ARI and $s(C)$ values are reported in Table 7.7 (refer to the first eight rows) over ten runs of each algorithm considered here. The table also contains the results produced by the average linkage, SOM and CRC algorithms. The number of clusters found in a majority of the runs is reported. It is evident from the table that the VGA (and hence the algorithms that use VGA in the first stage) has correctly evaluated the number of clusters to be ten, whereas IFCM and CRC wrongly found 11 and eight clusters in the dataset, respectively. Average linkage and SOM have been executed with ten clusters. It also appears from the table that irrespective of any algorithm in the two stages, application of second-stage clustering improves the *ARI* and $s(C)$ values, and the SiMM-TS (VGA-MOGA) technique provides the best of those values compared to any other combinations. Similar results have been obtained for other datasets.

These results indicate that for this dataset the SiMM-TS clustering method with VGA and MOGA in the first and second stages, respectively, performs better than any other combination of algorithms in the two stages, and is also better than the other algorithms considered for comparison. This is because VGA is capable of evolving the number of clusters automatically due to its variable string length and associated operators. Also, the use of Genetic Algorithm instead of IFCM in the first stage, enables the algorithm to come out of the local optima. Finally, the use of MOGA in the second stage allows the method to suitably balance different characteristics of clustering, unlike single objective techniques, which concentrate only on one characteristic by totally ignoring the others. Henceforth, SiMM-TS will indicate the VGA-MOGA combination only.

Table 7.7 Comparison of SiMM-TS (VGA-MOGA) with other combinations of algorithms in the two stages in terms of *ARI* and $s(C)$ for AD400_10_10 data

Algorithm	K	ARI	$s(C)$
SiMM-TS (VGA-MOGA)	10	0.6299	0.4523
IFCM	11	0.5054	0.2845
VGA	10	0.5274	0.3638
IFCM-FCM	11	0.5554	0.3692
IFCM-SGA	11	0.5835	0.3766
IFCM-MOGA	11	0.5854	0.3875
VGA-FCM	10	0.5894	0.4033
VGA-SGA	10	0.6045	0.4221
Average linkage	10	0.3877	0.2674
SOM	10	0.5147	0.3426
CRC	8	0.5803	0.3885

For the purpose of illustration, Figure 7.8 shows the boxplots representing the *ARI* scores produced by each algorithm for the AD400_10_10 dataset over ten runs. It is evident from the figure that the *ARI* index scores produced by the SiMM-TS (VGA-MOGA) clustering are better than those produced by the other algorithms.

7.5.4 Results for Yeast Sporulation Data

Table 7.8 shows the Silhouette index values for the algorithms SiMM-TS, IFCM, VGA, average linkage, SOM and CRC. It is evident from the table that VGA has determined the number of clusters as six, IFCM finds it to be seven, whereas the CRC algorithm gives $K = 8$. Note that among these three methods, VGA provides the best $s(C)$ value. Thus, for average linkage and SOM, K is chosen to be six. From the $s(C)$ values, it can be noticed that the SiMM-TS clustering performs the best, IFCM and average linkage perform poorly, whereas SOM and CRC perform reasonably well.

Figure 7.9 shows the boxplots representing the $s(C)$ index scores produced by each algorithm for the Yeast Sporulation dataset over ten runs of the algorithms. It

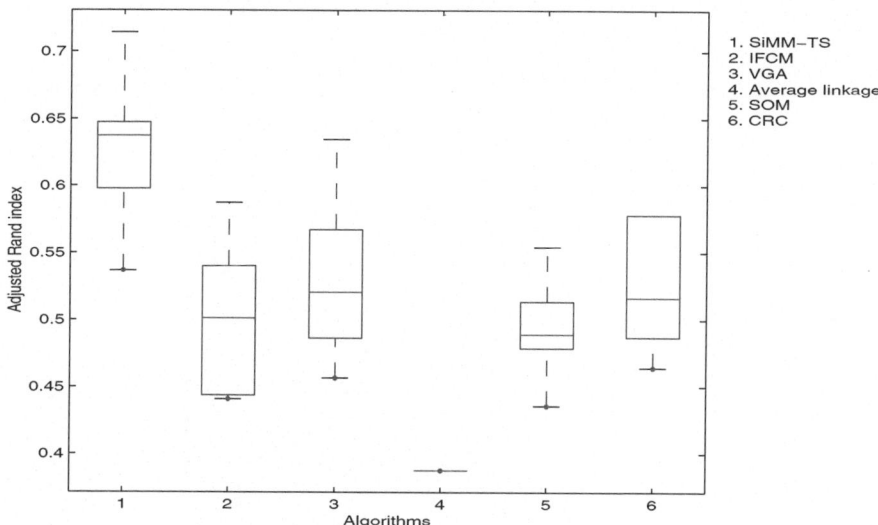

Fig. 7.8 Boxplots of the *ARI* index values produced by different algorithms for the AD400_10_10 dataset over ten runs.

Table 7.8 $s(C)$ values for real-life gene expression datasets

Algorithm	Sporulation		Serum		CNS	
	K	$s(C)$	K	$s(C)$	K	$s(C)$
SiMM-TS	6	0.6247	6	0.4289	6	0.5239
IFCM	7	0.4755	8	0.2995	5	0.4050
VGA	6	0.5703	6	0.3443	6	0.4486
Average linkage	6	0.5007	6	0.3092	6	0.3684
SOM	6	0.5845	6	0.3235	6	0.4122
CRC	8	0.5622	10	0.3174	4	0.4423

is evident from the figure that the $s(C)$ index scores produced by SiMM-TS (VGA-MOGA) clustering are better than those produced by the other algorithms.

For the purpose of illustration, the Eisen plot and the cluster profile plots for the clustering solution found in one of the runs of SiMM-TS have been shown in Figure 7.10. The six clusters are very much clear from the Eisen plot (Figure 7.10(a)). It is evident from the figure that the expression profiles of the genes of a cluster are similar to each other and produce similar color patterns. Cluster profile plots (Figure 7.10(b)) also demonstrate how the cluster profiles for the different groups of genes differ from each other, while the profiles within a group are reasonably similar.

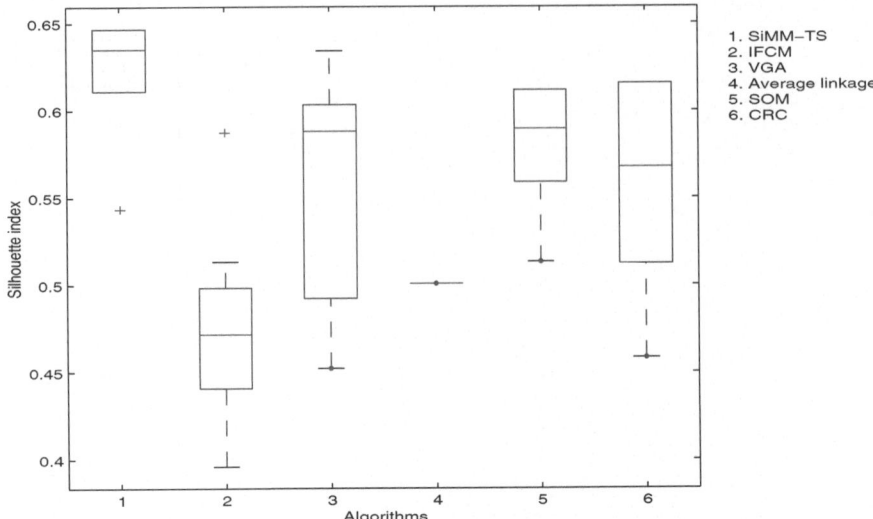

Fig. 7.9 Boxplots of the $s(C)$ index values produced by different algorithms for the Yeast Sporulation dataset over ten runs

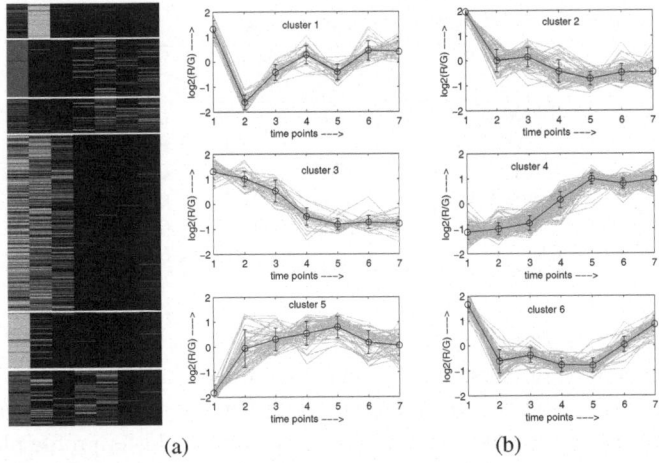

Fig. 7.10 Yeast Sporulation data clustered using the SiMM-TS clustering method. (a) Eisen plot, (b) cluster profile plots

7.5.5 Results for Human Fibroblasts Serum Data

The $s(C)$ values obtained by the different clustering algorithms on the Human Fibroblasts Serum data are shown in Table 7.8. Again, the VGA algorithm determines the number of clusters as six, whereas IFCM and CRC found eight and ten clusters, respectively, with the VGA providing the best value of $s(C)$ from among the three. The average linkage and SOM algorithms are therefore executed with six clusters. For this dataset also, the SiMM-TS clustering method provides a much better value of $s(C)$ than all the other algorithms, including VGA, while IFCM and average linkage perform poorly. It is also evident from the boxplots of $s(C)$ index scores for Serum data (Figure 7.11) that SiMM-TS (VGA-MOGA) produces a better $s(C)$ index scores that the other algorithms. Figure 7.12 shows the Eisen plot and the cluster profile plots of the clustering solution obtained in one of the runs of SiMM-TS algorithm on this data.

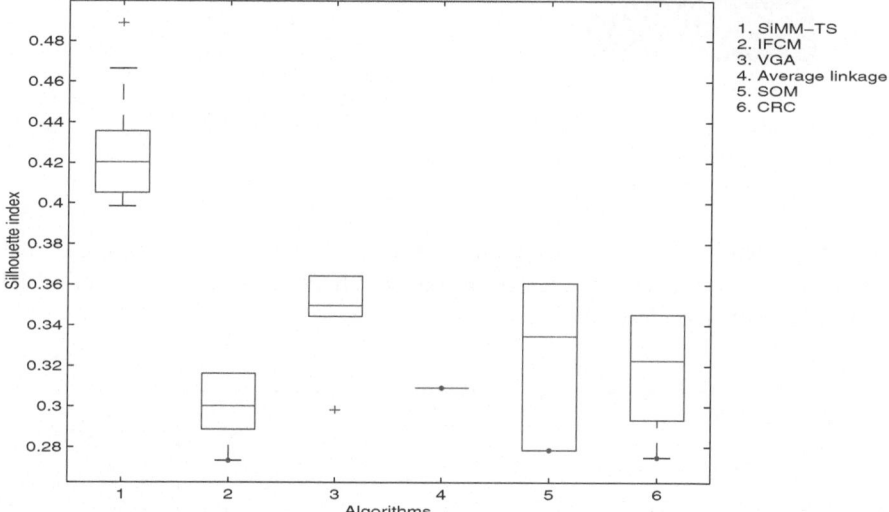

Fig. 7.11 Boxplots of the $s(C)$ index values produced by different algorithms for the Human Fibroblasts Serum dataset over ten runs

7.5.6 Results for Rat CNS Data

Table 7.8 reports the $s(C)$ values for the clustering results obtained by the different algorithms on Rat CNS data. VGA gives the number of clusters as six, which is the same as that found in [433]. IFCM and CRC found five and four clusters in the dataset, respectively, but with poorer $s(C)$ values than VGA. Thus, for average linkage and SOM, $K = 6$ is assumed. For this dataset also, the SiMM-TS method outperforms all the other algorithms in terms of $s(C)$. Figure 7.13 indicates that the $s(C)$ index scores produced by SiMM-TS (VGA-MOGA) clustering are better than

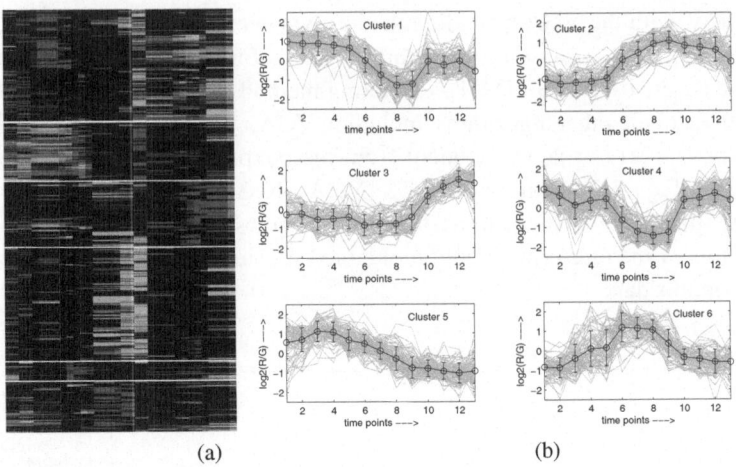

(a)   (b)

Fig. 7.12 Human Fibroblasts Serum data clustered using the SiMM-TS clustering method. (a) Eisen plot, (b) cluster profile plots

those produced by the other algorithms. Figure 7.14 illustrates the Eisen plot and the cluster profile plots of the clustering solution obtained in one of the runs of the SiMM-TS method on Rat CNS data.

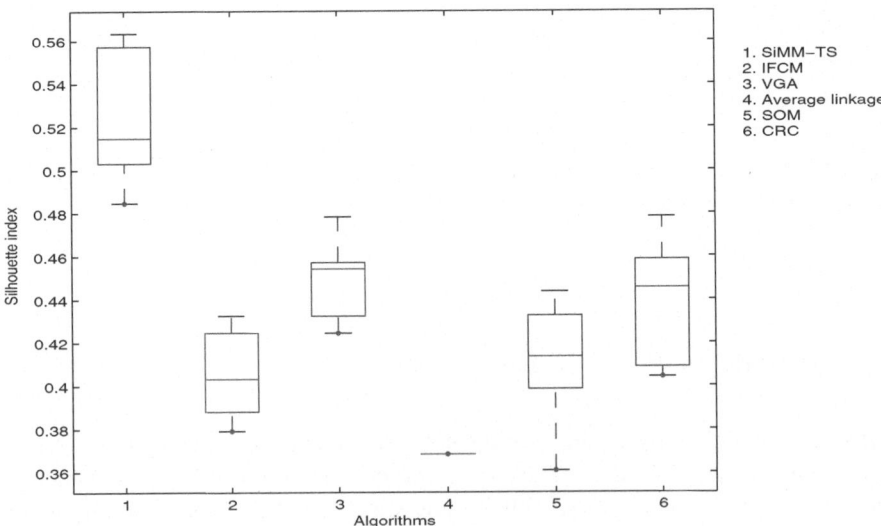

Fig. 7.13 Boxplots of the $s(C)$ index values produced by different algorithms for the Rat CNS dataset over ten runs

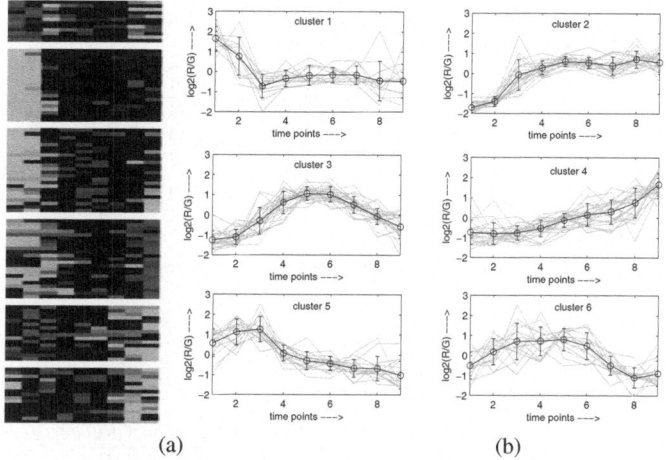

Fig. 7.14 Rat CNS data clustered using the SiMM-TS clustering method. (a) Eisen plot, (b) cluster profile plots

The results indicate significant improvement in clustering performance using the SiMM-TS clustering approach as compared to the other algorithms. In all the cases considered here, VGA is found to outperform IFCM in determining both the appropriate number of clusters and the partitioning in terms of the *ARI* and $s(C)$ indices. To justify that the solutions found by SiMM-TS are statistically superior to those of any other algorithm applied separately, statistical significance tests have been conducted and the results are reported in the next section.

7.5.7 Statistical Significance Test

A non-parametric statistical significance test called Wilcoxon's rank sum test for independent samples [215] has been conducted at the 5% significance level. Six groups, corresponding to six algorithms (SiMM-TS, VGA, IFCM, average linkage, SOM, CRC), have been created for each dataset. Each group consists of the performance scores (*ARI* for the artificial data and $s(C)$ for the real-life data) produced by ten consecutive runs of the corresponding algorithm. The median values of each group for all the datasets are shown in Table 7.9.

It is evident from Table 7.9 that the median values for SiMM-TS are better than those for other algorithms. To establish that this goodness is statistically significant, Table 7.10 reports the p-values produced by Wilcoxon's rank sum test for comparison of two groups (the group corresponding to SiMM-TS and a group corresponding to some other algorithm) at a time. As a null hypothesis, it is assumed that there is no significant difference between the median values of two groups. The alternative

Table 7.9 Median values of performance parameters (*ARI* for artificial and $s(C)$ for real-life datasets) over ten consecutive runs of different algorithms

Algorithms	AD400_10_10	Sporulation	Serum	Rat CNS
SiMM-TS	0.6370	0.6353	0.4203	0.5147
IFCM	0.5011	0.4717	0.3002	0.4032
VGA	0.5204	0.5880	0.3498	0.4542
Avg. Link	0.3877	0.5007	0.3092	0.3684
SOM	0.4891	0.5892	0.3345	0.4134
CRC	0.5162	0.5675	0.3227	0.4455

Table 7.10 The p-values produced by Wilcoxon's rank sum test comparing SiMM-TS (VGA-MOGA) with other algorithms

Datasets	p-values				
	IFCM	VGA	Avg. Link	SOM	CRC
AD400_10_10	2.1E-4	1.5E-3	5.2E-5	4.3E-5	1.1E-3
Sporulation	2.2E-5	2.1E-3	1.1E-5	1.2E-2	5.2E-3
Serum	1.5E-4	1.5E-4	5.4E-5	1.4E-4	1.4E-4
Rat CNS	1.1E-5	1.5E-4	5.4E-5	1.4E-4	1.7E-4

hypothesis is that there is significant difference in the median values of the two groups. All the p-values reported in the table are less than 0.05 (5% significance level). For example, the rank sum test between algorithms SiMM-TS and IFCM for Sporulation dataset provides a p-value of 2.2E–5, which is very small. This is strong evidence against the null hypothesis, indicating that the better median value of the performance metrics produced by SiMM-TS is statistically significant and has not occurred by chance. Similar results are obtained for all other datasets and for all other algorithms compared to SiMM-TS, establishing the significant superiority of the SiMM-TS algorithm.

7.5.8 Biological Relevance

The biological interpretation of a cluster of genes can be best assessed by studying the functional annotation of the genes of that cluster. Here, biological interpretations of the clusters are determined in terms of Gene Ontology (GO) [27]. For this purpose, a Web-based Gene Ontology tool FatiGO [9] (http://http://www.babelomics. org/) has been utilized. FatiGO extracts the Gene Ontology terms for a query and a reference set of genes. In our experiment, a query is the set of genes of a cluster whose biological relevance is to be measured. The union of the genes from the other clusters is taken as the reference set. The GO level is fixed at 7.

For illustration, similar clusters obtained using different algorithms for Yeast Sporulation data have been analyzed. Cluster *selectivity* is assessed for validating the clusters. The cluster selectivity denotes the proportion of genes with a certain annotation (a biological process in our case) in the cluster relative to all genes in the data with this annotation. The selectivity is computed as the difference between the percentage of genes with a certain annotation in the query and reference set. A high selectivity thus indicates that these genes are well distinguished in terms of their expression profiles, from all the genes.

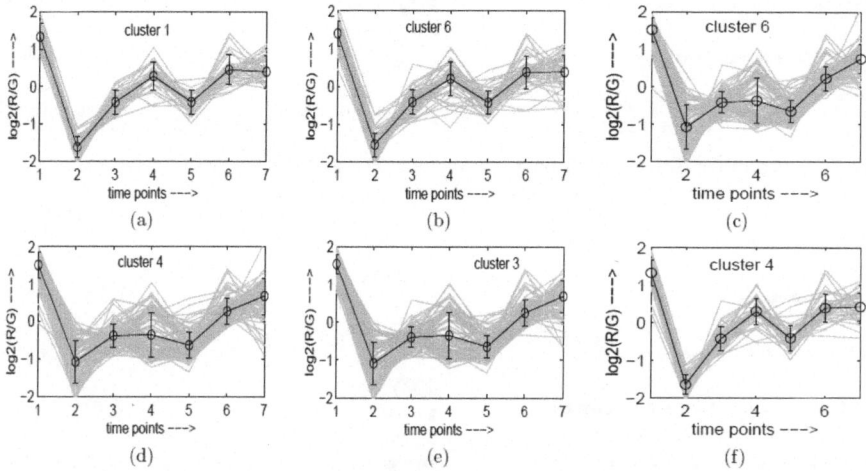

Fig. 7.15 Similar clusters found on Sporulation data by different algorithms: (a) cluster 1 of SiMM-TS (39 genes), (b) cluster 6 of VGA (44 genes), (c) cluster 6 of IFCM (87 genes), (d) cluster 4 of average linkage (92 genes), (e) cluster 3 of SOM (89 genes), (f) cluster 4 of CRC (35 genes)

Biological process. Level: 7	0 20 40 60 80 100	p-values(*)	
rRNA processing	62.50% / 1.14%	<1e-05	<1e-05
rRNA modification	12.50% / 0 %	0.00051	0.03236
M phase of meiotic cell cycle	0 % / 20.45%	0.01104	0.30141
nucleocytoplasmic transport	12.50% / 2.27%	0.03071	0.69876
nuclear transport	12.50% / 2.27%	0.03071	0.69876
sporulation (sensu Fungi)	0 % / 15.91%	0.03196	0.70194

Fig. 7.16 Part of the FatiGo results on Yeast Sporulation for cluster 1 of SiMM-TS

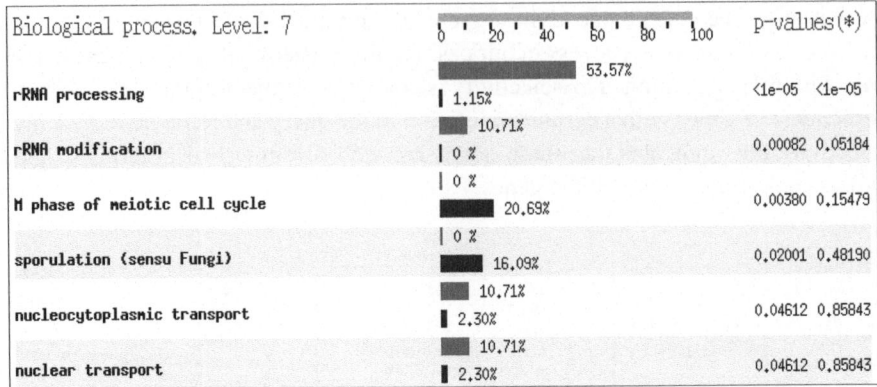

Fig. 7.17 Part of the FatiGo results on Yeast Sporulation data for cluster 6 of VGA

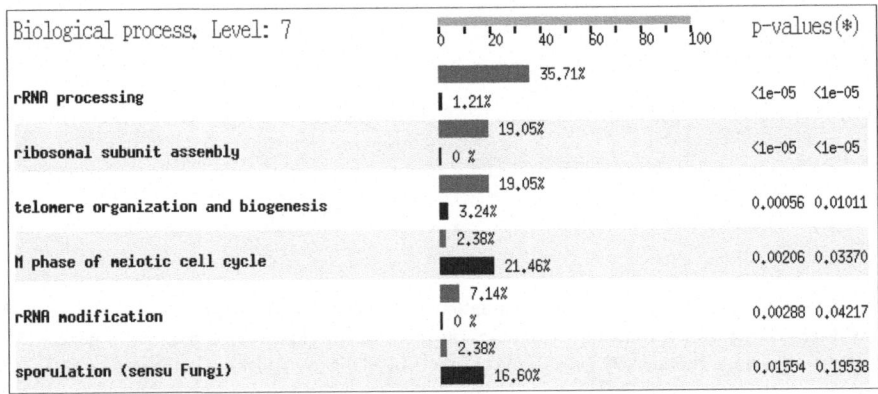

Fig. 7.18 Part of the FatiGo results on Yeast Sporulation data for cluster 6 of IFCM

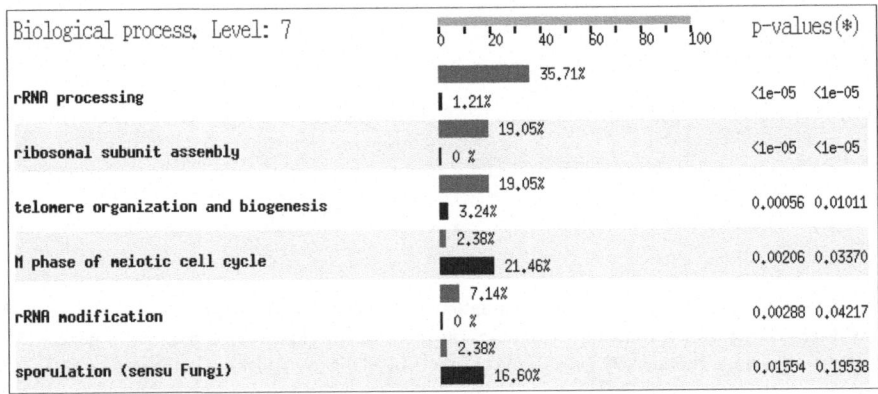

Fig. 7.19 Part of the FatiGo results on Yeast Sporulation data for cluster 4 of average linkage

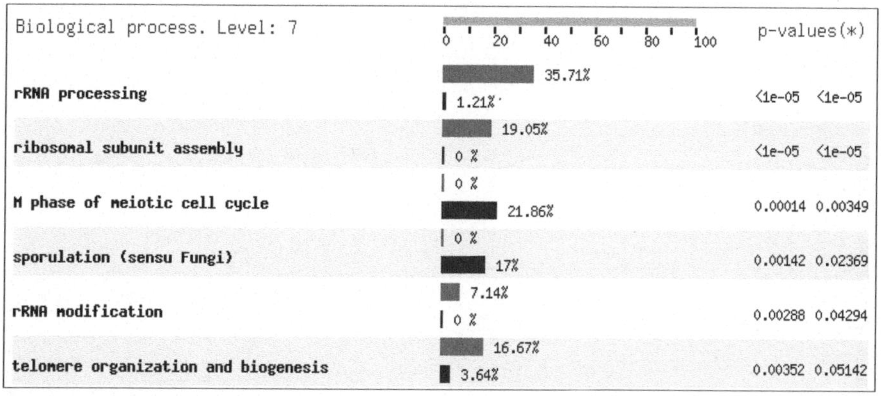

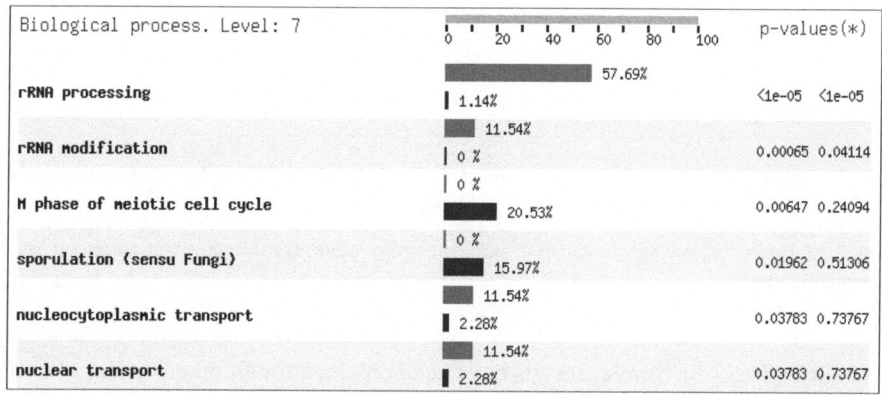

Fig. 7.20 Part of the FatiGo results on Yeast Sporulation data for cluster 3 of SOM

Fig. 7.21 Part of the FatiGo results on Yeast Sporulation data for cluster 4 of CRC

The performance of six algorithms, viz., SiMM-TS, VGA, IFCM, average linkage, SOM and CRC, on the Yeast Sporulation data are examined. From Figure 7.15, it is evident that cluster 1, obtained by SiMM-TS (VGA-MOGA) clustering, corresponds to cluster 6 of IFCM, cluster 6 of VGA, cluster 4 of average linkage, cluster 3 of SOM and cluster 4 of CRC clustering solutions because they provide similar expression patterns. Figures 7.16–7.21 show a part of the FatiGO results (first six most selective annotations) of the clusters mentioned for the six algorithms applied on Sporulation data. Here, biological processes have been used for annotation. The figure indicates that for most of the annotations, the SiMM-TS algorithm provides better selectivity than the other algorithms. All the algorithms, except average linkage, show maximum selectivity to the genes involved in rRNA processing. Average linkage shows maximum selectivity to the M phase of mitotic cell cycle. However, SiMM-TS shows the best maximum selectivity of 61.36% (62.5%–1.14%) compared to VGA (52.42%), IFCM (34.5%), average linkage (5.36%), SOM (34.5%)

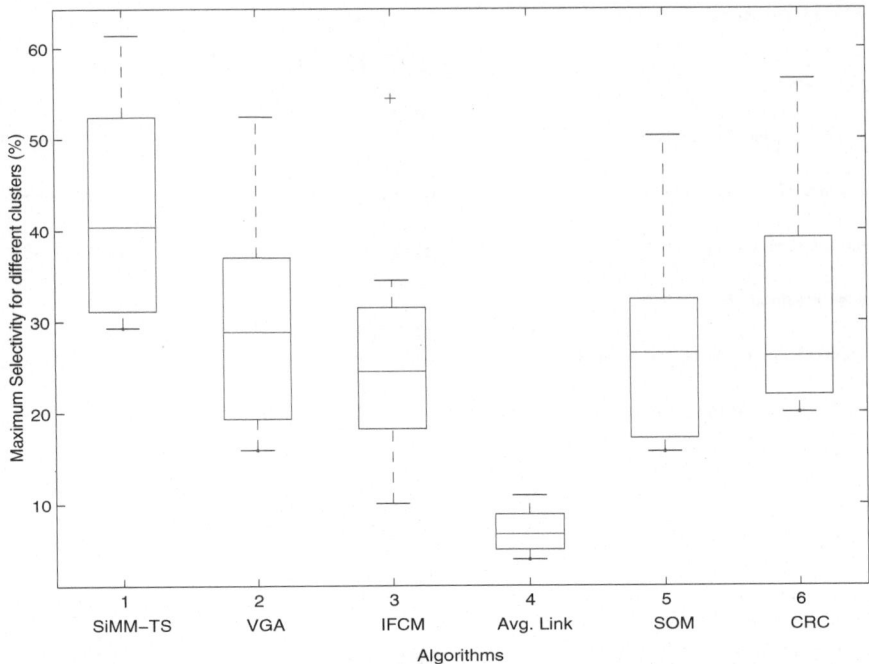

Fig. 7.22 Boxplots showing the range of best selectivity values for different algorithms on Yeast Sporulation data

and CRC (56.55%). Similar results have been obtained for all other similar clusters produced by different algorithms.

For the purpose of illustration, Figure 7.22 shows the boxplots representing the maximum selectivity for the first six most selective annotations for the clusters produced by different algorithms for the Yeast Sporulation data. It is evident from the figure that the range of selectivity values for the SiMM-TS (VGA-MOGA) clustering is better than that of the other algorithms. Overall, the results indicate the superiority of the two-stage clustering method.

7.6 Summary

This chapter describes a recently proposed two-stage clustering algorithm (SiMM-TS) using the idea of points having significant membership in multiple classes (SiMM points). A VGA-based clustering scheme and a MOGA-based clustering technique are utilized in the process. The number of clusters in a dataset is automatically evolved in the SiMM-TS clustering technique. The performance of the SiMM-TS clustering method has been compared with that of the other combinations of algorithms in the two stages as well as with the average linkage, SOM, CRC, VGA and IFCM clustering algorithms to show its effectiveness on remote sensing images and gene expression datasets.

In general it is found that the SiMM-TS clustering scheme outperforms all the other clustering methods significantly. Moreover, it is seen that VGA performs reasonably well in determining the appropriate value of the number of clusters. The clustering solutions are evaluated both quantitatively and visually. Statistical tests have also been conducted to establish the statistical significance of the results produced by the SiMM-TS technique. For gene expression datasets, a biological interpretation of the clustering solutions has been provided.

Chapter 8
Clustering Categorical Data in a Multiobjective Framework

8.1 Introduction

Most of the clustering algorithms are designed for such datasets where the dissimilarity between any two points of the dataset can be computed using standard distance measures such as the Euclidean distance. However, many real-life datasets are categorical in nature, where no natural ordering can be found among the elements in the attribute domain. In such situations, the clustering algorithms, such as K-means [238] and fuzzy C-means (FCM) [62], cannot be applied. The K-means algorithm computes the center of a cluster by computing the mean of the set of feature vectors belonging to that cluster. However, as categorical datasets do not have any inherent distance measure, computing the mean of a set of feature vectors is meaningless. A variation of the K-means algorithm, namely Partitioning Around Medoids (PAM) or K-medoids [243], has been proposed for such kinds of datasets. In PAM, instead of the cluster center, the cluster medoid, i.e., the most centrally located point in a cluster, is determined. Unlike the cluster center, a cluster medoid must be an actual data point. Another extension of K-means is the K-modes algorithm [222,223]. Here the cluster centroids are replaced with cluster modes (described later). A fuzzy version of the K-modes algorithm, i.e., fuzzy K-modes, is also proposed in [224]. Recently, a Hamming Distance (HD) vector-based categorical data clustering algorithm (CCDV, or Clustering Categorical Data based on Distance Vectors) has been developed in [462]. Hierarchical algorithms, such as average linkage [238], are also widely used for clustering categorical data. Some other developments in this area are available in [168, 175, 189]. However, all these algorithms rely on optimizing a single objective to obtain the partitioning. A single objective function may not work uniformly well for different kinds of categorical data. Hence it is natural to consider multiple objectives that need to be optimized simultaneously.

In this chapter, the problem of fuzzy partitioning of categorical datasets is modeled as one of multiobjective optimization (MOO), where search is performed over a number of often conflicting objective functions. Multiobjective Genetic Algorithms

(MOGAs) are used to determine the appropriate cluster centers (modes) and the corresponding partition matrix. Non-dominated Sorting GA-II (NSGA-II) [130] is used as the underlying optimization method. The two objective functions, the global fuzzy compactness of the clusters and fuzzy separation, are optimized simultaneously.

This chapter describes a fuzzy multiobjective algorithm for clustering categorical data. Unlike in the works [41] and [44], where chromosomes encode the cluster means (centers), here the chromosomes encode the cluster modes and hence differ in the chromosome updating process. Two fuzzy objective functions, viz., fuzzy compactness and fuzzy separation, have been simultaneously optimized, resulting in a set of non-dominated solutions. Subsequently, a technique based on majority voting among the non-dominated Pareto-optimal solutions followed by k-nn classification is used to obtain the final clustering solution from the set of non-dominated solutions. Thus, the requirement of the third cluster validity measure for selecting the final solution from the Pareto-optimal set and the resulting bias are eliminated.

Experiments have been carried out for four synthetic and four real-life categorical datasets. Comparison has been made among different algorithms such as fuzzy K-modes, K-modes, K-medoids, average linkage, CCDV, the single objective GA- (SGA-)based clustering algorithms and the NSGA-II-based multiobjective fuzzy clustering scheme. The superiority of the multiobjective algorithm has been demonstrated both quantitatively and visually. Also statistical significance tests are conducted to confirm that the superior performance of the multiobjective technique is significant and does not occur by chance.

8.2 Fuzzy Clustering of Categorical Data

This section describes the fuzzy K-modes clustering algorithm [224] for categorical datasets. The fuzzy K-modes algorithm is the extension of the well-known fuzzy C-means [62] algorithm in the categorical domain. Let $X = \{x_1, x_2, \ldots, x_n\}$ be a set of n objects having categorical attribute domains. Each object x_i, $i = 1, 2, \ldots, n$, is described by a set of p attributes $A_1, A_2, \ldots, A_p$. Let $DOM(A_j)$, $1 \leq j \leq p$, denote the domain of the jth attribute and it consists of different q_j categories such as $DOM(A_j) = \{a_j^1, a_j^2, \ldots, a_j^{q_j}\}$. Hence the ith categorical object is defined as $x_i = [x_{i1}, x_{i2}, \ldots, x_{ip}]$, where $x_{ij} \in DOM(A_j)$, $1 \leq j \leq p$.

The cluster centers in the fuzzy C-means are replaced by cluster modes in the fuzzy k-modes clustering. A mode is defined as follows: Let C_i be a set of categorical objects belonging to cluster i. Each object is described by attributes $A_1, A_2, \ldots, A_p$. The mode of C_i is a vector $m_i = [m_{i1}, m_{i2}, \ldots, m_{ip}]$, $m_{ij} \in DOM(A_j)$, $1 \leq j \leq p$, such that the following criterion is minimized:

$$\mathscr{D}(m_i, C_i) = \sum_{x \in C_i} D(m_i, x). \tag{8.1}$$

Here $D(m_i, x)$ denotes the dissimilarity measure between m_i and x. Note that m_i is not necessarily an element of set C_i.

The fuzzy K-modes algorithm partitions the dataset X into K clusters by minimizing the following criterion:

$$J_m(U, Z : X) = \sum_{k=1}^{n} \sum_{i=1}^{K} u_{ik}^m D(z_i, x_k). \tag{8.2}$$

For probabilistic fuzzy clustering, the following conditions must hold while minimizing J_m:

$$0 \leq u_{ik} \leq 1, \quad 1 \leq i \leq K, \quad 1 \leq k \leq n, \tag{8.3}$$

$$\sum_{i=1}^{K} u_{ik} = 1, \quad 1 \leq k \leq n, \tag{8.4}$$

and

$$0 < \sum_{k=1}^{n} u_{ik} < n, \quad 1 \leq i \leq K, \tag{8.5}$$

where m is the fuzzy exponent. $U = [u_{ik}]$ denotes the $K \times n$ fuzzy partition matrix and u_{ik} denotes the membership degree of the kth categorical object to the ith cluster. $Z = \{z_1, z_2, \ldots, z_K\}$ represents the set of cluster centers (modes).

The fuzzy K-modes algorithm is based on an alternating optimizing strategy. This involves iteratively estimating the partition matrix followed by the computation of new cluster centers (modes). It starts with random initial K modes, and then at every iteration it finds the fuzzy membership of each data point to every cluster using the following equation [224]:

$$u_{ik} = \frac{1}{\sum_{j=1}^{K} \left(\frac{D(z_i, x_k)}{D(z_j, x_k)} \right)^{\frac{1}{m-1}}}, \quad \text{for } 1 \leq i \leq K; \ 1 \leq k \leq n, \tag{8.6}$$

Note that while computing u_{ik} using Equation 8.6, if $D(z_j, x_k)$ is equal to zero for some j, then u_{ik} is set to zero for all $i = 1, \ldots, K$, $i \neq j$, while u_{jk} is set equal to 1.

Based on the membership values, the cluster centers (modes) are recomputed as follows: If the membership values are fixed, then the locations of the modes that minimize the objective function in Equation 8.2 will be [224] $z_i = [z_{i1}, z_{i2}, \ldots, z_{ip}]$ where $z_{ij} = a_j^r \in DOM(A_j)$ and

$$\sum_{k, x_{kj}=a_j^r} u_{ik}^m \geq \sum_{k, x_{kj}=a_j^t} u_{ik}^m, \quad 1 \leq t \leq q_j, \ r \neq t. \tag{8.7}$$

The algorithm terminates when there is no noticeable improvement in the J_m value (Equation 8.2). Finally, each object is assigned to the cluster to which it has the maximum membership. The main disadvantages of the fuzzy K-modes clustering algorithms are (1) it depends much on the initial choice of the modes and (2) it often gets trapped into some local optimum.

8.3 Multiobjective Fuzzy Clustering for Categorical Data

In this section, the method of using NSGA-II for evolving a set of near-Pareto-optimal non-degenerate fuzzy partition matrices is described.

8.3.1 Chromosome Representation

Each chromosome is a sequence of attribute values representing the K cluster modes. If each categorical object has p attributes $\{A_1, A_2, \ldots, A_p\}$, the length of a chromosome will be $K \times p$, where the first p positions (or, genes) represent the p dimensions of the first cluster mode, the next p positions represent those of the second cluster mode, and so on. As an illustration let us consider the following example. Let $p = 3$ and $K = 3$. Then the chromosome

$$c_{11} \; c_{12} \; c_{13} \; c_{21} \; c_{22} \; c_{23} \; c_{31} \; c_{32} \; c_{33}$$

represents the three cluster modes (c_{11}, c_{12}, c_{13}), (c_{21}, c_{22}, c_{23}) and (c_{31}, c_{32}, c_{33}), where c_{ij} denotes the jth attribute value of the ith cluster mode. Also $c_{ij} \in DOM(A_j)$, $1 \leq i \leq K$, $1 \leq j \leq p$.

8.3.2 Population Initialization

The initial K cluster modes encoded in each chromosome are chosen as K random objects of the categorical dataset. This process is repeated for each of the P chromosomes in the population, where P is the population size.

8.3.3 Computation of Objective Functions

The global compactness π [410] of the clusters and the fuzzy separation Sep [410] have been considered as the two objectives that need to be optimized simultaneously. For computing the measures, the modes encoded in a chromosome are first extracted. Let these be denoted as $z_1, z_2, \ldots, z_K$. The membership values u_{ik}, $i = 1, 2, \ldots, K$ and $k = 1, 2, \ldots, n$ are computed as follows [224]:

$$u_{ik} = \frac{1}{\sum_{j=1}^{K} \left(\frac{D(z_i, x_k)}{D(z_j, x_k)}\right)^{\frac{1}{m-1}}} \quad \text{for } 1 \leq i \leq K, \ 1 \leq k \leq n, \tag{8.8}$$

where $D(z_i, x_k)$ and $D(z_j, x_k)$ are as described earlier. m is the weighting coefficient. (Note that while computing u_{ik} using Equation 8.8, if $D(z_j, x_k)$ is equal to zero for some j, then u_{ik} is set to zero for all $i = 1, \ldots, K$, $i \neq j$, while u_{jk} is set equal to 1.) Subsequently, each mode encoded in a chromosome is updated to $z_i = [z_{i1}, z_{i2}, \ldots, z_{ip}]$ where $z_{ij} = a_j^r \in DOM(A_j)$ [224] and

$$\sum_{k, x_{kj} = a_j^r} u_{ik}^m \geq \sum_{k, x_{kj} = a_j^t} u_{ik}^m, \quad 1 \leq t \leq q_j, \ r \neq t. \tag{8.9}$$

This means that the category of the attribute A_j of the cluster centroid z_i is set to the category value that attains the maximum value of the summation of u_{ij} (the degrees of membership to the ith cluster) over all categories. Accordingly the cluster membership values are recomputed as per Equation 8.8.

The variation σ_i and fuzzy cardinality n_i of the ith cluster, $i = 1, 2, \ldots, K$, are calculated using the following equations [410]:

$$\sigma_i = \sum_{k=1}^{n} u_{ik}^m D(z_i, x_k), \quad 1 \leq i \leq K, \tag{8.10}$$

and

$$n_i = \sum_{k=1}^{n} u_{ik}, \quad 1 \leq i \leq K. \tag{8.11}$$

The global compactness π of the solution represented by the chromosome is then computed as [410]

$$\pi = \sum_{i=1}^{K} \frac{\sigma_i}{n_i} = \sum_{i=1}^{K} \frac{\sum_{k=1}^{n} u_{ik}^m D(z_i, x_k)}{\sum_{k=1}^{n} u_{ik}}. \tag{8.12}$$

To compute the other fitness function fuzzy separation Sep, the mode z_i of the ith cluster is assumed to be the center of a fuzzy set $\{z_j | 1 \leq j \leq K, j \neq i\}$. Hence the membership degree of each z_j to z_i, $j \neq i$, is computed as [410]

$$\mu_{ij} = \frac{1}{\sum_{l=1, l \neq j}^{K} \left(\frac{D(z_j, z_i)}{D(z_j, z_l)}\right)^{\frac{1}{m-1}}}, \quad i \neq j. \tag{8.13}$$

Subsequently, the fuzzy separation is defined as [410]

$$Sep = \sum_{i=1}^{K} \sum_{j=1, j \neq i}^{K} \mu_{ij}^m D(z_i, z_j). \tag{8.14}$$

Note that in order to obtain compact clusters, the measure π should be minimized. In contrast, to get well-separated clusters, the fuzzy separation Sep should be maximized. As the multiobjective problem is posed as minimization of both the objectives, the objective is to minimize π and $\frac{1}{Sep}$ simultaneously.

As multiobjective clustering deals with the simultaneous optimization of more than one clustering objective, its performance depends highly on the choice of these objectives. A careful choice of objectives can produce remarkable results, whereas arbitrary or unintelligent objective selection can unexpectedly lead to bad situations. The selection of objectives should be such that they can balance each other critically and are possibly contradictory in nature. Contradiction in the objective functions is beneficial since it leads to global optimum solution. It also ensures that no single clustering objective is optimized leaving the other probable significant objectives unnoticed.

Although several cluster validity indices exist, a careful study reveals that most of these consider cluster compactness and separation in some form [194, 299]. Hence the global cluster variance π (reflective of cluster compactness) and the fuzzy separation Sep (reflective of cluster separation) are chosen to be optimized. However, an exhaustive study involving two or more other powerful fuzzy cluster validity indices will constitute an area of interesting future work.

8.3.4 Selection, Crossover and Mutation

The selection operation used here is the crowded binary tournament selection used in NSGA-II. After selection, the selected chromosomes are put in the mating pool. Conventional single point crossover depending on crossover probability μ_c has been performed for generating the new offspring solutions from the chromosomes selected in the mating pool. For performing the mutation, a mutation probability μ_m has been used. If a chromosome is selected to be mutated, the gene position that will undergo mutation is selected randomly. After that, the categorical value of that position is replaced by another random value chosen from the corresponding categorical domain. The most characteristic part of NSGA-II is its elitism operation, where the non-dominated solutions among the parent and child populations are propagated to the next generation. For details on the different genetic processes, the reader may refer to [125]. The near-Pareto-optimal strings of the last generation provide the different solutions to the clustering problem.

8.3.5 Selecting a Solution from the Non-dominated Set

As discussed earlier, the multiobjective GA-based categorical data clustering algorithm produces a near-Pareto-optimal non-dominated set of solutions in the final generation. Hence, it is necessary to choose a particular solution from the set of non-dominated solutions $\mathbb{N}$. The technique adopted here is used to search for the complete approximated Pareto front and apply post-processing to identify the solution that shares the most information provided by all the non-dominated solutions. In this approach all the non-dominated solutions have been given equal importance and the idea is to extract the combined clustering information. In this regard, a majority voting technique followed by a *k-nearest neighbor* (*k*-nn) classification has been adopted to select a single solution from the set of the non-dominated solutions.

First the clustering label vectors are computed from the unique non-dominated solutions produced by the multiobjective technique and are made consistent with each other as per the method described in Chapter 6, Section 6.2.1. After relabeling all the label vectors, majority voting is applied for each point. The points that are voted by at least 50% of the solutions to have a particular class label are now taken as the training set for the remaining points. The remaining points are assigned a class label according to *k*-nn classifier. That is, for each unassigned point, *k* nearest neighbors are computed and the point is assigned a class label that is obtained by the majority voting of the k-nearest neighbors. The value for *k* is selected as 5.

The application of majority voting followed by *k*-nn classification produces a new cluster label vector $\mathbb{X}$ that shares the clustering information of all the non-dominated solutions. Thereafter, the percentage of matching with $\mathbb{X}$ is computed for each of the label vectors corresponding to each non-dominated solution. The label vector of the non-dominated solution that matches best with $\mathbb{X}$ is chosen from the set of the non-dominated solutions.

8.4 Contestant Methods

This section describes the contestant clustering algorithms that are used for the purpose of performance comparison.

8.4.1 K-Medoids

Partitioning Around Medoids (PAM), also called the K-medoids clustering [243], is a variation of the K-means with the objective to minimize the within-cluster variance W(K):

$$W(K) = \sum_{i=1}^{K} \sum_{x \in C_i} D(m_i, x). \tag{8.15}$$

Here m_i is the medoid of cluster C_i and $D(.)$ denotes a dissimilarity measure. A cluster medoid is defined as the most centrally located point within the cluster, i.e., it is the point from which the sum of distances to the other points of the cluster is minimum. Thus, a cluster medoid always belongs to the set of input data points X. The resulting clustering of the dataset X is usually only a local minimum of $W(K)$. The idea of PAM is to select K representative points, or medoids, in X and assign the rest of the data points to the cluster identified by the closest medoid. Initial medoids are chosen randomly. Then, all points in X are assigned to the nearest medoid. In each iteration, a new medoid is determined for each cluster by finding the data point with minimum total dissimilarity to all other points of the cluster. Subsequently, all the points in X are reassigned to their clusters in accordance with the new set of medoids. The algorithm iterates until $W(K)$ does not change any more.

8.4.2 K-Modes

K-modes clustering [223] is the crisp version of the fuzzy K-modes algorithm. The K-modes algorithm works similarly to K-medoids with the only difference that here, modes, instead of medoids, are used to represent a cluster. The K-modes algorithm minimizes the following objective function:

$$TC(K) = \sum_{i=1}^{K} \sum_{x \in C_i} D(m_i, x). \tag{8.16}$$

Here m_i denotes the mode of the cluster C_i. The mode of a set of points P is a point (not necessarily belonging to P) whose jth attribute value is computed as the most frequent value of the jth attribute over all the points in P. If there is more than one most frequent values, one of them is chosen arbitrarily. The iteration steps are similar to those of K-medoids and only differ in the center (mode) updating process.

8.4.3 Hierarchical Agglomerative Clustering

Agglomerative clustering techniques [238] begin with singleton clusters and combine two least distant clusters at every iteration. Thus in each iteration two clusters are merged, and hence the number of clusters reduces by 1. This proceeds iteratively in a hierarchy, providing a possible partitioning of the data at every level. When the target number of clusters (K) is achieved, the algorithms terminate. *Single, average* and *complete* linkage agglomerative algorithms differ only in the linkage metric used, i.e., they differ in computing the distance between two clusters. For the single linkage algorithm, the distance between two clusters C_i and C_j is computed as the smallest distance between all possible pairs of data points x and y, where $x \in C_i$ and $y \in C_j$. For the average and complete linkage algorithms, the linkage metrics are taken as the average and largest distances respectively.

8.4.4 Clustering Categorical Data Based on Distance Vectors

Clustering Categorical Data based on Distance Vectors (CCDV) [462] is a recently proposed clustering algorithm for categorical attributes. CCDV sequentially extracts the clusters from a given dataset based on the Hamming Distance (HD) vectors, with automatic evolution of the number of clusters. In each iteration, the algorithm identifies only one cluster, which is then deleted from the dataset in the next iteration. This procedure continues until there are no more significant clusters in the remaining data. For the identification and extraction of a cluster, the cluster center is first located by using a Pearson chi-squared-type statistic on the basis of HD vectors. The output of the algorithm does not depend on the order of the input data points.

8.4.5 Single Objective GA-Based Fuzzy Clustering Algorithms

Three single objective GA- (SGA-)based clustering algorithms with different objective functions have been considered here. All the algorithms have the same chromosome representation as that of the multiobjective one, and similar genetic operators. The first SGA-based algorithm minimizes the objective function π and is thus called SGA(π). The second SGA-based clustering maximizes Sep and hence is called SGA(Sep). The last one minimizes the objective function $\frac{\pi}{Sep}$ and is called SGA(π,Sep).

8.4.6 Multiobjective GA-Based Crisp Clustering Algorithm

To establish the utility of incorporating fuzziness, a multiobjective crisp clustering algorithm (MOGA$_{crisp}$) for categorical data has been utilized. This algorithm uses an encoding technique and genetic operators similar to the multiobjective fuzzy clustering method. The only difference is that it optimizes the crisp versions of the objective functions described in Equations 8.12 and 8.14, respectively. The objective functions for MOGA$_{crisp}$ are as follows:

$$\pi_{crisp} = \sum_{i=1}^{K} \frac{\sigma_i}{n_i} = \sum_{i=1}^{K} \frac{\sum_{x_k \in C_i} D(z_i, x_k)}{\sum_{k=1}^{n} R(z_i, x_k)}, \tag{8.17}$$

$$Sep_{crisp} = \sum_{i=1}^{K} \sum_{j=1, j \neq i}^{K} D(z_i, z_j). \tag{8.18}$$

Here $R(z_i, x_k)$ is defined as follows:

$$R(z_i, x_k) = \begin{cases} 1 & \text{if } x_k \in C_i \\ 0 & \text{otherwise.} \end{cases} \quad (8.19)$$

Here C_i denotes the ith cluster, and all other symbols have the same meaning as before. The data points are assigned to particular clusters as per the nearest distance criterion. The final solution is selected from the generated non-dominated front following the procedure described in Section 8.3.5.

8.5 Experimental Results

The performance of the clustering algorithms has been evaluated on four synthetic datasets (Cat250_15_5, Cat100_10_4, Cat500_20_10 and Cat280_10_6) and four real-life datasets (Congressional Votes, Zoo, Soybean and Breast cancer).

8.5.1 Dissimilarity Measure

As stated earlier, there is no inherent distance or dissimilarity measure such as Euclidean distance that can be directly applied to compute the dissimilarity between two categorical objects. This is because there is no natural order among the categorical values of any particular attribute domain. Hence it is difficult to measure the dissimilarity between two categorical objects.

The following dissimilarity measure has been used for all the algorithms considered. Let $x_i = [x_{i1}, x_{i2}, \ldots, x_{ip}]$ and $x_j = [x_{j1}, x_{j2}, \ldots, x_{jp}]$ be two categorical objects described by p categorical attributes. The dissimilarity measure between x_i and x_j, $D(x_i, x_j)$, can be defined as the total number of mismatches between the corresponding attribute categories of the two objects. Formally,

$$D(x_i, x_j) = \sum_{k=1}^{p} \delta(x_{ik}, x_{jk}), \quad (8.20)$$

where

$$\delta(x_{ik}, x_{jk}) = \begin{cases} 0 & \text{if } x_{ik} = x_{jk} \\ 1 & \text{if } x_{ik} \neq x_{jk}. \end{cases} \quad (8.21)$$

Note that $D(x_i, x_j)$ gives equal importance to all the categories of an attribute. However, in most of the categorical datasets, the distance between two data vectors depends on the nature of the datasets. Thus, if a dissimilarity matrix is predefined for a given dataset, the algorithms can adopt this for computing the dissimilarities.

8.5.2 *Visualization*

For the visualization of the datasets, the well-known VAT (Visual Assessment of clustering Tendency) representation [64] is used. To visualize a clustering solution, first the points are reordered according to the class labels given by the solution. Thereafter the distance matrix is computed on this reordered data matrix. In the graphical plot of the distance matrix, the boxes lying on the main diagonal represent the clustering structure.

The plots of the Pareto frontier produced by the multiobjective fuzzy clustering algorithm have also been used for visualization of the results.

8.5.3 *Synthetic Datasets*

8.5.3.1 Cat250_15_5

This synthetic dataset has a one-layer clustering structure (Figure 8.1(a)) with 15 attributes and 250 points. It has five clusters of the same size (50 points in each cluster). Each cluster has random categorical values selected from $\{0,1,2,3,4,5\}$ in a distinct continuous set of 12 attributes, while the remaining attributes are set to 0.

8.5.3.2 Cat100_10_4

This is a synthetic dataset with 100 points and ten attributes (Figure 8.1(b)). The dataset has four clusters of same sizes (25 points each). For each cluster, two random attributes of the points of that cluster are zero-valued and the remaining attributes have values in $\{0,1,2,3,4,5\}$.

8.5.3.3 Cat500_20_10

This synthetic dataset is generated by using the data generator available at the Web site http://www.datgen.com. This generator provides various options to specify the number of attributes, the attribute domains and the number of tuples. The number of classes in the dataset is specified by the use of conjunctive rules of the form $(Attr_1 = a, Attr_2 = b, \ldots) \Longrightarrow class \; c_1$. The dataset contains 500 points and 20 attributes (Figure 8.1(c)). The points are clustered into ten clusters.

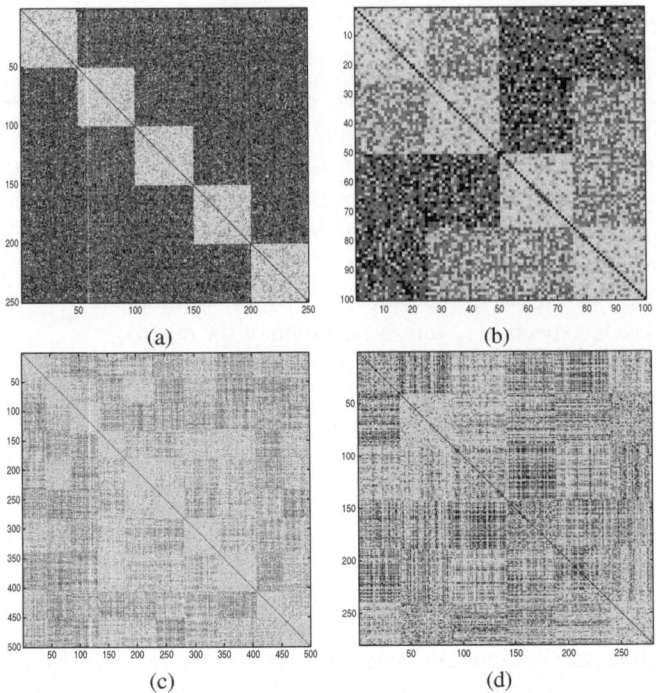

Fig. 8.1 True clustering of synthetic datasets using VAT representation: (a) Cat250_15_5 dataset, (b) Cat100_10_4 dataset, (c) Cat500_20_10 dataset, (d) Cat280_10_6 dataset

8.5.3.4 Cat280_10_6

This is another synthetic dataset obtained using the data generator. The dataset contains 280 points, ten attributes and six clusters (Figure 8.1(d)).

8.5.4 Real-Life Datasets

8.5.4.1 Congressional Votes

This dataset is the United States Congressional voting records in 1984 (Figure 8.2(a)). The total number of records is 435. Each row corresponds to one Congressman's votes on 16 different issues (e.g., education spending, crime). All the attributes are Boolean with a Yes (i.e., 1) or No (i.e., 0) value. A classification label of Republican or Democrat is provided with each data record. The dataset contains records for 168 Republicans and 267 Democrats.

8.5.4.2 Zoo

The Zoo dataset consists of 101 instances of animals in a zoo with 17 features (Figure 8.2(b)). The name of the animal constitutes the first attribute. This attribute is neglected. There are 15 boolean attributes corresponding to the presence of hair, feathers, eggs, milk, backbone, fins, tail, and whether airborne, aquatic, predator, toothed, breathing, venomous, domestic and cat-size. The character attribute corresponds to the number of legs lying in the set $\{0, 2, 4, 5, 6, 8\}$. The dataset consists of seven different classes of animals.

8.5.4.3 Soybean

The Soybean dataset contains 47 data points on diseases in soybeans (Figure 8.2(c)). Each data point has 35 categorical attributes and is classified as one of the four diseases, i.e., the number of clusters in the dataset is four.

8.5.4.4 Breast Cancer

This dataset has a total of 699 records, and nine attributes, each of which is described by ten categorical values (Figure 8.2(d)). The 16 rows which contain missing values are deleted from the dataset and the remaining 683 records are used. The dataset is classified into two classes, benign and malignant.

The real-life datasets are obtained from the UCI Machine Learning Repository (http://archive.ics.uci.edu/ml/datasets.html).

8.5.5 Comparison Procedure

The multiobjective fuzzy clustering technique and its crisp version search in parallel a number of solutions and finally a single solution is chosen from the set of non-dominated solutions as discussed before. The single objective GA-based algorithms also search in parallel and the best chromosome of the final generation is treated as the desired solution. In contrast, the iterated algorithms, such as fuzzy K-modes, K-modes, K-medoids and CCDV, try to improve a single solution iteratively. They depend a lot on the initial configuration and often get stuck at the local optima. In order to compare these algorithms with GA-based methods, the following procedure is adopted. Iterated algorithms are run $\mathcal{N}$ times where each run consists of $\mathcal{I}$

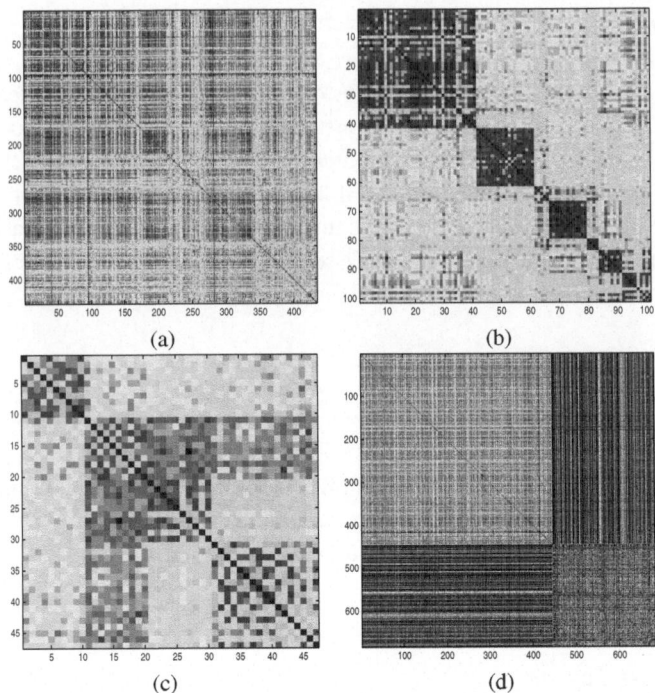

Fig. 8.2 True clustering of real-life datasets using VAT representation: (a) Congressional Votes dataset, (b) Zoo dataset, (b) Soybean dataset, (b) Breast cancer dataset

re-iterations, as follows:

for $i = 1$ **to** $\mathcal{N}$
    for $j = 1$ **to** $\mathcal{I}$
        $ARI[j] = ARI$ score obtained by running the algorithm
            with new random seed.
    end for
    $ARIB[i] = \max\{ARI[1], \dots, ARI[\mathcal{I}]\}.$
end for
$AvgARIB = \text{avg}\{ARIB[1], \dots, ARIB[\mathcal{N}]\}.$

In Tables 8.1 and 8.3 the average $ARIB$ scores ($AvgARIB$) for each algorithm are reported.

The GA-based algorithms have been run $\mathcal{N}$ times, with the number of generations as $\mathcal{I}$. The average of the best ARI scores for the GA-based algorithms is computed from the ARI scores of the $\mathcal{N}$ runs.

8.5.6 Input Parameters

The GA-based algorithms are run for 100 generations with population size 50. The crossover and mutation probabilities are fixed at 0.8 and $1/chromosome\ length$ respectively. These values are chosen after several experiments. The parameters $\mathcal{N}$ and $\mathcal{I}$ are taken as 50 and 100. Each re-iteration of the fuzzy K-modes, K-modes and K-medoids algorithms has been executed 500 times unless they converge before that. This means that each of these three iterative algorithms has been executed 50×100 times and each such execution is allowed for a maximum of 500 iterations. This is done for a fair comparison of these algorithms with the GA-based techniques, which explore a total of 50×100 combinations (since the number of generations and population size of the GA-based techniques are 100 and 50, respectively). The fuzzy exponent m has been chosen to be 2.

8.5.7 Results for Synthetic Datasets

The clustering results in terms of the average values of the $ARIB$ scores over 50 runs ($AvgARIB$) on the four synthetic datasets using different algorithms are reported in Table 8.1. The maximum values of $AvgARIB$ are shown in bold letters. From the table it can be observed that the multiobjective genetic fuzzy clustering algorithm gives the best $AvgARIB$ scores for all the datasets. It is also evident from the table that for all the synthetic datasets, the fuzzy version of a clustering method performs better than its crisp counterpart. For example, the $AvgARIB$ scores for the fuzzy K-modes and K-modes algorithms are 0.5092 and 0.4998, respectively for Cat280_10_6 data. This is also the case for multiobjective clustering. MOGA(π,Sep) and MOGA$_{crisp}$ provide $AvgARIB$ scores 0.5851 and 0.5442, respectively for this dataset. For all other datasets also, the fuzzy algorithms provide better results than the corresponding crisp versions. This establishes the utility of incorporating fuzziness into clustering categorical datasets.

Table 8.2 reports another interesting observation. Here the best $ARIB$ scores for single objective and multiobjective GA-based fuzzy algorithms have been shown for the Cat250_15_5 dataset. The final objective function values are also reported. As expected, SGA(π) produces the minimum π value (11.29), whereas SGA(Sep) gives the maximum Sep value (16.39). The MOGA(π,Sep) method provides a π value (11.34) greater than that provided by SGA(π), whereas a Sep value (15.38) smaller than that is provided by SGA(Sep). However, in terms of the $ARIB$ scores, the multiobjective fuzzy approach provides the best result ($ARIB = 1$). This signifies the importance of optimizing both π and Sep simultaneously instead of optimizing them separately and this finding is very similar to that in [200].

Figure 8.3 plots the Pareto fronts produced by one of the runs of the multiobjective fuzzy algorithm along with the best solutions provided by the other algorithms

Table 8.1 *AvgARIB* scores for synthetic datasets over 50 runs of different algorithms

Algorithm	Cat250	Cat100	Cat500	Cat280
Fuzzy K-modes	0.7883	0.5532	0.3883	0.5012
K-modes	0.7122	0.4893	0.3122	0.4998
K-medoids	0.7567	0.4977	0.3003	0.4901
Average linkage	**1.0000**	0.5843	0.2194	0.0174
CCDV	**1.0000**	0.5933	0.0211	0.5002
SGA (π)	0.8077	0.5331	0.4243	0.4894
SGA (*Sep*)	0.7453	0.4855	0.2954	0.4537
SGA(π,*Sep*)	**1.0000**	0.5884	0.4276	0.5264
MOGA$_{crisp}$	**1.0000**	0.5983	0.4562	0.5442
MOGA (π, *Sep*)	**1.0000**	**0.6114**	**0.4842**	**0.5851**

Table 8.2 Objective function values and the best *ARIB* scores for the Cat250_15_5 dataset

Algorithm	π	*Sep*	*ARI*
Single objective GA minimizing π	**11.29**	13.44	0.8119
Single objective GA maximizing *Sep*	11.57	**16.39**	0.7701
multiobjective GA optimizing π and *Sep*	11.34	15.38	**1.0000**

for the synthetic datasets. The figure also marks the selected solution from the non-dominated Pareto-optimal set. It appears that these selected solutions tend to fall at the knee regions of the Pareto fronts. Similar plots have been used for illustrations in [200] for showing the Pareto front generated by the multiobjective algorithm along with the solutions generated by other crisp clustering methods for continuous data. Here the solutions for both fuzzy and crisp clustering methods used for clustering categorical data are plotted. As expected, each of the fuzzy K-modes, K-modes, K-medoids and SGA(π) algorithms tends to minimize the objective π and thus gives smaller values for *Sep* (larger values for $1/Sep$). On the other hand, SGA(*Sep*) maximizes the objective *Sep* and hence gives larger values of the objective π. The algorithms CCDV, SGA(π,*Sep*) and MOGA$_{crisp}$ are found to come nearest to the selected solution in the Pareto front.

8.5.8 Results for Real-Life Datasets

Table 8.3 reports the *AvgARIB* scores over 50 runs of the different clustering algorithms on the real-life datasets. It is evident from the table that for all the datasets, the multiobjective clustering technique produces the best *AvgARIB* scores. Here it can also be noted that the fuzzy clustering procedures outperform the corresponding crisp versions for all the real-life datasets, indicating the utility of incorporating fuzziness.

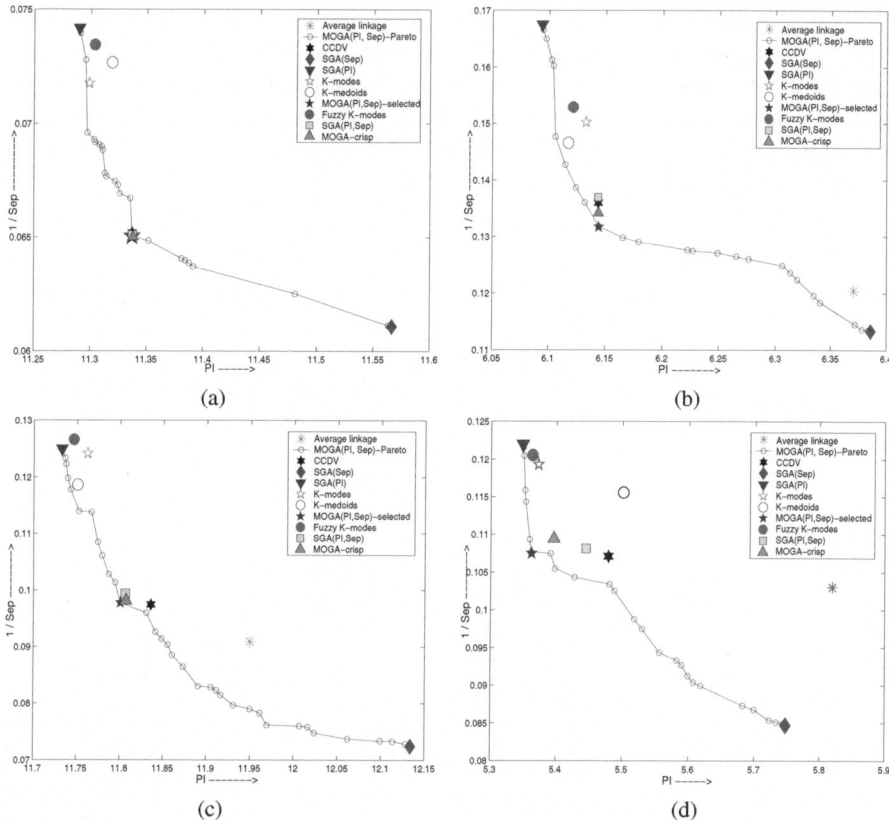

Fig. 8.3 Pareto-optimal fronts produced by the multiobjective fuzzy clustering technique for synthetic datasets along with the best results provided by other algorithms: (a) Cat250_15_5 dataset, (b) Cat100_10_4 dataset, (c) Cat500_20_10 dataset, (d) Cat280_10_6 dataset

Table 8.3 *AvgARIB* scores for real-life datasets over 50 runs of different algorithms

Algorithm	Votes	Zoo	Soybean	Cancer
Fuzzy K-modes	0.4983	0.6873	0.8412	0.7155
K-modes	0.4792	0.6789	0.5881	0.6115
K-medoids	0.4821	0.6224	0.8408	0.7124
Average linkage	0.5551	0.8927	**1.0000**	0.0127
CCDV	0.4964	0.8143	**1.0000**	0.7145
SGA (π)	0.4812	0.8032	0.8861	0.7268
SGA (Sep)	0.4986	0.8011	0.8877	0.2052
SGA(π,Sep)	0.5012	0.8348	0.9535	0.7032
MOGA$_{crisp}$	0.5593	0.8954	**1.0000**	0.7621
MOGA (π, Sep)	**0.5707**	**0.9175**	**1.0000**	**0.8016**

Figure 8.4 shows the Pareto fronts produced by one of the runs of the multi-objective technique along with the best solutions provided by the other algorithms for the real-life datasets. Here also the selected solutions from the Pareto front are mostly in the knee regions of the Pareto fronts. In the case of real datasets also, the best competitive algorithms are CCDV, SGA(π,Sep) and MOGA$_{crisp}$, which come closest to the selected non-dominated solution.

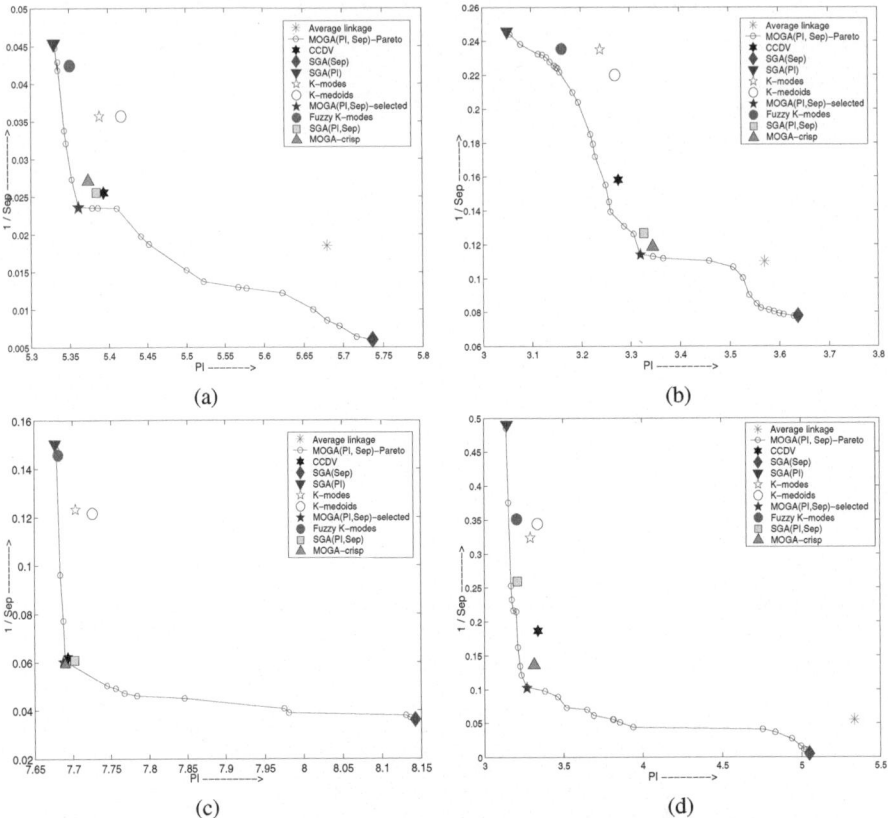

Fig. 8.4 Pareto-optimal fronts produced by the multiobjective fuzzy clustering technique for real-life datasets along with the best results provided by other algorithms: (a) Votes dataset, (b) Zoo dataset, (c) Soybean dataset, (d) Cancer dataset

For both synthetic and real-life and categorical datasets, it has been found that the fuzzy clustering methods perform better than their crisp counterparts. This is due to the fact that since the fuzzy membership functions allow a data point to belong to multiple clusters simultaneously with different degrees of membership, the incorporation of fuzziness makes the algorithm capable of handling the overlapping partitions better. For this reason, the fuzzy algorithms outperform their corresponding crisp versions. Also note that both the fuzzy and crisp versions of the multiobjective categorical data clustering methods use a similar encoding policy and a final solution selection based on majority voting followed by k-nn classification. Hence the better performance of MOGA(π,Sep) than MOGA$_{crisp}$ indicates that improvement in clustering results is solely due to the introduction of fuzziness in the objective function as well as in the clustering stage, and not due to the final solution selection strategy. This signifies the utility of incorporating fuzziness into the clustering techniques.

8.5.9 Execution Time

The execution times of the fuzzy clustering algorithms are usually more than those of corresponding crisp versions for computing and updating the fuzzy membership values. In this section, the time consumption of the fuzzy clustering algorithms has been compared. All the algorithms have been implemented in Matlab and executed on an Intel Core 2 Duo 2.0 GHz machine with the Windows XP operating system. On average, the MOGA(π,Sep) clustering executes in 990.43 seconds for the Cat250_15_5 dataset, whereas fuzzy K-modes, SGA(π), SGA(Sep) and SGA(π,Sep) take 610.32, 680.73, 630.44 and 890.81 seconds, respectively for this dataset. The execution times have been computed on the basis of the parameter setting discussed in Section 8.5.6. As expected, the execution time of the multiobjective fuzzy clustering technique is larger than that of the other single objective fuzzy clustering methods because of some additional operations necessitated by its multiobjective nature. However, as is evident from the results, the clustering performance of MOGA(π,Sep) is the best among all the methods for all the datasets considered here. It is also found during experimentation that even if the other algorithms used for comparison (both fuzzy and crisp) are allowed to run for the time taken by MOGA(π,Sep), they are not able to improve their clustering results.

The execution time of the MOGA(π,Sep) algorithm for the other datasets considered here are as follows: Cat100_10_4, 376.35 seconds; Cat500_20_10, 2045.58 seconds; Cat280_10_6, 1030.33 seconds; Votes, 780.28 seconds; Zoo, 530.47 seconds; Soybean, 120.49 seconds; and Cancer, 1,080.56 seconds. The timing requirements of the multiobjective fuzzy clustering technique can be reduced by using a stopping criterion based on some test of convergence for multiobjective evolutionary process.

8.6 Statistical Significance Test

Tables 8.1 and 8.3 report the *AvgARIB* scores produced by different algorithms over 50 consecutive runs for the synthetic and real-life datasets, respectively. It is evident from the table that the *AvgARIB* scores produced by the multiobjective fuzzy clustering technique are better than those produced by all other algorithms except for some of the datasets, where average linkage, CCDV and SGA(π,*Sep*) provide scores similar to that of the multiobjective fuzzy clustering technique. To establish that this better performance of the multiobjective fuzzy clustering algorithm is statistically significant, some statistical significance test is required. Here, the statistical significance of the clustering solutions is tested through a t-test [67] at the 5% significance level. Ten groups, corresponding to the ten algorithms (fuzzy K-modes, K-modes, K-medoids, average linkage, CCDV, SGA(π), SGA(*Sep*), SGA(π,*Sep*), MOGA(π,*Sep*) and MOGA$_{crisp}$) have been created for each dataset. Each group consists of the *ARIB* scores produced by 50 consecutive runs of the corresponding algorithm.

Table 8.4 reports the p-values produced by a t-test for the comparison of two groups (the group corresponding to MOGA(π,*Sep*) and a group corresponding to some other algorithm) at a time. As a null hypothesis, it is assumed that there are no significant differences between the *AvgARIB* scores of the two groups. The alternative hypothesis is that there are significant differences in the mean values of the two groups. All the p-values reported in the table are less than 0.05 (5% significance level). For example, the t-test between the algorithms MOGA(π,*Sep*) and fuzzy K-modes for the Votes dataset provides a p-value of $4.33E - 08$, which is much less than the significance level 0.05. This is strong evidence against the null hypothesis, indicating that the better *AvgARIB* scores produced by the multiobjective fuzzy clustering method are statistically significant and have not occurred by chance. Similar results are obtained for all other datasets and for all other algorithms compared to MOGA(π,*Sep*), establishing the significant superiority of the multiobjective fuzzy clustering algorithm.

8.7 Summary

In this chapter, the issues of fuzzy clustering for categorical data are discussed. In this context, a recently developed multiobjective Genetic Algorithm-based fuzzy clustering algorithm for clustering categorical datasets has been discussed. This method optimizes two objectives, namely the fuzzy compactness and the fuzzy separation of the clusters, simultaneously. The algorithm is based on NSGA-II.

The performance of the multiobjective fuzzy clustering technique, based on the adjusted Rand index, has been compared with that of the several other well-known categorical data clustering algorithms. Four synthetic and four real-life categorical datasets have been used for performing the experiments. The superiority of the multiobjective technique has been demonstrated and the use of multiple objectives rather than a single objective has been justified. Moreover, a statistical significance test based on t-statistic has been carried out in order to judge the statistical significance of the clustering solutions.

Table 8.4 The p-values produced by t-test comparing MOGA(π,Sep) with other algorithms.

Datasets					p-values					
	F	K-modes	K-medoids	Avg link	CCDV	SGA(π)	SGA(Sep)	SGA(π,Sep)	MOGA$_{crisp}$
Cat250	2.13E-07	2.13E-10	7.44E-07	same	same	5.45E-10	1.67E-20	same	same
Cat100	3.02E-10	4.07E-17	5.39E-11	1.46E-12	6.28E-07	5.11E-15	3.82E-12	4.88E-07	3.98E-06
Cat500	2.73E-08	1.02E-10	7.88E-10	5.07E-13	1.22E-24	2.12E-11	1.43E-08	2.93E-11	6.92E-08
Cat280	4.06E-10	9.32E-12	4.95E-12	2.69E-31	3.44E-09	4.63E-12	3.66E-14	1.82E-09	5.21E-09
Votes	4.33E-08	3.85E-11	2.56E-09	7.23E-07	4.84E-08	2.33E-10	7.83E-08	4.72E-08	3.73E-08
Zoo	4.57E-19	6.45E-19	7.48E-20	5.11E-09	3.54E-07	4.12E-13	8.44E-14	2.18E-07	4.66E-09
Soybean	5.62E-06	2.04E-18	8.55E-06	same	same	5.66E-05	3.18E-05	5.08E-05	same
Cancer	1.83E-11	6.17E-12	2.48E-10	2.33E-40	7.55E-09	6.03E-08	5.22E-26	2.66E-08	8.56E-08

Chapter 9
Unsupervised Cancer Classification and Gene Marker Identification

9.1 Introduction

Microarray technology has significant impact on cancer research. It is utilized in cancer diagnosis by means of classification of tissue samples. When microarray datasets are organized as gene vs. sample, they are very helpful in the classification of different types of tissues and the identification of those genes whose expression levels are good diagnostic indicators. The microarrays where tissue samples represent cancerous (malignant) and non-cancerous (benign) cells, their classification results in a binary cancer classification.

A microarray gene expression dataset consisting of g genes and s tissue samples is typically organized in a 2D matrix $E = [e_{ij}]$ of size $s \times g$. Each element e_{ij} gives the expression level of the jth gene for the ith tissue sample. Most of the researches in the area of cancer diagnosis have focussed on supervised classification of cancer datasets through training, validation and testing to classify the tumor samples as malignant or benign, or as their subtypes [13, 14, 124, 184, 246, 447]. However, unsupervised classification or clustering of tissue samples should also be studied since in many cases labeled tissue samples are not available. In this chapter, the application of the multiobjective genetic clustering for unsupervised classification of cancer data has been explored.

Usually, a fuzzy clustering solution is defuzzified by assigning each point to the cluster to which the point has the highest membership degree. In general, it has been observed that for a particular cluster, among the points that are assigned to it based on the maximum membership criterion, some have higher membership degree to that cluster, whereas the other points of the same cluster may have a lower membership degree. Thus the points in the latter case are not assigned to that cluster with high confidence. This observation motivates us to improve the clustering result obtained by the multiobjective fuzzy clustering method using some supervised classification tool, such as Support Vector Machine (SVM) [417], which is trained by the points with high membership degree in a cluster. The trained classifier thereafter can be used to classify the remaining points [301, 313].

A characteristic of any MOO approach is that evidently all the solutions in the Pareto-optimal set have some information that is inherently good for the problem at hand. Motivated by this observation, this chapter describes a method to obtain the final solution by combining the information of all the non-dominated solutions produced in the final generation. In this approach, first each solution (fuzzy membership matrix) of the final non-dominated front is considered and each is improved using the SVM classifier [417] as discussed above. Finally, to combine the clustering information of the non-dominated solutions, a cluster ensemble method based on the majority voting technique is applied to obtain the final optimal clustering solution.

Furthermore, the clustering solution produced by the MOGASVMEN clustering technique (multiobjective GA-based clustering improved with SVM and cluster ensemble) has been used to identify the gene markers that are mostly responsible for distinguishing between the two classes of tissue samples. SNR statistic-based gene ranking followed by multiobjective feature selection has been utilized for this purpose.

The performance of the MOGASVMEN clustering technique has been reported for three publicly available benchmark cancer datasets, viz., Leukemia, Colon cancer, and Lymphoma. The superiority of the MOGASVMEN technique, as compared to K-means clustering [237], FCM [62], single objective GA minimizing the XB index (SGA(XB)) [300], hierarchical average linkage clustering and Self Organizing Map (SOM) clustering [396], is demonstrated both quantitatively and visually. The superiority of the MOGASVMEN clustering technique has also been proved to be statistically significant through statistical significance tests. Finally, it has been demonstrated how the MOGASVMEN clustering result can be used for identifying the relevant gene markers.

9.2 MOGASVMEN Clustering Technique

This section describes the scheme for improving the non-dominated set of solutions produced by MOGA clustering through SVM classification and then combining them via the majority voting ensemble technique to yield the final optimal clustering solution. As discussed before, the procedure is motivated by two things: one is to utilize the points with high membership degree as the training points, and the other is to combine the clustering information possessed by each non-dominated solution. The different steps of the MOGASVMEN technique are described below in detail.

Procedure MOGASVMEN

1. Apply MOGA clustering (described in Chapter 5) on the given dataset to obtain a set $S = \{s_1, s_2, \ldots, s_N\}$, $N \leq P$ (P is the population size), of non-dominated solution strings consisting of cluster centers.
2. Using Equation 5.4, compute the fuzzy membership matrix $U^{(i)}$ for each of the non-dominated solutions s_i, $1 \leq i \leq N$.
3. For each non-dominated solution $s_i \in S$, repeat the following:

   a. Assign each point k, ($k = 1, \ldots, n$), to some cluster j ($1 \leq j \leq K$) such that $u_{jk}^{(i)} = \max_{l=1,\ldots,K} \{u_{lk}\}$.
   b. For each cluster l ($l = 1, \ldots, K$), select all the points k of that cluster for which $u_{lk}^{(i)} \geq \mathscr{T}$, where $\mathscr{T}$ ($0 < \mathscr{T} < 1$) is a threshold value of the membership degree. These points act as training points for cluster l. Combine the training points of all the clusters to form the complete training set. Keep the remaining points as the test set.
   c. Train the SVM classifier by the selected training points.
   d. Predict the class labels for the remaining points (test points) using the trained SVM classifier.
   e. Combine the label vectors corresponding to the training and testing points to obtain the label vector λ_i for the complete dataset corresponding to the non-dominated solution s_i.

4. Reorganize the membership matrices to make them consistent with each other following the method discussed in Chapter 6, Section 6.2.1.
5. Apply majority voting on the label vectors λ_i, $i = 1, \ldots, N$, to obtain the final clustering label vector λ. The majority voting is done as follows: assign each point $k = 1, \ldots, n$ to the cluster j where the label j appears the maximum number of times among all the labels for the point k in all the λ_is.

The parameter $\mathscr{T}$ acts as a threshold value to separate the training set from the test set. If $\mathscr{T}$ is small, then the membership degrees of a large number of points exceed $\mathscr{T}$. Hence the size of the training set increases; however, the confidence of the training set will be low since many training points will have low membership degrees. On the other hand, if $\mathscr{T}$ is increased, then the membership degrees of a smaller number of points will exceed the threshold and thus the size of the training set will decrease. However, the training set will then contain only the points having high membership degrees to their respective clusters. Hence the training set will have more confidence. To balance the size and confidence of the training set, the threshold $\mathscr{T}$ has been set to 0.5 after several experiments.

9.3 Datasets and Preprocessing

Three publicly available benchmark cancer datasets, viz., Leukemia, Colon cancer and Lymphoma, have been used for experiments. The datasets and their preprocessing are described in this section.

9.3.1 Leukemia Data

The Leukemia dataset [184] consists of 72 tissue samples. The samples consists of two types of leukemia, 25 of AML and 47 of ALL. The samples are taken from 63 bone marrow samples and nine peripheral blood samples. There are 7,129 genes in the dataset. The dataset is publicly available at http://www.genome.wi.mit.edu/MPR.

The dataset is subjected to a number of preprocessing steps to find the genes with most variability. As this work considers the problem of unsupervised classification, the initial gene selection steps followed here are also completely unsupervised. However, more sophisticated methods for gene selection could have been applied. First the genes whose expression levels fall between 100 and 15,000 are selected. From the resulting 1,015 genes, the 100 genes with the largest variation across the samples are selected, and the remaining expression values are log-transformed. The resultant dataset is of dimension 72×100.

9.3.2 Colon Cancer Data

The Colon cancer dataset [14] consists of 62 samples of colon epithelial cells from colon cancer patients. The samples consist of tumor biopsies collected from tumors (40 samples) and normal biopsies collected from the healthy part of the colons (22 samples) of the same patients. The number of genes in the dataset is 2,000. The dataset is publicly available at http://microarray.princeton.edu/oncology.

This dataset is preprocessed as follows: first the genes whose expression levels fall between 10 and 15,000 are selected. From the resulting 1,756 genes, the 100 genes with the largest variation across samples are selected, and the remaining expression values are log-transformed. The resultant dataset is of dimension 62×100.

9.3.3 Lymphoma Data

The diffuse large B-cell lymphoma (DLBCL) dataset [13] contains expression measurements of 96 normal and malignant lymphocyte samples, each measured using a specialized cDNA microarray containing 4,026 genes that are preferentially expressed in lymphoid cells or which are of known immunological or oncological importance. There are 42 DLBCL and 54 other cancer disease samples. The dataset is publicly available at http://genome-www.stanford.edu/lymphoma.

The preprocessing steps for this dataset are as follows: As the dataset contains some missing values, only those genes which do not contain any missing value are selected. This results in 854 genes. Next, each gene is normalized to have an expression value between 0 and 1. Thereafter, the top 100 genes with respect to variance are selected. Hence the dataset contains 96 samples, each described by 100 genes.

9.4 Experimental Results

The performance of the MOGASVMEN clustering has been compared with that of K-means clustering [237, 401], FCM [62], single objective genetic clustering scheme that minimizes the XB validity measure (SGA (XB)) [300], hierarchical average linkage clustering [408] and Self Organizing Map (SOM) clustering [396].

For evaluating the performance of the clustering algorithms on the three cancer datasets, adjusted Rand index (ARI) [449] and silhouette index ($s(C)$) [369] are used. As a distance measure, the Pearson correlation-based distance is used.

9.4.1 Input Parameters

The different parameters of MOGA and single objective GAs are taken as follows: number of generations = 100, population size = 50, crossover probability = 0.8 and mutation probability = 0.01. The value of the parameters $\mathscr{T}$ is taken as 0.5. The parameters have been set after several experiments. The fuzzy exponent m is chosen as in [132, 249], and the values of m for the datasets Leukemia, Colon cancer and Lymphoma are set to 1.19, 1.53 and 1.34, respectively. The K-means and the fuzzy C-means algorithms have been run for 200 iterations unless they converge before that. The SVM classifier uses the RBF kernel.

9.4.2 Results

Each algorithm considered here has been executed for 50 consecutive runs and Tables 9.1 and 9.2 report the average ARI index scores and the average $s(C)$ index scores over these 50 runs, respectively, for the Leukemia, Colon cancer and

Table 9.1 Average *ARI* scores produced by 50 consecutive runs of different algorithms for Leukemia, Colon cancer and Lymphoma datasets

Algorithms	Leukemia	Colon cancer	Lymphoma
MOGASVMEN	0.8424	0.6657	0.4086
K-means	0.7093	0.4264	0.3017
FCM	0.7019	0.4794	0.3414
SGA (*XB*)	0.7842	0.5789	0.3203
Avg. linkage	0.6284	-0.0432	0.3973
SOM	0.7409	0.5923	0.3367

Table 9.2 Average $s(C)$ scores produced by 50 consecutive runs of different algorithms for Leukemia, Colon cancer and Lymphoma datasets

Algorithms	Leukemia	Colon cancer	Lymphoma
MOGASVMEN	0.3129	0.4357	0.3253
K-means	0.2252	0.1344	0.2291
FCM	0.2453	0.1649	0.2239
SGA (*XB*)	0.2612	0.3024	0.2575
Avg. linkage	0.1934	0.2653	0.2235
SOM	0.2518	0.3153	0.2665

Lymphoma datasets. As is evident from the tables, the MOGASVMEN clustering produces the best average *ARI* and $s(C)$ index scores compared to the other algorithms. For example, for the leukemia dataset, MOGASVMEN produces an average *ARI* index score of 0.8424, whereas the next best *ARI* score is 0.7842, provided by SGA(*XB*). Similar results are obtained for all the datasets and for the $s(C)$ index. In general, it can be observed that MOGASVMEN consistently outperforms the other algorithms for each of the Leukemia, Colon cancer and Lymphoma datasets in terms of both *ARI* and $s(C)$.

From the tables, it appears that MOGASVMEN also outperforms its single objective counterparts FCM and SGA(*XB*). FCM minimizes the J_m measure, whereas SGA(*XB*) minimizes the *XB* validity index. MOGASVMEN minimizes both of these objectives simultaneously. As MOGASVMEN performs better in terms of both the internal and the external validity indices, it is established that optimizing multiple criteria simultaneously can yield better clustering than optimizing the objectives separately.

For the purpose of illustration, Figures 9.1 and 9.2 show the boxplots representing the *ARI* and $s(C)$ index values, respectively, over 50 runs of the algorithms for the three datasets considered here. It is evident from the figures that the boxplots corresponding to the MOGASVMEN clustering method are situated in the upper half of the figures, which indicates that MOGASVMEN produces higher *ARI* and $s(C)$ index values compared to the other algorithms in all the runs.

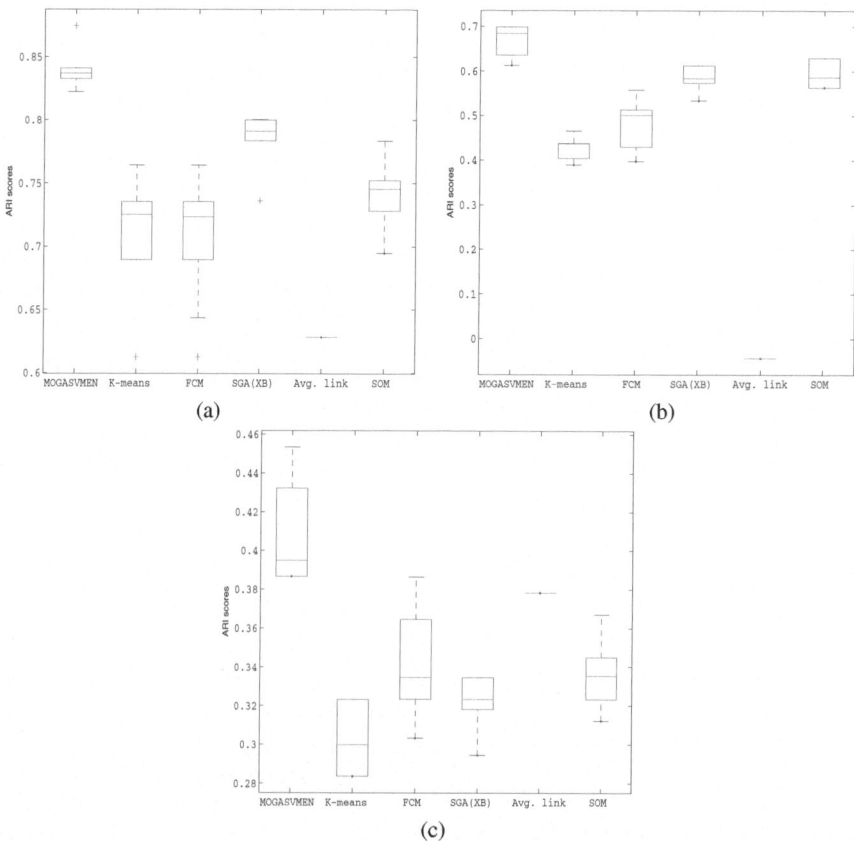

Fig. 9.1 Boxplots showing the *ARI* index scores produced by different algorithms over 50 consecutive runs for (a) Leukemia, (b) Colon cancer and (c) Lymphoma datasets

9.4.3 Statistical Significance Test

To establish that MOGASVMEN is significantly superior to the other algorithms, the statistical significance test t-test has been conducted at the 5% significance level. Six groups, corresponding to the six algorithms (MOGASVMEN, K-means, FCM, SGA(XB), average linkage and SOM), have been created for each dataset. Each group consists of the *ARI* index scores produced by 50 consecutive runs of the corresponding algorithm.

As is evident from Table 9.1, the average values of *ARI* scores for MOGASVMEN are better than those for the other algorithms. To establish that this goodness is statistically significant, Table 9.3 reports the p-values produced by the t-test for the comparison of two groups (the group corresponding to MOGASVMEN and a group corresponding to some other algorithm) at a time. As a null hypothesis, it is

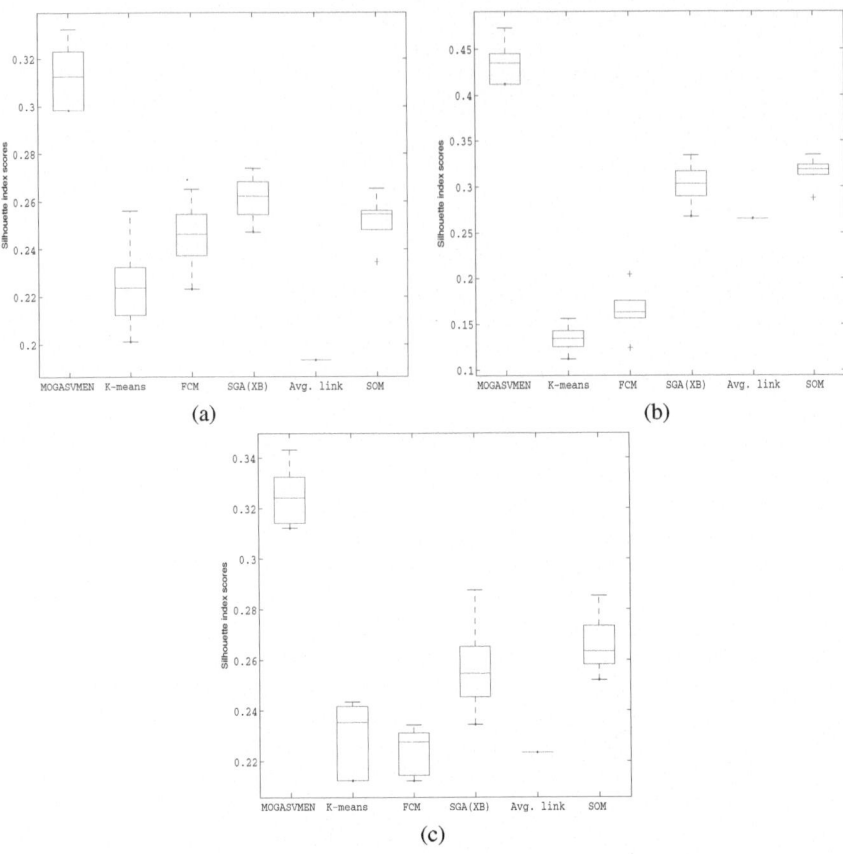

Fig. 9.2 Boxplots showing the $s(C)$ index scores produced by different algorithms over 50 consecutive runs for (a) Leukemia, (b) Colon cancer and (c) Lymphoma datasets

Table 9.3 p-values produced by t-test comparing MOGASVMEN with other algorithms.

Datasets	p-values (comparing *ARI* means of MOGASVMEN with other algorithms)				
	K-means	FCM	SGA(*XB*)	Average Linkage	SOM
Leukemia	3.1E-07	2.17E-07	2.41E-03	1.08E-06	6.5E-05
Colon cancer	2.21E-05	1.67E-08	3.4E-05	4.52E-12	1.44E-04
Lymphoma	3.42E-05	7.43E-08	5.8E-05	2.7E-07	2.1E-05

assumed that there is no significant difference between the mean values of the two groups. The alternative hypothesis is that there is significant difference in the mean values of the two groups. All the p-values reported in the table are less than 0.05 (5% significance level). This is strong evidence against the null hypothesis, indicating that the better mean values of the *ARI* index produced by MOGASVMEN are statistically significant and have not occurred by chance.

9.5 Identification of Gene Markers

In this section, it is demonstrated how the clustering result of the MOGASVMEN clustering technique can be used to identify the gene markers mostly responsible for distinguishing between the two classes of tissue samples. This has been done as follows.

First, MOGASVMEN is applied to cluster the samples of the preprocessed datasets into two classes, labeled 1 and 2, respectively. Thereafter, for each of the genes, a statistic called Signal-to-Noise Ratio (SNR) [184] is computed. SNR is defined as

$$SNR = \frac{\mu_1 - \mu_2}{\sigma_1 + \sigma_2}, \tag{9.1}$$

where μ_i and σ_i, $i \in \{1,2\}$ denote the mean and standard deviation of class i for the corresponding gene. Note that the larger value of SNR for a gene indicates that the gene's expression level is high in one class and low in another. Hence this bias is very useful in identifying the genes that are expressed differently in the two classes of samples.

After computing the SNR statistic for each gene, the genes are sorted in descending order of their SNR values. From the sorted list, the genes whose SNR values are greater than the average SNR value are selected. The number of genes selected in this way for the Leukemia, Colon cancer and Lymphoma datasets are 38, 31 and 36, respectively. These genes are mostly responsible for distinguishing between the two sample classes.

The list of genes obtained is further reduced by a *multiobjective feature selection* technique. In this technique, each chromosome is a binary string of length equal to the number of genes selected through the SNR method. For a chromosome, bit '1' indicates that the corresponding gene is selected, and bit '0' indicates that the corresponding gene is not selected. Here also NSGA-II is used as the underlying optimization tool. The two objective functions are *classification accuracy* and *number of selected genes*. The classification accuracy is computed by using Leave One Out Cross Validation (LOOCV) with an SVM (RBF) classifier. Note that the cross validation with SVM is done for the subset of genes encoded in the chromosome and the classification accuracy is computed using the clustering result obtained using the MOGASVMEN method. The goal is to maximize the first objective while minimizing the second one simultaneously, as the minimum number of possible gene markers are to be selected. From the final non-dominated front produced by the multiobjective feature selection algorithm, the solution with the maximum classification accuracy is selected and the corresponding gene subset is selected as the final set of gene markers.

Table 9.4 reports the number of gene markers obtained as above for the three datasets. The numbers of gene markers for the three datasets are 11, 8 and 9, respectively. The table also reports the *ARI* index scores obtained by MOGASVMEN clustering on the complete preprocessed datasets (with 100 genes) and on the reduced dataset consisting of the marker genes only. It is evident from the table that the performance of MOGASVMEN clustering improves when applied to the dataset with the identified marker genes only. This indicates the ability of the gene markers to distinguish between the two types of samples in all the datasets.

Table 9.4 Number of gene markers selected for different datasets and performance of MOGASV-MEN on the set of all 100 genes and on the set of marker genes in terms of *ARI* index scores

Dataset	# gene markers	MOGASVMEN *ARI* on all 100 genes	MOGASVMEN *ARI* on marker genes
Leukemia	11	0.8424	0.9436
Colon cancer	8	0.6657	0.6947
Lymphoma	9	0.4086	0.5182

9.5.1 Gene Markers for Leukemia Data

In Figure 9.3, the heatmap of the gene vs. sample matrix, where the rows correspond to the top 30 genes in terms of SNR statistic scores and the columns correspond to the ALL and AML samples as obtained by MOGASVMEN clustering, is shown. The cells of the heatmap represent the expression levels of the genes in terms of colors. The shades of red represent high expression levels, the shades of green represent low expression levels and the darker colors represent the absence of differential expression values. The 11 gene markers identified are placed at the top 11 rows of the heatmap. It is evident from the figure that these 11 gene markers clearly distinguish the AML samples from the ALL ones. The characteristics of the gene markers are as follows: The genes M92287_at, HG1612-HT1612_at, X51521_at, Z15115_at, U22376_cds2_s_at and X67951_at are upregulated in the ALL samples and downregulated in the AML samples. On the other hand, the genes M63138_at, X62320_at, HG2788-HT2896_at, U46751_at and L08246_at are downregulated in the ALL samples and upregulated in the AML samples. In Table 9.5, the 11 gene markers are reported along with their description and associated Gene Ontological (GO) terms. It is evident from the table that all the 11 genes share most of the GO terms, which indicates that these genes have similar molecular functions (mainly related to cell, cell part and organelle).

9.5.2 Gene Markers for Colon Cancer Data

Figure 9.4 shows the heatmap of the gene vs. sample matrix for the top 30 gene markers of the Colon Cancer data. The eight gene markers identified are placed at the top eight rows of the heatmap. It is evident from visual inspection that these eight gene markers partition the tumor samples from the normal ones. The characteristics of the gene markers are as follows: The genes M63391 and Z24727 are downregulated in the tumor samples and upregulated in the normal samples. In contrast, the genes T61609, T48804, T57619, M26697, T58861 and T52185 are upregulated in the tumor samples and downregulated in the normal samples. In Table 9.6, the eight gene markers are described along with the associated Gene Ontological (GO) terms. It is evident from the table that all the eight genes mainly take part in metabolic process, cellular process and gene expression and share most of the GO terms. This indicates that these genes have similar molecular functions.

9.5.3 Gene Markers for Lymphoma Data

In Figure 9.3, the heatmap for the top 30 gene markers for the Lymphoma data is shown. The topmost nine gene markers selected are placed at the top 9 rows of the heatmap. Visual inspection reveals that these nine gene markers efficiently distinguish the DLBCL samples from the non-DLBCL ones. The characteristics of the gene markers are as follows: The genes 19335, 19338, 20344, 18344, 19368, 20392 and 16770 are upregulated in the non-DLBCL samples and downregulated in the DLBCL samples. On the other hand, the genes 13684 and 16044 are downregulated in the non-DLBCL samples and upregulated in the DLBCL samples. In Table 9.7, the nine gene markers along with their description and associated Gene Ontological (GO) terms have been reported. It is evident from the table that all the nine genes share most of the GO terms (mainly related to different kinds of binding functions), which indicates that these genes have similar molecular functions.

Table 9.5 The description and associated Gene Ontological (GO) terms for the 11 gene markers identified in Leukemia data

AFFY_ID	Gene description	Gene function (GO terms)
M92287_at	cyclin d3	cell, macromolecular complex, organelle, cell part
HG1612-HT1612_at	marcks-like 1	cell, cell part
X51521_at	villin 2 (ezrin)	cell, organelle, organelle part, cell part
Z15115_at	topoisomerase (dna) ii beta 180kda	cell, membrane-enclosed lumen, organelle, organelle part, cell part
U22376_cds2_s_at	v-myb myeloblastosis viral oncogene homolog (avian)	cell, membrane-enclosed lumen, macromolecular complex, organelle, organelle part, cell part
X67951_at	peroxiredoxin 1	cell, organelle, cell part
M63138_at	cathepsin d (lysosomal aspartyl peptidase)	extracellular region, cell, organelle, cell part
X62320_at	granulin	extracellular region, cell, organelle, extracellular region part, cell part
HG2788-HT2896_at	s100 calcium binding protein a6 (calcyclin)	cell, envelope, organelle, organelle part, cell part
U46751_at	sequestosome 1	cell, organelle, cell part
L08246_at	myeloid cell leukemia sequence 1 (bcl2-related)	cell, envelope, organelle, organelle part, cell part

Table 9.6 The description and associated Gene Ontological (GO) terms for the eight gene markers identified in Colon Cancer data

Gene.ID	Gene description	Gene function (GO terms)
M63391	desmin	cellular process, multicellular organismal process, biological regulation
Z24727	tropomyosin 1 (alpha)	cellular process, multicellular organismal process, localization biological regulation
T61609	ribosomal protein sa	metabolic process, cellular process, gene expression, biological adhesion
T48804	ribosomal protein s24	metabolic process, cellular process, gene expression
T57619	ribosomal protein s6	metabolic process, cellular process, gene expression, biological regulation
M26697	nucleophosmin (nucleolar phosphoprotein b23, numatrin)	metabolic process, cellular process, gene expression, developmental process, response to stimulus, localization, establishment of localization, biological regulation
T58861	ribosomal protein 130	metabolic process, cellular process, gene expression
T52185	ribosomal protein s19	immune system process, metabolic process, cellular process, gene expression, multicellular organismal process, developmental process, locomotion, response to stimulus, localization, estabshment of localization, biological regulation

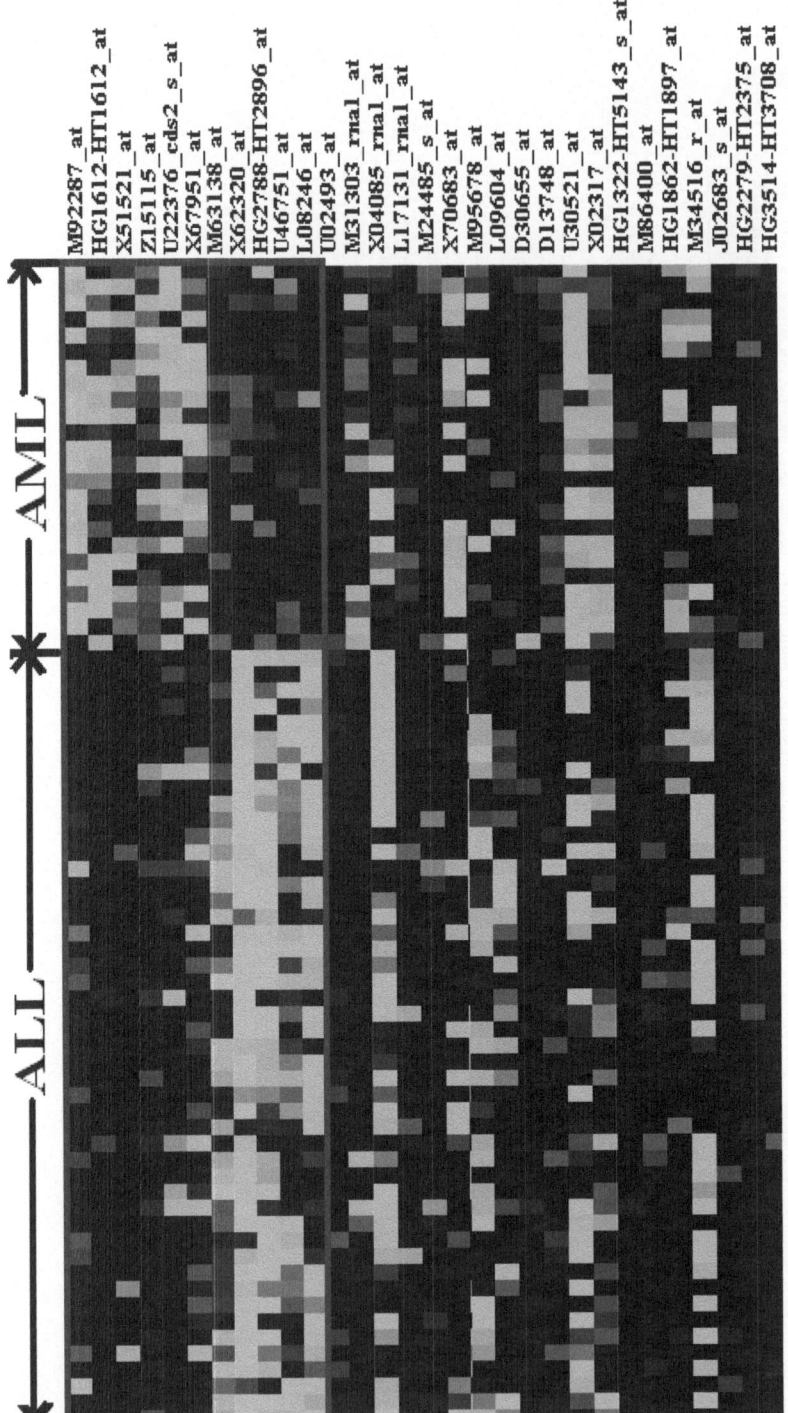

Fig. 9.3 Heatmap of the expression levels of the top 30 gene markers distinguishing the AML and ALL samples of Leukemia data. Red/green represent up/down regulation relative to black. The top 11 markers are put in a blue box

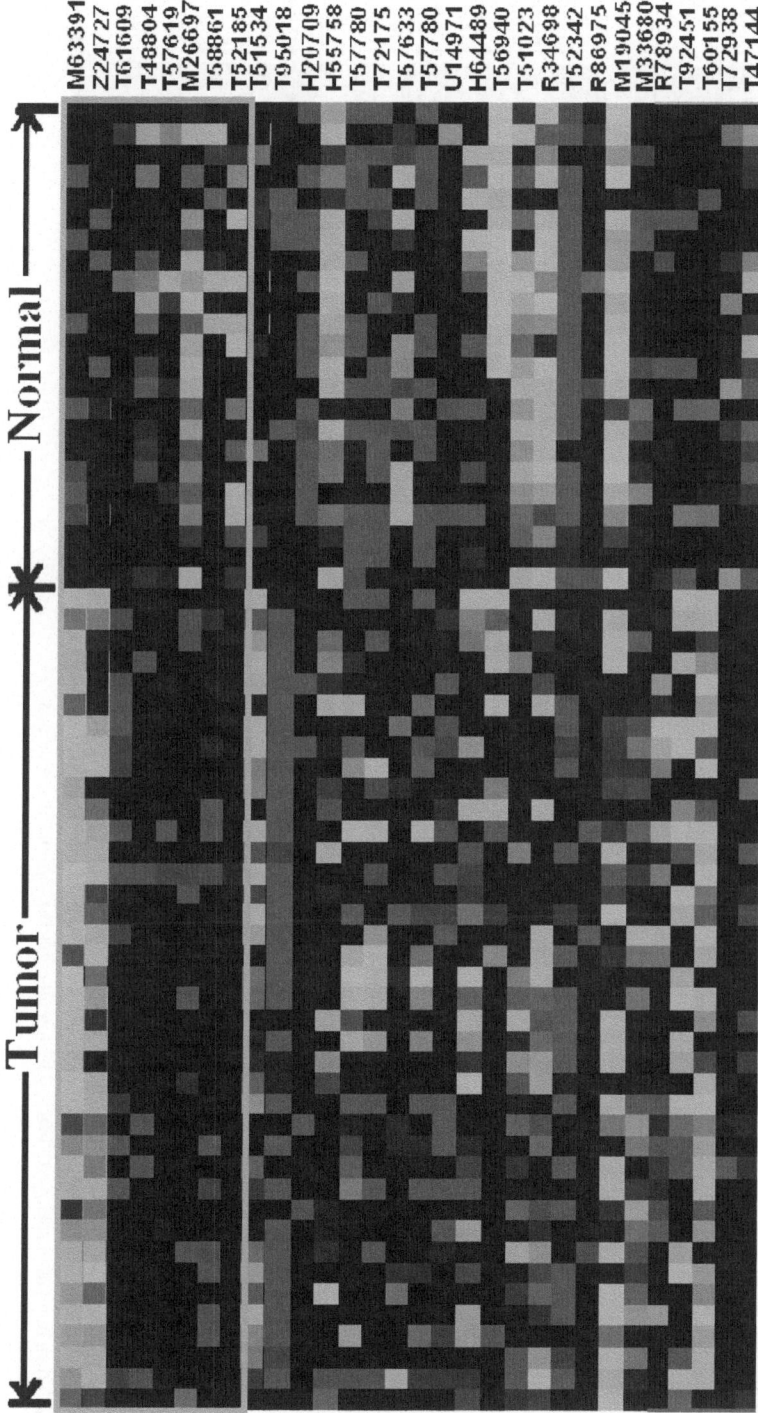

Fig. 9.4 Heatmap of the expression levels of the top 30 gene markers distinguishing the Tumor and Normal samples of Colon Cancer data. Red/green represent up/down regulation relative to black. The top eight markers are put in a blue box

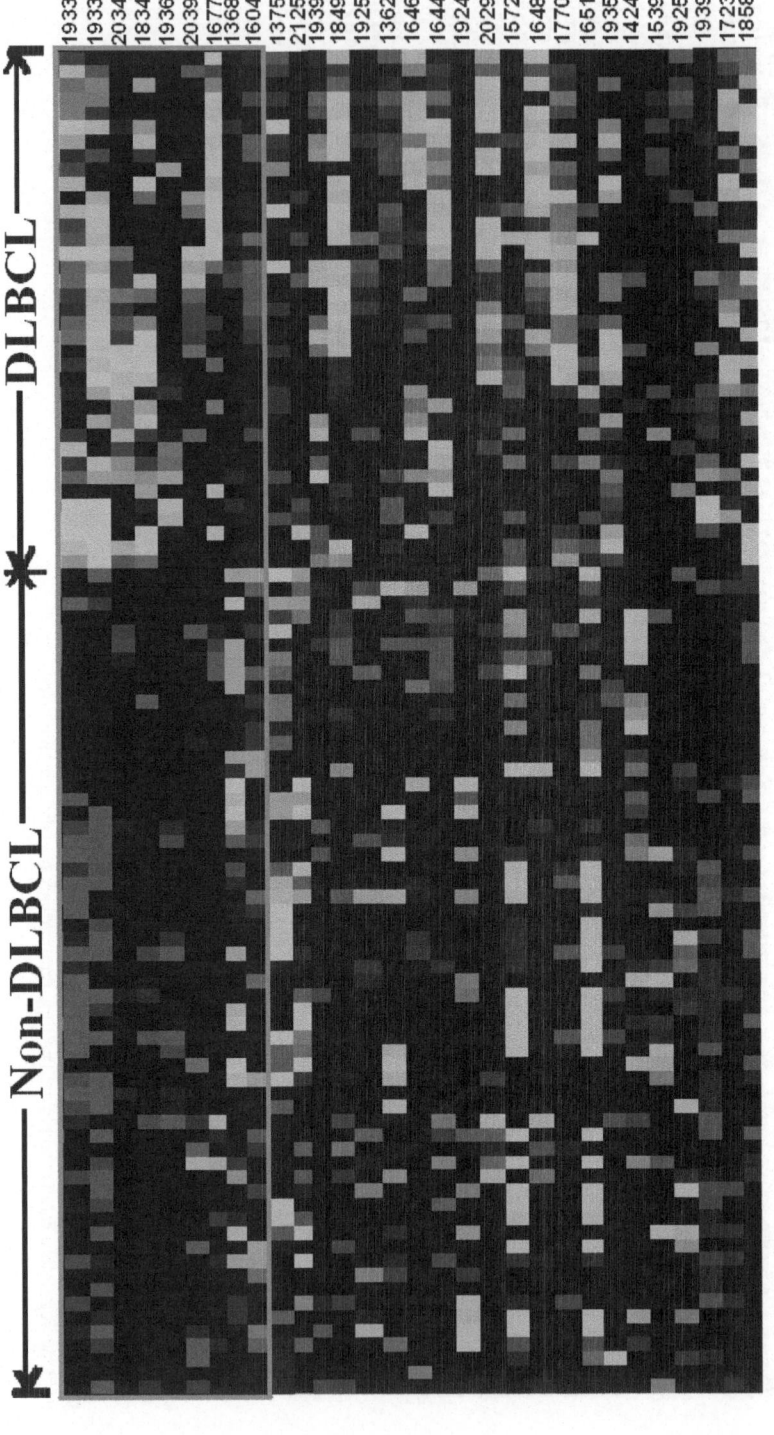

Fig. 9.5 Heatmap of the expression levels of the top 30 gene markers distinguishing the Tumor and Normal samples of Colon Cancer data. Red/green represent up/down regulation relative to black. The top nine markers are put in a blue box

Table 9.7 The description and associated Gene Ontological (GO) terms for the nine gene markers identified in Lymphoma data

Gene_ID	Gene description	Gene function (GO terms)
19335	rab23, member ras oncogene family	nucleotide binding, binding, GTP binding, purine nucleotide binding, guanyl nucleotide binding, ribonucleotide binding, purine ribonucleotide binding, guanyl ribonucleotide binding
19338	rab33b, member of ras oncogene family	nucleotide binding, binding, GTP binding, purine nucleotide binding, guanyl nucleotide binding, ribonucleotide binding, purine ribonucleotide binding, guanyl ribonucleotide binding
20344	selectin, platelet	glycoprotein binding, binding, protein binding, sugar binding, carbohydrate binding, sialic acid binding, monosaccharide binding, calcium-dependent protein binding
18344	olfactory receptor 45	rhodopsin-like receptor activity, signal transducer activity, receptor activity, transmembrane receptor activity, G-protein coupled receptor activity, olfactory receptor activity, molecular transducer activity
19368	retinoic acid early transcript 1, alpha	receptor binding, binding, protein binding, phospholipid binding, lipid binding, phosphoinositide binding, natural killer cell lectin-like receptor binding, GPI anchor binding
20392	sarcoglycan, epsilon	binding, calcium ion binding, protein binding, ion binding, cation binding, metal ion binding,
16770	lactalbumin, alpha	catalytic activity, lactose synthase activity, binding, calcium ion binding, UDP-glycosyltransferase activity, galactosyltransferase activity, transferase activity, transferase activity, transferring glycosyl groups, transferase activity, transferring hexosyl groups, UDP-galactosyltransferase activity, ion binding, cation binding, metal ion binding,
13684	eukaryotic translation initiation factor 4e	nucleic acid binding, RNA binding, translation initiation factor activity, binding, protein binding, translation factor activity, nucleic acid binding, translation regulator activity
16044	immunoglobulin heavy chain sa2	antigen binding, binding

9.6 Summary

In this chapter, the NSGA-II-based multiobjective fuzzy clustering technique that optimizes the XB and J_m validity indices is used for unsupervised cancer data classification. The non-dominated solutions produced at the final generation are improved through SVM classification. A technique for producing the final clustering result by combining the clustering information of the non-dominated solutions through a majority voting ensemble method is described.

Results on three publicly available benchmark cancer datasets, viz., Leukemia, Colon cancer and Lymphoma, have been demonstrated. The performance of MO-GASVMEN has been compared with that of K-means, FCM, SGA(XB), average linkage and SOM clustering methods. The results have been demonstrated both quantitatively and visually. The MOGASVMEN clustering technique consistently outperformed the other algorithms considered here. Also, statistical superiority has been established through statistical significance test.

Finally, relevant gene markers are identified using the clustering result of MO-GASVMEN. For this purpose, SNR statistic-based gene selection followed by a multiobjective feature selection technique is utilized. The identified gene markers are found to share many GO terms and molecular functions.

Chapter 10
Multiobjective Biclustering in Microarray Gene Expression Data

10.1 Introduction

An important goal in microarray data analysis is to identify sets of genes with similar expression profiles. Clustering algorithms have been applied on microarray data either to group the genes across experimental conditions/samples [44, 302, 355, 396, 448] or group the samples across the genes [317, 339, 384, 400]. Clustering techniques, which aim to find the clusters of genes over all experimental conditions, may fail to discover the genes having similar expression patterns over a subset of conditions. Similarly, a clustering algorithm that groups conditions/samples across all the genes may not capture the group of samples having similar expression values for a subset of genes. It is often the case that a subset of genes is co-regulated and co-expressed across a subset of experimental conditions and have almost different expression patterns over the remaining conditions. Traditional clustering methods are not able to identify such local patterns, usually termed *biclusters*. Thus biclustering [318] can be thought as simultaneously clustering genes and conditions instead of clustering them separately. The aim of biclustering algorithms is to discover a subset of genes that are co-regulated over a subset of experimental conditions. Hence they provide a better reflection of the biological reality.

Although biclustering is a relatively new approach applied to gene expression data, it has a fast-growing literature. In this chapter, we discuss several issues of biclustering, including a comprehensive review of recent literature. Thereafter a recently proposed multiobjective biclustering algorithm (MOGAB) and its application in both simulated and real-life microarray datasets is described. Finally, fuzzification of the MOGAB algorithm is discussed and demonstrated with some experiments.

10.2 Biclustering Problem and Definitions

Given a $\mathscr{G} \times \mathscr{C}$ microarray data matrix $\mathscr{A}(G,C)$ consisting of a set of $\mathscr{G}$ genes $G = \{I_1, I_2, \ldots, I_{\mathscr{G}}\}$ and a set of $\mathscr{C}$ conditions $C = \{J_1, J_2, \ldots, J_{\mathscr{C}}\}$, a bicluster can be defined as follows:

Definition 10.1. (Bicluster) A bicluster is a submatrix $\mathscr{M}(I,J) = [m_{ij}]$, $i \in I, j \in J$, of matrix $\mathscr{A}(G,C)$, where $I \subseteq G$ and $J \subseteq C$, and the subset of genes in the bicluster is similarly expressed over the subset of conditions and vice versa.

The problem of biclustering is thus to identify a set of biclusters from a given data matrix depending on some coherence criterion to evaluate the quality of the biclusters. In general, the complexity of a biclustering problem depends on the coherence criterion used. However, in almost all cases, the biclustering problem is known to be NP-complete. Therefore, a number of approaches use heuristics for discovering biclusters from a gene expression matrix.

Depending on how the genes in a bicluster are similar to each other under the experimental conditions, biclusters can be categorized into different types. The following subsection provides the definitions of different types of biclusters.

10.2.1 Types of Biclusters

There are mainly six types of biclusters, viz., (1) biclusters with constant values, (2) biclusters with constant rows, (3) biclusters with constant columns, (4) biclusters with additive patterns, (5) biclusters with multiplicative patterns and (6) biclusters with both additive and multiplicative patterns. The additive and multiplicative patterns are also called shifting and scaling patterns, respectively [6]. The different types of biclusters are defined as follows:

Definition 10.2. (Biclusters with Constant Values) In a bicluster $\mathscr{M}(I,J) = [m_{ij}]$, $i \in I, j \in J$, with constant values, all the elements have the same value, i.e.,

$$m_{ij} = \pi, \quad i \in I, j \in J. \tag{10.1}$$

Definition 10.3. (Biclusters with Constant Rows) In a bicluster $\mathscr{M}(I,J) = [m_{ij}]$, $i \in I, j \in J$, with constant rows, all the elements of each row of the bicluster have the same value. Hence, in this type of bicluster, each element is represented using one of the following notations:

$$m_{ij} = \pi + a_i, \quad i \in I, j \in J, \tag{10.2}$$

$$m_{ij} = \pi b_i, \quad i \in I, j \in J, \tag{10.3}$$

$$m_{ij} = \pi b_i + a_i, \quad i \in I, j \in J. \tag{10.4}$$

Here π is a constant value for a bicluster, a_i is the additive (shifting) factor for row i and b_i is the multiplicative (scaling) factor for row i.

Definition 10.4. (Biclusters with Constant Columns) In a bicluster $\mathcal{M}(I,J) = [m_{ij}]$, $i \in I, j \in J$, with constant columns, all the elements of each column of the bicluster have the same value. Hence, in this type of bicluster, each element is represented using one of the following notations:

$$m_{ij} = \pi + p_j, \quad i \in I, j \in J, \tag{10.5}$$

$$m_{ij} = \pi q_j, \quad i \in I, j \in J, \tag{10.6}$$

$$m_{ij} = \pi q_j + p_j, \quad i \in I, j \in J. \tag{10.7}$$

Here π is a constant value for a bicluster, p_j is the additive (shifting) factor for column j and q_j is the multiplicative (scaling) factor for column j.

Definition 10.5. (Biclusters with Additive Pattern) In a bicluster $\mathcal{M}(I,J) = [m_{ij}]$, $i \in I, j \in J$, with additive (shifting) pattern, each column and row have only some additive (shifting) factors. Hence, in this type of bicluster, each element is represented as

$$m_{ij} = \pi + a_i + p_j, \quad i \in I, j \in J. \tag{10.8}$$

Definition 10.6. (Biclusters with Multiplicative Pattern) In a bicluster $\mathcal{M}(I,J) = [m_{ij}]$, $i \in I, j \in J$, with multiplicative (scaling) pattern, each column and row have only some multiplicative (scaling) factors. Hence, in this type of bicluster, each element is represented as

$$m_{ij} = \pi b_i q_j, \quad i \in I, j \in J. \tag{10.9}$$

Definition 10.7. (Biclusters with both Additive and Multiplicative Patterns) In a bicluster $\mathcal{M}(I,J) = [m_{ij}]$, $i \in I, j \in J$ with both additive (shifting) and multiplicative (scaling) patterns, each column and row has both additive (shifting) and multiplicative (scaling) factors. Hence in this type of bicluster, each element is represented as:

$$m_{ij} = \pi b_i q_j + a_i + p_j, \quad i \in I, j \in J. \tag{10.10}$$

Note that these biclusters are the most general form of biclusters. All other types of biclusters are special cases of these biclusters. Figure 10.1 shows examples of different types of biclusters.

10.2.2 Some Important Definitions

Here we discuss some important terms regarding biclusters and the biclustering problem.

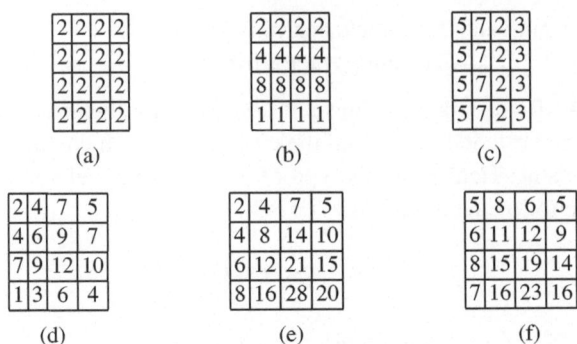

Fig. 10.1 Examples of different types of biclusters: (a) constant, (b) row-constant, (c) column-constant, (d) additive pattern, (e) multiplicative pattern, (f) both additive and multiplicative patterns

Definition 10.8. (Bicluster Variance) Bicluster variance $VARIANCE(I,J)$ of a bicluster $\mathcal{M}(I,J)$ is defined as follows:

$$VARIANCE(I,J) = \sum_{i \in I, j \in J} (m_{ij} - m_{IJ})^2, \qquad (10.11)$$

where $m_{IJ} = \frac{1}{|I||J|} \sum_{i \in I, j \in J} m_{ij}$, i.e., the mean of the elements in the bicluster.

Definition 10.9. (Residue) The residue r_{ij} of any element m_{ij} of a bicluster $\mathcal{M}(I,J)$ is defined as

$$r_{ij} = m_{ij} - m_{iJ} - m_{Ij} + m_{IJ}, \qquad (10.12)$$

where m_{iJ} is the mean of the ith row, i.e., $m_{iJ} = \frac{1}{|J|} \sum_{j \in J} m_{ij}$, m_{Ij} is the mean of the jth column, i.e., $m_{Ij} = \frac{1}{|I|} \sum_{i \in I} m_{ij}$, and m_{IJ} is the mean of all the elements in the bicluster, i.e., $m_{IJ} = \frac{1}{|I||J|} \sum_{i \in I, j \in J} m_{ij}$.

Definition 10.10. (Mean Squared Residue) The mean squared residue ($MSR(I,J)$) of a bicluster $\mathcal{M}(I,J)$ is defined as

$$MSR(I,J) = \frac{1}{|I||J|} \sum_{i \in I, j \in J} r_{ij}^2. \qquad (10.13)$$

The mean squared residue score of a bicluster represents the level of coherence among the elements of the bicluster. A lower residue score indicates greater coherence and thus a better quality of the bicluster.

Definition 10.11. (Row Variance) The row variance $VAR(I,J)$ of a bicluster $\mathcal{M}(I,J)$ is defined as

$$VAR(I,J) = \frac{1}{|I||J|} \sum_{i \in I, j \in J} (m_{ij} - m_{iJ})^2. \qquad (10.14)$$

A high row variance indicates that the rows (genes) of the biclusters have large variance across the conditions. Sometimes high row variance is desirable in order to escape from trivial constant biclusters.

10.3 Biclustering Algorithms

In recent years, a large number of biclustering algorithms are being proposed for gene expression data analysis. In this section, we discuss some popular biclustering algorithms in different categories, such as iterative greedy search, randomized greedy search, evolutionary techniques, graph-based algorithms, and fuzzy methods, etc.

10.3.1 Iterative Greedy Search

The concept of biclustering was first introduced by Hartigan in [206] in the form of direct clustering. As a coherence measure of a bicluster $\mathscr{M}(I,J)$, bicluster *variance* ($VARIANCE(I,J)$) was used (Equation 10.11). The goal of the algorithm was to extract K biclusters from the given dataset while minimizing the sum of the bicluster variances of the K biclusters. In each iteration, the algorithm partitions the data matrix into a set of submatrices, each of which is considered as a bicluster. As can be observed, for a constant bicluster, $VARIANCE(I,J)$ is zero. As each element of the data matrix satisfies the zero variance criterion, to avoid this, the algorithm was executed until the data matrix was partitioned into K submatrices. Hartigan's algorithm was able to detect constant biclusters only. However, Hartigan proposed to use other homogeneity criteria to detect other types of biclusters.

Cheng and Church first introduced the biclustering problem in the case of microarray gene expression data [94]. The coherence measure called *mean squared residue (MSR)* was introduced by them (Equation 10.13). Cheng and Church proposed a greedy search heuristic that searches for the largest possible bicluster keeping MSR under a threshold δ (called as δ-bicluster). The algorithm has two phases. In the first phase, starting with the complete data matrix, they first delete rows and columns in order to bring the MSR score below δ. In this regard, Cheng and Church suggested a greedy heuristic to rapidly converge to a locally maximal submatrix with the MSR score below δ. In the second phase, the rows and columns are added as long as the MSR score does not increase. The same procedure is executed for K iterations in order to discover K δ-biclusters. At each iteration, the bicluster found in the previous iteration is masked with random values in order to avoid overlaps. Since the MSR score is zero for the biclusters with constant values, constant rows, constant columns and additive patterns, the Cheng and Church algorithm is able to detect these kinds of biclusters only. However, the algorithm is known to get stuck at local optima often and also suffers from random interference due to the masking of biclusters with random values.

In [446], the authors extended the concept of δ-bicluster to cope with the problem of masking the missing values as well as masking the biclusters found in the previous iteration with random values. In this algorithm, the residue of a specified (non-missing) element in a bicluster is taken as per Equation 10.12, but the residue of an unspecified (missing) element is taken to be zero. This algorithm allows the biclusters to overlap and thus is termed FLexible Overlapped biClustering (FLOC). The FLOC algorithm begins with an initial set of biclusters (seeds) and iteratively improves the overall quality of the biclustering. At each iteration, each row and column are moved among the biclusters to yield a better biclustering in terms of the lower *MSR*. The best biclustering obtained during an iteration is used as the initial biclustering seed in the next iteration. The algorithm terminates automatically when the current iteration fails to improve the overall biclustering quality. Thus FLOC is able to evolve k biclusters simultaneously. However, this algorithm also can only identify constant and additive patterns, and fails to detect multiplicative patterns.

In [59], an algorithm called Order Preserving SubMatrix (OPSM) is proposed. Here a bicluster is defined as a submatrix where the order of the selected conditions is preserved for all of the selected genes. Hence, the expression values of the genes within a bicluster induce an identical linear ordering across the selected conditions. The authors proposed a deterministic iterative algorithm to find large and statistically significant biclusters. The time complexity of this technique is $O(\mathcal{G}\mathcal{C}^3 k)$ where $\mathcal{G}$ and $\mathcal{C}$ are the number of genes and conditions of the input dataset, respectively, and k is the number of biclusters found. Thus OPSM does not scale well for high-dimensional datasets.

In [229] and [230], the authors proposed the Iterative Signature Algorithm (ISA) where a bicluster is considered to be a transcription module, i.e., a set of co-regulated genes together with the associated set of regulating conditions. The algorithm starts with an initial set of genes and all samples are scored with respect to this gene set. The samples for which the score exceeds a predefined threshold are chosen. Similarly, all genes are scored regarding the selected samples and a new set of genes is selected based on another user-defined threshold. This procedure is iterated until the set of genes and the set of samples converge, i.e., do not change anymore. ISA can discover more than one bicluster by starting with different initial gene sets. The choice of the initial reference gene set plays an important role in ISA in obtaining good quality results. ISA is highly sensitive to the threshold values and often tends to identify a strong bicluster many times.

In xMotif biclustering [319], the biclusters which contain genes that are almost constantly expressed across the selected conditions are identified. At first, each gene is assigned a set of statistically significant states which define the set of valid biclusters. In xMotif, a bicluster is considered to be a submatrix where each gene is in exactly the same state for all the selected conditions. The aim is to identify the largest bicluster. To identify the largest valid biclusters, an iterative search method is proposed that is run on different initial random seeds. It should be noted that the xMotif framework requires pre-identification of the classes of biclusters present in the data, which may not be feasible for most of the real-life datasets.

In general, greedy search algorithms scale well in large datasets. However, they mainly suffer from the problem of getting stuck at local optima depending on the initial configuration.

10.3.2 Two-Way Clustering

In [173], the authors present a coupled two-way clustering (CTWC) approach to gene microarray data analysis. The main idea is to identify subsets of the genes and samples, such that when one of these is used to cluster the other, stable and significant partitions emerge. They present an algorithm, based on iterative clustering, that performs such a search. This two-way clustering algorithm repeatedly performs one-way clustering on the rows and columns of the data matrix, using stable clusters of rows as attributes for column clustering and vice-versa. Although the authors used hierarchical clustering, any reasonable choice of clustering method and definition of stable cluster can be used within the framework of CTWC. As a preprocessing step, they used normalization, which allowed them to capture biclusters with constant columns also.

Interrelated Two-Way Clustering (ITWC) [399], an algorithm similar to CTWC, combines the results of one-way clustering both dimensions of the gene expression matrix for producing biclusters. As a preprocessing step, the rows of the data matrix are first normalized. Thereafter, the vector-angle cosine value between each row and a predefined stable pattern are computed to determine whether the row values vary much among the columns. The rows with very little variation are then removed. After that, the correlation coefficient is used to measure the strength of the linear relationship between two rows or two columns, to perform the two-way clustering. As the correlation coefficient is independent of the magnitude and only depends on the pattern, ITWC is able to detect both additive and multiplicative biclusters.

Double Conjugated Clustering (DCC) [78] is a node-driven algorithm that unifies the two view points of microarray clustering, viz., clustering the samples taking the genes as the features and clustering the genes taking the samples as the features. DCC performs the two tasks simultaneously to achieve a unified clustering where the sample clusters are distinguished by subsets of genes. The clustering in the sample space and gene space are synchronized by a projection of nodes between the spaces mapping the sample clusters to the corresponding gene clusters. The method may utilize any relevant clustering technique such as SOM and K-means. The data does not scatter across all offered nodes due to the projection between the two clustering spaces. The DCC algorithm can provide sharp clusters and empty nodes even in the case of the number of nodes exceeding the number of clusters. However, DCC can only find constant biclusters from the input dataset.

The two-way clustering algorithms in general cluster the dataset from both the dimensions (rows and columns) and finally try to combine the clustering of the two dimensions in order to obtain the biclusters. However, there is no standard rule for the choice of the number of clusters in both the gene and condition dimensions.

10.3.3 Evolutionary Search

Evolutionary algorithms, such as Genetic Algorithms (GAs) [181] and Simulated Annealing (SA) [252], have been used extensively in the biclustering problem. Some of these algorithms are described below.

10.3.3.1 GA-Based Biclustering

In [69], a Genetic Algorithm-based biclustering framework has been developed. As an encoding strategy, the authors use a binary string of length $\mathscr{G} + \mathscr{C}$, where $\mathscr{G}$ and $\mathscr{C}$ denote the number of genes and number of conditions/samples/time points, respectively. If a bit position is '1', the corresponding gene or condition is selected in the bicluster and if a bit position is '0', the corresponding gene or condition is not selected in the bicluster. Hence, each chromosome encodes one possible bicluster. The following fitness function $\mathscr{F}$ is minimized:

$$\mathscr{F} = \begin{cases} \frac{1}{|I||J|} & \text{if } MSR(I,J) \leq \delta \\[2ex] \frac{MSR(I,J)}{\delta} & \text{otherwise.} \end{cases} \tag{10.15}$$

Hence, if the *MSR* of the bicluster encoded in a chromosome is less than or equal to the threshold δ (i.e., a δ-bicluster), the objective is to maximize the volume. Otherwise, the objective is to minimize the *MSR*. The algorithm employs a special selection operator called environment selection to maintain the diversity of the population in order to identify a set of biclusters at one run. A local search strategy is used to expedite the rate of convergence. As the local search, one iteration of the Cheng and Church node deletion and addition algorithm is executed before computing the fitness value of a chromosome. Also, the chromosome is updated with the new bicluster obtained after the local search. Standard uniform crossover and bit-flip mutation operators are adopted for producing the next generation.

A similar GA-based biclustering approach can be found in [88]. Here, instead of using the Cheng and Church algorithm as a local search strategy in each step of fitness computation, it is used only once initially. The initial population consists of bicluster seeds generated through K-means clustering in both dimensions and the combined gene and sample clusters. Thereafter these seeds are grown through the Cheng and Church algorithm. Subsequently, the normal GA process follows. As the fitness function, the authors minimized the ratio of *MSR* to the volume of the biclusters in order to capture large yet coherent biclusters.

Another GA-based biclustering, called Sequential Evolutionary BIclustering (SEBI), is proposed in [141]. In this work also, the authors use binary chromosomes as discussed above. SEBI minimizes the following fitness function:

$$\mathcal{F} = \frac{MSR(I,J)}{\delta} + \frac{1}{VAR(I,J)} + w_d + penalty, \qquad (10.16)$$

where $w_d = w_V(w_r \frac{\delta}{|I|} + w_c \frac{\delta}{|J|})$. Here w_V, w_r and w_c represent weights on volume, number of rows and number of columns in the bicluster, respectively. Also, $penalty = \sum_{i \in I, j \in J} w_p(m_{ij})$, where $w_p(m_{ij})$ is a weight associated with each element m_{ij} of the bicluster, and it is defined as

$$w_p(m_{ij}) = \begin{cases} 0 & \text{if } |COV(m_{ij})| = 0 \\ \frac{\sum_{k \in I, l \in J} e^{-|COV(m_{kl})|}}{e^{-|COV(m_{ij})|}} & \text{if } |COV(m_{ij})| > 0. \end{cases} \qquad (10.17)$$

Here $|COV(m_{ij})|$ denotes the number of biclusters containing m_{ij}. The weight $w_p(m_{ij})$ is used to control the amount of overlap among the biclusters. Binary tournament selection is used. Three crossover operators, one-point, two-point and uniform crossover, have been studied. Also, three mutation operators, namely standard bit-flip mutation, mutation by adding a row and mutation by adding a column are used for study. SEBI does not use any local search strategy for updating the chromosomes.

10.3.3.2 SA-Based Biclustering

There are many instances in the literature that use Simulated Annealing (SA) for the biclustering problem. A standard representation of a configuration in SA is equivalent to a binary string used in GA-based biclustering. In [76], this representation is used. Here the initial configuration consists of all '1's, i.e., it encodes the complete dataset. The perturbation is equivalent to the bit-flipping mutation used in GA. The energy to be minimized is taken as the *MSR* score of the encoded bicluster.

A similar approach is found in [87], where instead of starting from the complete data matrix, the author first creates a seed bicluster by clustering the genes and samples and combining them. Thereafter SA is used to grow the seed. Here also, *MSR* is used as the coherence measure. The perturbation includes only the addition of a random gene and/or condition.

10.3.3.3 Hybrid Approaches

In [441], a hybrid Genetic Algorithm-Particle Swarm Optimization (GA-PSO) approach, which uses binary strings to encode the biclusters, is proposed. The GA and PSO have their own populations that evolve through the standard GA and PSO processes, respectively. At each iteration, a random set of individual solutions is exchanged between the two population. As the fitness function, it uses the one described in Equation 10.15.

In general, evolutionary algorithms are known for their strength in avoiding locally optimum solutions. Specially when they are equipped with some local search, they can converge fast toward the global optimum. However, the algorithms which optimize *MSR* as an objective function fail to discover the multiplicative patterns. Also, evolutionary algorithms are inherently slower compared to the greedy iterative algorithms and depend a lot on different parameters such as population size, number of generations, crossover and mutation rates, and annealing schedule, etc. But in general, it has been found that evolutionary algorithms work better than the greedy search strategies in terms of performance.

10.3.4 Fuzzy Biclustering

Microarray gene expression data usually contains noise. It is also highly likely that a gene or condition may belong to different biclusters simultaneously with different degrees of belongingness. Therefore a lot of uncertainty is involved in microarray datasets. Hence the concept of fuzzy set theory is useful for discovering such overlapping biclusters from the noisy background. Some recent biclustering algorithms employ fuzzy set theory in developing biclustering algorithms in order to capture overlapping biclusters. In [407], a flexible fuzzy co-clustering algorithm which incorporates feature-cluster weighting in the formulation is proposed. The algorithm, called Flexible Fuzzy Co-clustering with Feature-cluster Weighting (FFCFW), allows the number of object clusters to be different from the number of feature clusters. A feature-cluster weighting scheme is incorporated for each object cluster generated by FFCFW so that the relationships between the two types of clusters are manifested in the feature-cluster weights. This enables FFCFW to generate a more accurate representation of fuzzy co-clusters. FFCFW uses an iterative optimization procedure.

In [157], a GA-based possibilistic fuzzy biclustering algorithm GFBA is proposed. In GFBA, instead of a binary chromosome, the authors use real-valued chromosome of length $\mathcal{G} + \mathcal{C}$. Each position in the chromosome has a value between 0 and 1, representing the degree of membership of the corresponding gene or condition to the encoded bicluster. They fuzzified the different coherence and quality metrics such as *MSR*, *VAR* and the volume of the biclusters as follows: The means of each row (m_{iJ}), each column (m_{Ij}) and all the elements (m_{IJ}) of a bicluster are redefined as

$$m_{iJ} = \frac{\sum_{j=1}^{\mathscr{C}} f_J(j)^\mu . m_{ij}}{\sum_{j=1}^{\mathscr{C}} f_J(j)^\mu}, \tag{10.18}$$

$$m_{Ij} = \frac{\sum_{i=1}^{\mathscr{G}} f_I(i)^\mu . m_{ij}}{\sum_{i=1}^{\mathscr{G}} f_I(i)^\mu}, \tag{10.19}$$

and

$$m_{iJ} = \frac{\sum_{i=1}^{\mathscr{G}} \sum_{j=1}^{\mathscr{C}} f_I(i)^\mu . f_J(j)^\mu . m_{ij}}{\sum_{i=1}^{\mathscr{G}} \sum_{j=1}^{\mathscr{C}} f_I(i)^\mu . f_J(j)^\mu}, \tag{10.20}$$

where $f_I(i)$ and $f_J(j)$ denote the membership degree of the ith gene and jth condition to the bicluster, respectively and μ is the fuzzy exponent. Hence the fuzzy mean squared residue $FMSR$ is defined as

$$FMSR(I,J) = \frac{1}{|I||J|} \sum_{i=1}^{\mathscr{G}} \sum_{j=1}^{\mathscr{C}} f_I(i)^\mu . f_J(j)^\mu (m_{ij} - m_{iJ} - m_{Ij} + m_{IJ})^2, \tag{10.21}$$

where $|I| = \sum_{i=1}^{\mathscr{G}} f_I(i)$ and $|J| = \sum_{j=1}^{\mathscr{C}} f_J(j)$. The objective function to be minimized is selected as

$$\begin{aligned} \mathscr{F} = & \ FMSR(I,J) \\ & + \frac{\sum_{i=1}^{\mathscr{G}} \eta . FMSR(I,J).(1 - f_I(i))^\mu}{\mathscr{G} - |I|} \\ & + \frac{\sum_{j=1}^{\mathscr{C}} \xi . FMSR(I,J).(1 - f_J(j))^\mu}{\mathscr{C} - |J|}, \end{aligned} \tag{10.22}$$

where η and ξ are parameters provided to satisfy different requirements on the incoherence and the size of the biclusters. Conventional roulette wheel selection and single point crossover followed by mutation (increasing or decreasing a membership value) have been used. GFBA also uses a bicluster optimization technique at each generation for faster convergence.

In [198], another fuzzy biclustering algorithm called Fuzzy Biclustering for Microarray Data Analysis (FBMDA) is proposed. The method employs a combination of the Nelder-Mead and min-max algorithms to construct hierarchically structured biclustering; thus it can represent the biclustering information at different levels. FB-MDA uses multiobjective optimization that optimizes volume, variance and fuzzy entropy simultaneously. The Nelder-Mead algorithm is used to compute a single objective optimal solution, and the min-max algorithm is used to trade-off between multiple objectives. FBMDA is not subject to the convexity limitations, and need not use derivative information. FBMDA ensures that the current local optimal solution is removed and that higher precision is reached.

The incorporation of fuzziness into biclustering algorithms enables them to deal with noisy data and overlapping biclusters efficiently. But as most of the aforementioned fuzzy algorithms use evolutionary techniques as the underlying optimization strategy, they suffer from the fundamental disadvantages of evolutionary methods. Furthermore, computation of fuzzy membership degrees takes additional time, which adds to the time taken by the fuzzy biclustering methods.

10.3.5 Graph Theoretic Approaches

Graph-theoretic concepts and techniques have been utilized in detecting biclusters. In [398], the authors introduced SAMBA (Statistical-Algorithmic Method for Bicluster Analysis), a graph-theoretic approach to biclustering, in combination with a statistical data model. In SAMBA the expression matrix is modeled as a bipartite graph consisting of two sets of vertices corresponding to genes and conditions. A bicluster is defined as a subgraph, and a likelihood score is used to assess the significance of observed subgraphs. SAMBA repeatedly finds the maximal highly connected subgraph in the bipartite graph. Then it performs local improvement by adding or deleting a single vertex until no further improvement is possible. SAMBA's time complexity is $O(N2^d)$, where d is the upper bound on the degree of each vertex.

The Binary inclusion-Maximal (Bimax) biclustering algorithm proposed in [352] identifies all biclusters in the input matrix. The Bimax algorithm works on a binary matrix. The input matrix is first discretized to 0s and 1s according to a user-specified threshold. Based on this binary matrix, Bimax identifies all maximal biclusters, where a bicluster is defined as a submatrix E containing all 1s. An inclusion-maximal bicluster means that this bicluster is not completely contained in any other bicluster. The authors used an divide-and-conquer algorithm to find the inclusion-maximal biclusters, exploiting the fact that the matrix E induces a bipartite graph. As Bimax works with a binary matrix, it is suitable only for detecting constant biclusters.

In [7], the optimal biclustering problem is posed as a problem of maximal crossing number reduction (minimization) in a weighted bipartite graph. In this regard, an algorithm called cHawk is proposed that employs a barycenter heuristic and local search technique. There are three main steps of the algorithm, viz., the construction of a bipartite graph from the input matrix, the bipartite graph crossing minimization and finally, the bicluster identification. This approach reorders the matrix so that all rows and columns belonging to the same bicluster are brought into the vicinity of each other. cHawk is able to detect constant, additive and overlapped noisy biclusters.

The graph-based biclustering algorithms usually model the input dataset as a bipartite graph with two sets of nodes corresponding to the genes and conditions, respectively. The edges of the graph represent the level of overexpression and underexpression of a gene under certain conditions. A bicluster is a subgraph of the bipartite graph, where the genes have coherence across the selected conditions. In these types of algorithms, the genes and conditions are partitioned into the same number of clusters, which may be impractical. Moreover, the input dataset has to be discretized properly before applying graph-based algorithms.

10.3.6 Randomized Greedy Search

In [24], a greedy random walk search technique for the biclustering problem that is enriched by a local search strategy to escape local optima has been presented. The algorithm begins with an initial random solution and searches for a locally optimal solution by successive transformations (including random moves depending on some probability) to improve a gain function defined as a combination of the mean squared residue, the expression profile variance and the volume of the biclusters. The algorithm iterates k times to generate k biclusters.

In [136], the basic concepts of the metaheuristics Greedy Randomized Adaptive Search Procedure (GRASP) construction and local search phases are reviewed. Also, a method which is a variant of GRASP, called Reactive Greedy Randomized Adaptive Search Procedure (Reactive GRASP), is proposed to detect significant biclusters from large microarray datasets. The method has two major steps. First, high-quality bicluster seeds are generated by using the K-means clustering from both dimensions and combining the clusters. In the second step, these seeds are grown using Reactive GRASP. In Reactive GRASP, the basic parameter that defines the restrictiveness of the candidate list is self-adjusted, depending on the quality of the solutions found previously.

Randomized greedy search algorithms try to combine the advantages of greedy search and randomization, so that they execute fast and not get stuck at local optima. However, these algorithms still heavily depend on the initial choice of the solution and there is no clear way to get out of a poor choice.

10.3.7 Other Recent Approaches

There are a number of biclustering algorithms that have appeared in recent literature that follow new methodologies. Some of them are described here.

In [272], the authors introduces the plaid model as a statistical model assuming that the expression value m_{ij} in a bicluster is the sum of the main effect π, the gene effect p_i, the condition effect q_j, and the noise term ε_{ij}:

$$m_{ij} = \pi + p_i + q_j + \varepsilon_{ij}. \tag{10.23}$$

Also, it is assumed that the expression values of two overlapping biclusters are the sum of the two module effects. In the plaid model, a greedy search strategy is used; hence errors can accumulate easily. Moreover, in the case of multiple clusters, the clusters identified by the method tend to overlap to a great extent.

In [386], a biclustering algorithm is proposed based on probabilistic Gibbs sampling. Gibbs sampling does not suffer from the problem of local minima that often characterizes expectation maximization. However, when the microarray data is organized in patient vs. gene fashion, and the number of patients is much lower than the number of genes, the algorithm faces computational difficulties. Moreover, the algorithm is only able to identify biclusters with constant columns.

In [253], the authors developed a spectral biclustering method that simultaneously clusters genes and conditions, finding distinctive checkerboard patterns in matrices of gene expression data, if they exist. The method is based on the observation that checkerboard structures can be found in the eigenvectors corresponding to the characteristic expression patterns across the genes or conditions. In addition, these eigenvectors can be readily identified by commonly used linear algebra approaches such as singular value decomposition (SVD) coupled with closely integrated normalization steps.

In [452], the authors proposed a biclustering method that employs dynamic programming and a divide-and-conquer technique, as well as efficient data structures such as the trie and Zero-suppressed Binary Decision Diagrams (ZBDDs). The use of ZBDDs extends the stability of the method substantially.

In [464], the authors developed MicroCluster, a deterministic biclustering method. In MicroCluster, only the maximal biclusters satisfying certain homogeneity criteria are considered. The clusters can be arbitrarily positioned anywhere in the input data matrix, and they can have arbitrary overlapping regions. MicroCluster uses a flexible definition of a cluster that lets it mine several types of biclusters. Moreover, MicroCluster can delete or merge biclusters that have large overlaps. So, it can tolerate some noise in the dataset and let the users focus on the most important clusters. As MicroCluster relies on extracting maximal cliques from the constructed range multigraph, it is computationally demanding. Moreover, there are several input parameters that are to be tuned properly in order to find suitable biclusters.

A method based on application of the non-smooth nonnegative matrix factorization technique for discovering local structures (biclusters) from gene expression datasets is developed in [83]. This method utilizes nonnegative matrix factorization with non-smoothness constraints for identifying biclusters in gene expression data for a given factorization rank.

In [403], biclustering algorithms using basic linear algebra and arithmetic tools have been developed. The proposed biclustering algorithms can be used to search for all biclusters with constant values, biclusters with constant values on rows, biclusters with constant values on columns, and biclusters with coherent values from a set of data in a timely manner and without solving any optimization problem.

In [405], the authors proposed a biclustering method by alternatively sorting the genes and conditions using the dominant set. By using the weighted correlation co-efficient, they emphasize the similarities across a subset of the genes or conditions. Additionally, a coherence measure called Average Correlation Value (ACV) is proposed; it is effective in determining both additive and multiplicative patterns. Some special preprocessing of the input dataset is needed for detecting additive and multiplicative biclusters. To detect different types of biclusters, different runs are needed.

In [158], a biclustering algorithm that adopts a bucketing technique to find a raw submatrix is proposed. The algorithm refines and extends the raw submatrix into a bicluster. The algorithm is called the Bucketing and Extending Algorithm (BEA).

A Bayesian BiCustering (BBC) model is proposed in [188] that uses Gibbs sampling. For a single bicluster, the same model as that in the plaid model is assumed. For multiple biclusters, the overlapping of biclusters is allowed either in genes or conditions. Moreover, the authors use a flexible error model, which permits the error term of each bicluster to have a different variance.

In [137] the authors presented a rigorous approach to biclustering, which is based on the Optimal RE-Ordering (OREO) of the rows and columns of a data matrix so as to globally minimize the dissimilarity metric. The physical permutations of the rows and columns of the data matrix can be modeled as either a network flow problem or a traveling salesman problem. Cluster boundaries in one dimension are used to partition and reorder the other dimensions of the corresponding submatrices to generate the biclusters. The reordering of the rows and the columns for large datasets can be computationally demanding.

The authors in [290] propose an algorithm that finds and reports all maximal contiguous column coherent (CCC) biclusters in time linear in the size of the expression matrix. The linear time complexity of CCC biclustering relies on the use of a discretized matrix and efficient string processing techniques based on suffix trees. This algorithm can only detect biclusters with columns arranged contiguously.

In [291], an iterative density-based biclustering algorithm called BIDENS is proposed. BIDENS is able to detect a set of k possibly overlapping biclusters simultaneously. The algorithm is similar to FLOC, but instead of having the residue as the objecting function, it tries to maximize the overall density of the biclusters. The input dataset is needed to be discretized before the application of the BIDENS algorithm.

10.4 Multiobjective Biclustering

As the biclustering problem requires several objectives to be optimized, such as *MSR*, volume, and row variance, there are some approaches that pose the biclustering problem as multiobjective optimization [125]. In [140], the authors extended their work of [141] to the multiobjective case. The algorithm is called as Sequential

Multi-Objective Biclustering (SMOB). Here also they used a binary encoding strategy. Three objective functions, viz., mean squared residue, volume and row variance are optimized simultaneously. In [281], a Crowding distance-based Multi-Objective Particle Swarm Optimization Biclustering (CMOPSOB) algorithm is proposed that uses binary encoding. The algorithm optimizes the *MSR*, volume and *VAR* simultaneously. In [156], a hybrid multiobjective biclustering algorithm that combines NSGA-II and Estimation of Distribution Algorithm (EDA) [268] for searching biclusters is proposed. The volume and *MSR* of the biclusters are optimized simultaneously. A multiobjective artificial immune system capable of performing a multi-population search, named MOM-aiNet, is proposed in [105].

All the above algorithms use chromosomes of length equal to the number of genes plus the number of conditions. Thus the chromosomes are very large if the dataset is large. This may cause the other operators, such as crossover and mutation, to take longer and thus slow down the convergence. Moreover, there are several objectives that a biclustering algorithm needs to optimize. Two important objectives are to minimize the mean squared residue (MSR) and maximize the row variance in order to obtain highly coherent non-trivial biclusters. A recently proposed multiobjective biclustering algorithm (MOGAB) [303] takes these issues into account. The following section describes different steps of MOGAB that employ a variable string length encoding scheme, where each string encodes a number of possible biclusters. Non-dominated Sorting Genetic Algorithm-II (NSGA-II) [130], is used as the underlying optimization strategy. The biclustering problem has been viewed as the simultaneous clustering of genes and conditions. Two objective functions, mean squared residue and row variance of the biclusters, are optimized simultaneously. The performance of the MOGAB algorithm has been compared with that of some well-known biclustering algorithms for both simulated and real-life microarray datasets. Also, a biological significance test based on Gene Ontology (GO) [352, 398, 453] has been carried out to establish that the biclusters discovered by MOGAB are biologically relevant.

10.5 MOGAB Biclustering Algorithm

This section discusses in detail the aforementioned multiobjective GA-based biclustering (MOGAB) technique. MOGAB uses NSGA-II as the underlying multiobjective optimization technique.

10.5.1 Chromosome Representation

Each chromosome has two parts: one for clustering the genes, and another for clustering the conditions. The first M positions represent the M cluster centers for the genes, and the remaining N positions represent the N cluster centers for the conditions. Thus a string looks like $\{gc_1 \ gc_2 \ \ldots \ gc_M \ cc_1 \ cc_2 \ \ldots \ cc_N\}$, where each gc_i, $i = 1 \ldots M$, represents the index of a gene acting as a cluster center of a set of genes, and each cc_j, $j = 1 \ldots N$, represents the index of a condition acting as a cluster center of a set of conditions. For a dataset having n points, it is usual to assume that the dataset will contain at most $\sqrt{n}$ clusters [44] in the absence of any domain-specific information. Taking this into account, for a $\mathscr{G} \times \mathscr{C}$ microarray, the values of the maximum number of gene clusters and the maximum number of condition clusters are $\lceil \sqrt{\mathscr{G}} \rceil$ and $\lceil \sqrt{\mathscr{C}} \rceil$, respectively. Therefore the value of M varies from 2 to $\lceil \sqrt{\mathscr{G}} \rceil$ and the value of N varies from 2 to $\lceil \sqrt{\mathscr{C}} \rceil$. Hence the length of the strings will vary from 4 to $\lceil \sqrt{\mathscr{G}} \rceil + \lceil \sqrt{\mathscr{C}} \rceil$. However, if some domain-specific information is available, it is always better to incorporate that to estimate the upper bound on the number of clusters. The first M positions can have values from $\{1, 2, \ldots, \mathscr{G}\}$ and the next N positions can have values from $\{1, 2, \ldots, \mathscr{C}\}$. Hence the gene and condition cluster centers are represented by the indices of the genes and conditions, respectively.

A string that encodes A gene clusters and B condition clusters represents a set of $A \times B$ biclusters, one for each pair of gene and condition cluster. Each pair $< gc_i, cc_j >$, $i = 1, \ldots, M$, $j = 1, \ldots, N$, represents a bicluster that consists of all genes of the gene cluster centered at gene gc_i, and all conditions of the condition cluster centered at condition cc_j.

10.5.2 Initial Population

The initial population contains randomly generated individuals. Each gene or condition is equally likely to become the center of a gene or a condition cluster, respectively.

10.5.3 Fitness Computation

Given a valid string (i.e., the string contains no repetition of gene or condition indices, and $2 \leq M \leq \lceil \sqrt{\mathscr{G}} \rceil$, $2 \leq N \leq \lceil \sqrt{\mathscr{C}} \rceil$), first all the gene and condition clusters encoded in it are extracted and each gene and condition are assigned to the respective least distant cluster centers. Subsequently each cluster center (for both genes and conditions) is updated by selecting the most centrally located point, from which the summation of the distances of other points of that cluster is minimum. Accordingly, the strings are updated. As most of the distance functions are known to perform equally on normalized data [381], any distance function such as Euclidean, Pearson Correlation, or Manhattan can be used here. The Euclidean distance measure has been adopted here.

Next, all the δ-biclusters denoted by some *<gene cluster, condition cluster>* pair, encoded in the updated string are found. The two objective functions are *mean squared residue (MSR)* (Equation 10.13) and *row variance (VAR)* (Equation 10.14). *MSR* is to be minimized and *VAR* is to be maximized to have good quality biclusters. As MOGAB is posed as a minimization algorithm, the two objective functions (f_1 and f_2) of a bicluster $B(I,J)$ are taken as follows:

$$f_1(I,J) = \frac{MSR(I,J)}{\delta}, \quad \text{and} \tag{10.24}$$

$$f_2(I,J) = \frac{1}{1 + VAR(I,J)}. \tag{10.25}$$

The denominator of f_2 is chosen in such a way to avoid the divide by zero condition when *row variance* = 0. Both f_1 and f_2 are to be minimized to obtain highly coherent yet "interesting" biclusters. For each encoded δ-bicluster, the fitness vector $\overline{f} = \{f_1, f_2\}$ is computed. The fitness vector of a string is then the mean of the fitness vectors of all encoded δ-biclusters in it. Due to the randomness of the genetic operators, invalid strings may arise at any point in the algorithm. The invalid strings are given the fitness vector $\overline{f} = \{X, X\}$, where X is an arbitrary large number. Thus the invalid strings will be automatically out of the competition in subsequent generations.

From the non-dominated solutions produced in the final population, all the δ-biclusters are extracted from each non-dominated string to produce the final biclusters.

10.5.4 Selection

MOGAB uses the crowded binary tournament selection method as used in NSGA-II. The detailed description of the selection process is available in [130].

10.5.5 Crossover

In MOGAB, single point crossover is used. Each part (gene part and condition part) of the string undergoes crossover separately. For the crossover in the gene part, two crossover points are chosen on the two parent chromosomes, respectively, and the portions of the chromosomes (the gene parts) around these crossover points are exchanged. Similarly, the crossover is performed for the condition parts. Performing the crossover may generate invalid strings, i.e., strings with repeated gene or condition indices or with an invalid number of gene or condition clusters (less than two or greater than the maximum number of clusters). If such strings are generated, they are given the fitness vector $\overline{f} = \{X, X\}$, where X is an arbitrary large number. Thus the invalid strings will be automatically out of the competition in subsequent generations. Crossover probability p_c is taken as 0.8. Figure 10.2 illustrates the crossover operation.

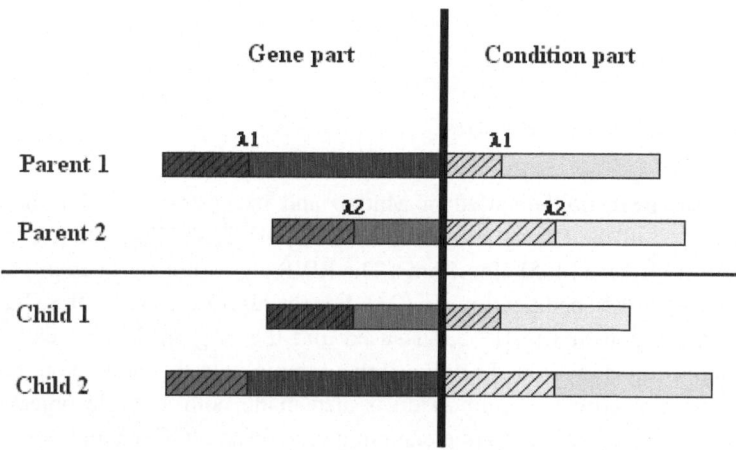

Fig. 10.2 The illustration of the crossover operation in MOGAB

10.5.6 Mutation

Suppose $\mathcal{G}$ and $\mathcal{C}$ are the total number of genes and conditions, respectively. The mutation is done as follows. A random position is chosen from the first M positions and its value is replaced with an index randomly chosen from $\{1, 2, \ldots, \mathcal{G}\}$. Similarly, to mutate the condition portion of the string, a random position is selected from the next N positions and its value is substituted with a randomly selected index from $\{1, 2, \ldots, \mathcal{C}\}$. A string undergoes mutation depending on a mutation probability $p_m = 0.1$.

As an example, consider the following string:

$$Parent: \ 8 \ 11 \ 7 \ 18 \ 21 \ 3 \ 12 \mid 13 \ 2 \ 8 \ 4 \ 15.$$

Assume that $\mathscr{G} = 40$ and $\mathscr{C} = 15$. If the string undergoes mutation, suppose the fourth gene index and the third condition index are selected to undergo mutation. Also assume that gene 20 and condition 12 have been selected randomly to replace the gene index and the condition index chosen to undergo mutation. Hence, after mutation, the resultant string will be

$$Child : 8\ 11\ 7\ \mathbf{20}\ 21\ 3\ 12\ |\ 13\ 2\ \mathbf{12}\ 4\ 15.$$

The mutated positions are shown in boldfaces.

Elitism has been incorporated into MOGAB to track the non-dominated solutions found so far. MOGAB has been executed for 100 generations with a population size 50.

10.6 Experimental Results

MOGAB has been implemented in Matlab and its performance has been compared with that of the Cheng and Church (CC) [94], RWB [23], OPSM [59], ISA [229, 230] and BiVisu [405] algorithms. The RWB algorithm has been implemented in Matlab as per the specifications in [23]. For the algorithms CC, OPSM and ISA, the implementation in BicAT [53] is used and the original Matlab code for the BiVisu algorithm is utilized. The algorithms are run with the parameters suggested in the respective articles. Comparison is also made with a single objective GA-based algorithm that uses the same encoding scheme as MOGAB and optimizes the product of the objective functions f_1 (Equation 10.24) and f_2 (Equation 10.25). A simulated dataset SD_200_100 and three real-life datasets, i.e., Yeast, Human and Leukemia are used for performance comparison. All the algorithms have been executed on an Intel Core 2 Duo 2.0 GHz machine with the Windows XP operating system.

10.6.1 Datasets and Data Preprocessing

10.6.1.1 Simulated Data

A simulated dataset SD_200_100 having 200 genes and 100 conditions is generated. To construct the simulated data, first the 200×100 background matrix filled with uniform random numbers ranging from 0 to 100 is generated randomly. Thereafter, to generate a $p \times q$ bicluster, the matrix values in a reference row $(r_1, r_2, \ldots, r_q)$ are generated randomly according to the standard normal distribution with mean 0 and standard deviation 1.2. To get a row $(r_{i1}, r_{i2}, \ldots, r_{iq})$ in the bicluster, a distance d_i (based on the standard normal distribution) is generated randomly and we set

$r_{ij} = a_j + d_i$, for $j = 1, 2, \ldots q$. To get the $p \times q$ bicluster, p such rows are generated. This way, 12 biclusters are generated and planted in the background matrix. The number of rows of the biclusters varies from ten to 60 and the number of columns of the biclusters varies from seven to 27. The maximum mean squared residue among the biclusters is 2.18. Hence the δ value for this dataset is chosen as 2.5.

10.6.1.2 Yeast Cell Cycle Data

This dataset contains 2,884 genes and 17 conditions [94]. Each entry has a value in $\{0, 1, \ldots, 600\}$. There are two rows with missing values denoted by -1. These rows are omitted from the dataset to form a data matrix of size $2,882 \times 17$. The dataset is publicly available at http://arep.med.harvard.edu/biclustering.

10.6.1.3 Human Large B-cell Lymphoma Data

There are 4,026 genes and 96 conditions in this dataset [94] as described in Section 9.3.3. This data matrix has elements with expression values ranging from -750 to 650. There are 47,639 missing values represented by 999. The rows with missing values have been removed to reduce the data matrix to a smaller size of 854×96. This dataset is also publicly available at http://arep.med.harvard.edu/biclustering.

10.6.1.4 ALL-AML Leukemia Data

This dataset provides the RNA value of 7,129 probes of human genes for 72 acute leukemia patients [14] as described in Section 9.3.1. The patients are classified as having either Acute Lymphoblastic Leukemia (ALL) (47 patients) or Acute Myeloid Leukemia (AML) (25 patients). This dataset is freely available at the following Web site: http://sdmc.lit.org.sg/GEDatasets/Datasets.html.

The values of δ for the above four datasets are taken to be 2.5, 300 [94], 1,200 [94] and 500 [23], respectively. Here the missing values have been omitted to avoid random interference (using random values for missing values). However, MOGAB can be modified to adopt any missing value estimation technique. All the datasets are normalized so that each row has mean 0 and variance 1. The supplementary Web site provides the details of the datasets and preprocessing.

10.6.2 Performance Indices

10.6.2.1 Match Score

For the simulated data, the *match score* as defined in [352] is used to compare the performance of different algorithms. Suppose $\mathcal{M}_1(I_1,J_1)$ and $\mathcal{M}_2(I_2,J_2)$ are two biclusters. The gene match score $S_I(I_1,I_2)$ is defined as

$$S_I(I_1,I_2) = \frac{|I_1 \cap I_2|}{|I_1 \cup I_2|}. \tag{10.26}$$

Similarly, the condition match score $S_J(J_1,J_2)$ is defined as

$$S_J(J_1,J_2) = \frac{|J_1 \cap J_2|}{|J_1 \cup J_2|}. \tag{10.27}$$

Note that the gene and condition match scores are symmetric and range from 0 (when two sets are disjoint) to 1 (when two sets are identical).

In order to evaluate the similarity between two sets of biclusters, the average gene match score and average condition match score can be computed. Let B_1 and B_2 be two sets of biclusters. The average gene match score of B_1 with respect to B_2 can be defined as

$$S_I^*(B_1,B_2) = \frac{1}{|B_1|} \sum_{(I_1,J_1)\in B_1} \max_{(I_2,J_2)\in B_2} S_I(I_1,I_2). \tag{10.28}$$

$S_I^*(B_1,B_2)$ represents the average of the maximum gene match scores for all the biclusters in B_1 with respect to the biclusters in B_2. Note that $S_I^*(B_1,B_2)$ is not symmetric and yields different values if B_1 and B_2 are exchanged. Similarly, the average condition match score can be defined as

$$S_J^*(B_1,B_2) = \frac{1}{|B_1|} \sum_{(I_1,J_1)\in B_1} \max_{(I_2,J_2)\in B_2} S_J(J_1,J_2). \tag{10.29}$$

Overall average match score of B_1 with respect to B_2 can now be defined as

$$S^*(B_1,B_2) = \sqrt{(S_I^*(B_1,B_2) \times S_J^*(B_1,B_2))}. \tag{10.30}$$

Now, if B_{im} denotes the set of implanted biclusters and B is the set of biclusters provided by some biclustering method, then $S^*(B_{im},B)$ represents how well each of the true biclusters is recovered by the biclustering algorithm. This ranges from 0 to 1, and takes the maximum value of 1 when $B_{im} = B$.

10.6.2.2 $\mathscr{BI}$ Index

This is used for the real datasets. The objective of the biclustering technique is to find δ-biclusters having large row variances. In this regard, a Biclustering Index ($\mathscr{BI}$) [303] is used here to measure the goodness of the biclusters. Suppose a bicluster has mean squared residue $\mathscr{H}$ and row variance $\mathscr{R}$. The biclustering index $\mathscr{BI}$ for that bicluster is then defined as

$$\mathscr{BI} = \frac{\mathscr{H}}{(1+\mathscr{R})}. \tag{10.31}$$

As the objective is to minimize $\mathscr{H}$ and maximize $\mathscr{R}$, a lower value of the $\mathscr{BI}$ index implies highly coherent and non-trivial biclusters.

10.6.3 Results on the Simulated Dataset

A single run of MOGAB on the simulated dataset takes about 40 seconds. Figure 10.3 shows the six non-trivial biclusters found in a single run of MOGAB on the simulated data. The figure indicates that the biclusters obtained using MOGAB are highly coherent (i.e., have low *MSR*) and non-trivial (i.e., have high row variance *VAR*). Hence the biclusters are "interesting" in nature. In terms of the $\mathscr{BI}$ index, bicluster (b) is the best; it has the minimum $\mathscr{BI}$ score of 0.2878. Hence these results demonstrate the effectiveness of the MOGAB algorithm.

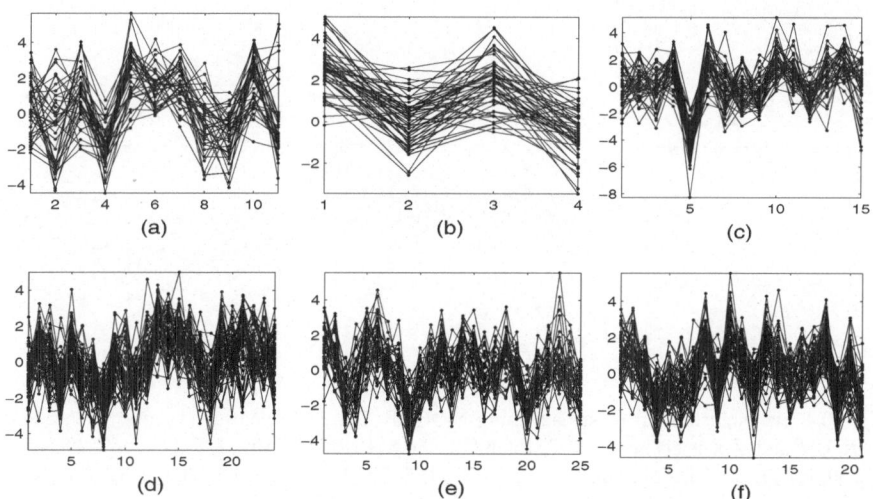

Fig. 10.3 Six biclusters of SD_200_100 data using MOGAB. For each bicluster, #genes, #conditions, *MSR*, *VAR* and $\mathscr{BI}$ is given: (a) 38, 11, 2.1003, 3.7407, 0.4430, (b) 52, 4, 1.0224, 2.5519, 0.2878, (c) 42, 15, 1.6321, 3.5932, 0.3553, (d) 59, 24, 1.5124, 2.5529, 0.4257, (e) 41, 25, 1.3584, 2.4389, 0.3950, (f) 58, 21, 1.3206, 2.4333, 0.3846

Table 10.1 reports the percentage match score for the simulated data for all the algorithms. It appears that MOGAB produces the maximum match score of 91.3%. This establishes that MOGAB is able to discover the implanted biclusters from the background reasonably well.

Table 10.1 Percentage match scores provided by different algorithms for the SD_200_100 dataset

MOGAB	SGAB	CC	RWB	Bimax	OPSM	ISA	BiVisu
91.3	80.2	33.6	61.3	33.1	59.5	17.1	55.2

10.6.4 Results on the Yeast Dataset

A single run of MOGAB on the Yeast dataset takes about 140 seconds. In Figure 10.4, six example biclusters found by MOGAB on Yeast data are shown. Visual inspection reveals that MOGAB discovers "interesting" biclusters having high row variance and low $\mathscr{BI}$ scores.

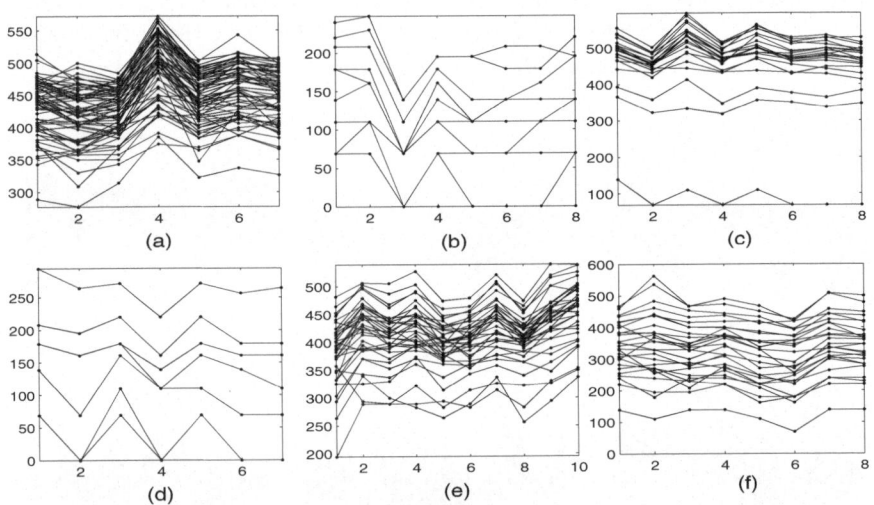

Fig. 10.4 Six biclusters of Yeast data using MOGAB. For each bicluster, #genes, #conditions, *MSR*, *VAR* and $\mathscr{BI}$ is given: (a) 71, 7, 165.6892, 766.8410, 0.2158, (b) 11, 8, 223.3569, 996.4375, 0.2239, (c) 21, 8, 125.9566, 506.4115, 0.2482, (d) 8, 7, 222.3437, 669.3858, 0.2492, (e) 36, 10, 222.7392, 892.7143, 0.3317, (f) 27, 8, 269.7345, 645.2176, 0.4174

Figure 10.5 shows the plot of $\mathscr{BI}$ scores of the 100 best biclusters (in terms of $\mathscr{BI}$) produced by all the algorithms. The biclusters of each algorithm are sorted in ascending order of their $\mathscr{BI}$ scores. Note that OPSM and ISA provided only 16 and 23 biclusters, respectively. The figure indicates that MOGAB has the most stable behavior, as for most of the biclusters, it provides lower $\mathscr{BI}$ values than the other algorithms. A comparative study of all the algorithms is reported in Table 10.2. It appears from the table that on average, the biclusters discovered by MOGAB have lower *MSR* and higher row variance simultaneously. Also, MOGAB provides the lowest average $\mathscr{BI}$ score. Moreover, the standard deviations (shown within brackets) of the range of values are on the lower side and this means all the biclusters are equally interesting. Though some algorithms, such as CC, Bimax, OPSM and ISA, beat MOGAB in terms of the lowest $\mathscr{BI}$ index score, Figure 10.5 shows that MOGAB provides a low $\mathscr{BI}$ index score for the most of the biclusters. This means that MOGAB produces a set of biclusters with similar quality, whereas the other algorithms generate some uninteresting biclusters with higher $\mathscr{BI}$ scores.

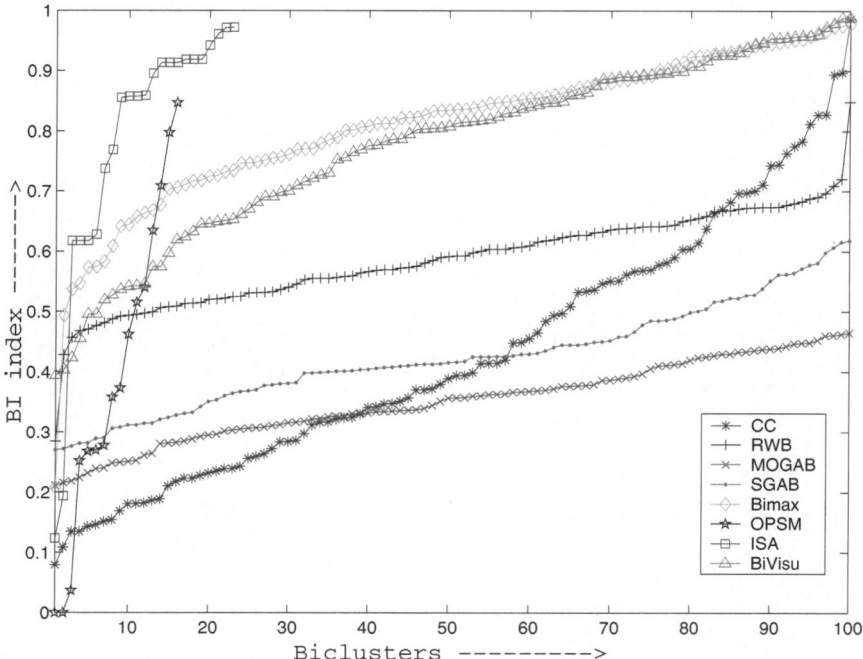

Fig. 10.5 Plot of $\mathscr{BI}$ index values of 100 best biclusters of Yeast data for different algorithms

Table 10.2 Comparison of the biclusters of different algorithms for Yeast data. Bracketed values represent the standard deviations

Algorithm	MSR		Row variance		Volume		$\mathcal{BI}$ index	
	Average	Best (min)	Average	Best (max)	Average	Best (max)	Average	Best (min)
MOGAB	185.85 (40.21)	116.34	932.04 (120.78)	3823.46	329.93 (93.50)	1848	0.3329 (0.061)	0.2123
SGAB	198.88 (40.79)	138.75	638.23 (126.93)	3605.93	320.72 (110.63)	1694	0.4026 (0.087)	0.2714
CC	204.29(42.78)	163.94	801.27 (563.19)	3726.22	1576.98 (2178.46)	10523	0.3822 (0.189)	0.0756
RWB	295.81 (42.96)	231.28	528.97 (198.52)	1044.37	1044.43 (143.34)	5280	0.5869 (0.078)	0.2788
Bimax	32.16 (26.23)	5.73	39.53 (35.81)	80.42	60.52 (55.78)	256	0.4600 (0.126)	0.2104
OPSM	320.39 (187.82)	118.53	1083.24 (1002.57)	3804.56	1533.31 (2005.51)	3976	0.3962 (0.267)	0.0012
ISA	281.59 (165.11)	125.76	409.29 (287.12)	1252.34	79.22 (32.95)	168	0.7812 (0.229)	0.1235
BiVisu	290.59 (26.57)	240.96	390.73 (100.81)	775.41	2136.34 (1293.43)	4080	0.7770 (0.149)	0.3940

10.6.5 Results on the Human Dataset

A single run of MOGAB on the Human dataset takes around 180 seconds. Figure 10.6 shows six example biclusters generated by MOGAB for Human data. It is evident from the figure that all of the six biclusters are highly coherent and have a high value of row variance. The $\mathscr{BI}$ scores are also very low for the biclusters.

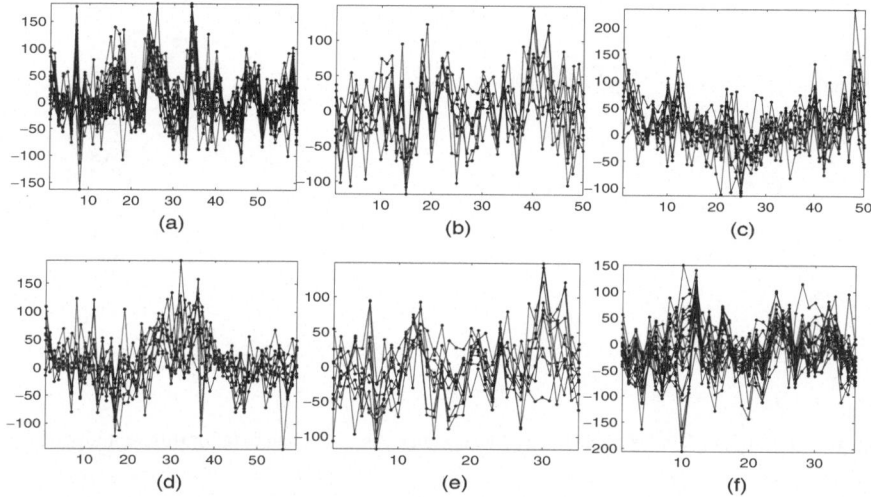

Fig. 10.6 Six biclusters of Human data using MOGAB. For each bicluster, #genes, #conditions, *MSR*, *VAR* and $\mathscr{BI}$ is given: (a) 19, 59, 996.8422, 2315.0660, 0.4304, (b) 8, 50, 1005.4557, 1976.0596, 0.5086, (c) 12, 50, 995.8271, 1865.1578, 0.5336, (d) 10, 59, 1006.7489, 1825.6937, 0.5511, (e) 10, 35, 1054.5312, 1895.1600, 0.5561, (f) 23, 36, 1194.3213, 1917.9915, 0.6224

Figure 10.7 plots the $\mathscr{BI}$ scores for all 100 biclusters obtained by various algorithms for Human data. OPSM, ISA and BiVisu produced 14, 65 and 93 biclusters, respectively. It is evident that for most of the biclusters, MOGAB provides smaller $\mathscr{BI}$ scores than the other algorithms. The values in Table 10.3 also confirm that the biclusters produced by MOGAB have lower average residue and higher average variance. Also, the average $\mathscr{BI}$ score of MOGAB is the best. Furthermore, lower standard deviation proves that the scores do not vary much from one bicluster to another. This establishes that MOGAB provides a set of equal quality biclusters.

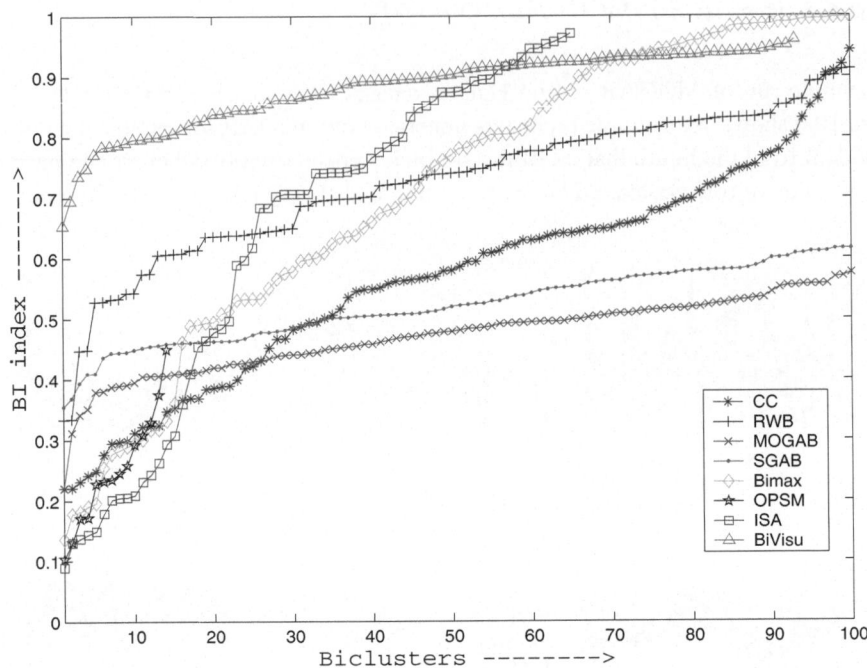

Fig. 10.7 Plot of $\mathcal{BI}$ index values of 100 best biclusters of Human data for different algorithms

Table 10.3 Comparison of the biclusters of different algorithms for Human data. Bracketed values represent the standard deviations

Algorithm	MSR		Row variance		Volume		$\mathscr{BI}$ index	
	Average	Best (min)	Average	Best (max)	Average	Best (max)	Average	Best (min)
MOGAB	801.37 (138.84)	569.23	2378.25 (380.43)	5377.51	276.48 (160.93)	1036	0.4710 (0.059)	0.2283
SGAB	855.36 (232.44)	572.54	2222.19 (672.93)	5301.83	242.93 (171.23)	996	0.5208 (0.063)	0.3564
CC	1078.38 (143.85)	779.71	2144.14 (895.32)	5295.42	388.86 (1118.42)	9100	0.5662 (0.174)	0.2298
RWB	1185.69 (11.59)	992.76	1698.99 (386.09)	3575.40	851.54 (169.98)	1830	0.7227 (0.120)	0.3386
Bimax	387.71 (210.53)	96.98	670.83 (559.56)	3204.35	64.34 (15.31)	138	0.7120 (0.250)	0.1402
OPSM	1249.73 (652.63)	43.07	6374.64 (2439.28)	11854.63	373.5472 (432.55)	1299	0.2520 (0.094)	0.1024
ISA	2006.83 (1242.93)	245.28	4780.65 (4713.62)	14682.47	69.35 (35.95)	220	0.6300 (0.278)	0.0853
BiVisu	1680.23 (70.51)	1553.43	1913.24 (137.31)	2468.63	1350.24 (1243.15)	17739	0.8814 (0.062)	0.6537

10.6.6 Results on the Leukemia Dataset

MOGAB takes about 570 seconds for a single run on the Leukemia data. As evident from Figure 10.8, six biclusters found by MOGAB for Leukemia data have low residue and high row variance, and thus have low $\mathscr{BI}$ scores.

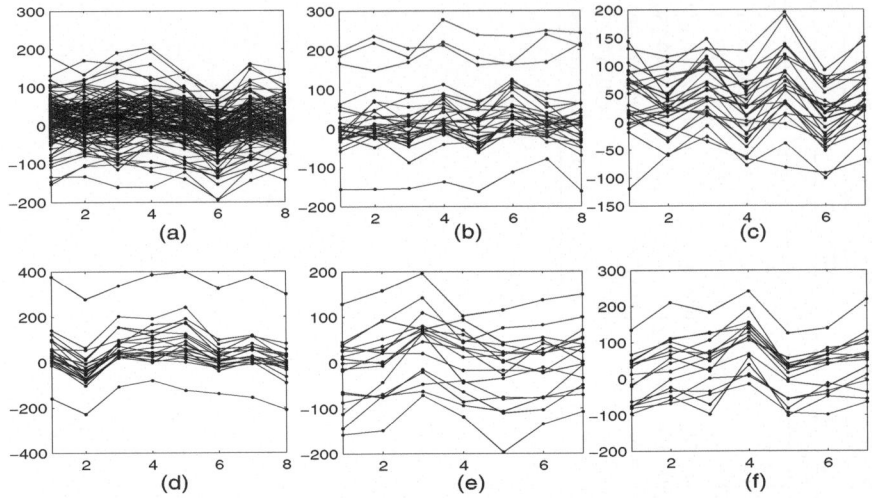

Fig. 10.8 Six biclusters of Leukemia data using MOGAB. For each bicluster, #genes, #conditions, *MSR*, *VAR* and $\mathscr{BI}$ is given: (a) 126, 8, 428.9371, 1865.9349, 0.2298, (b) 29, 8, 432.0511, 1860.1433, 0.2321, (c) 27, 7, 396.8000, 2462.4681, 0.1611, (d) 21, 8, 441.5094, 5922.4800, 0.0745, (e), 17, 7, 358.1648, 2378.7116, 0.1505, (f), 16, 7, 377.4619, 4561.3842, 0.0827

The plots of $\mathscr{BI}$ scores is shown in Figure 10.9 for the 100 best biclusters for different algorithms for Leukemia data. OPSM, ISA and BiVisu provide 16, 21 and 54 biclusters, respectively. The figure establishes that for most of the biclusters, MOGAB provides lower $\mathscr{BI}$ scores than the other algorithms. Also, the performance of MOGAB is more stable. It appears from Table 10.4 that the general nature of the biclusters produced by MOGAB is lower residue and higher row variance, producing lower $\mathscr{BI}$ scores.

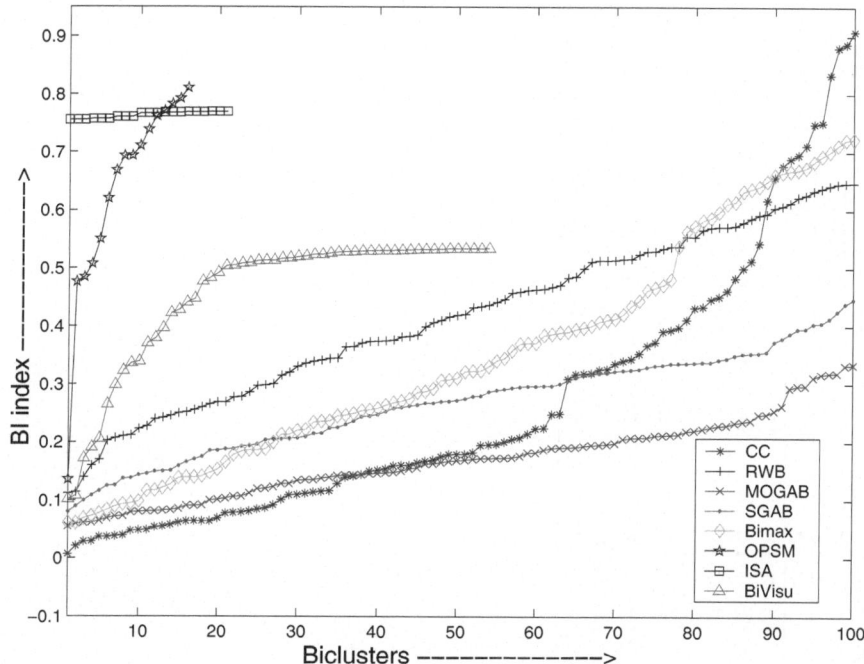

Fig. 10.9 Plot of $\mathscr{BI}$ index values of 100 best biclusters of Leukemia data for different algorithms

Table 10.4 Comparison of the biclusters of different algorithms for Leukemia data. Bracketed values represent the standard deviations

Algorithm	MSR		Row variance		Volume		$\mathcal{BI}$ index	
	Average	Best (min)	Average	Best (max)	Average	Best (max)	Average	Best (min)
MOGAB	206.55 (61.15)	59.84	2852.76 (1846.73)	27688.93	330.56 (166.47)	1008	0.1698 (0.069)	0.0553
SGAB	243.54 (72.43)	79.41	2646.25 (1938.83)	21483.74	323.54 (182.49)	840	0.2663 (0.087)	0.0803
CC	335.81 (83.60)	109.49	2794.36 (3291.42)	24122.64	230.90 (233.13)	1124	0.2648 (0.227)	0.0071
RWB	583.42 (18.35)	561.73	1137.62 (84.33)	2937.76	388.65 (103.66)	420	0.4156 (0.142)	0.1123
Bimax	1243.45 (2946.32)	703.45	1834.54 (2037.63)	13842.65	95.25 (22.44)	130	0.3475 (0.195)	0.0694
OPSM	3.58E6 (2.63E6)	5.34E4	5.03E6 (4.50E6)	1.73E7	3176.34 (4309.32)	12226	0.6375 (0.175)	0.1359
ISA	1.12E7 (1.30E7)	1.03E6	1.13E7 (1.30E7)	3.18E7	731.86 (357.28)	1398	0.7638 (0.006)	0.7556
BiVisu	1580.45 (838.61)	587.40	3006.15 (1563.92)	12547.74	2380.13 (197.61)	3072	0.4540 (0.119)	0.1018

10.6.7 Statistical Significance Test

A statistical significance test based on the t-statistic is conducted at a 5% significance level to establish that the better average $\mathcal{BI}$ scores produced by MOGAB
compared to the other algorithms are statistically significant. Table 10.5 shows the
p-values provided by the t-test by comparing the mean $\mathcal{BI}$ scores of MOGAB
with those of the other algorithms for the real datasets. It appears that the p-values
are less than 0.05 (5% significance level). There is only one exception where t-test
gives p-value larger than 0.05. This is the case of comparing MOGAB with OPSM
for the Human dataset. However, this is acceptable as OPSM produces only 14 biclusters for this dataset. Hence this test proves that the performance of MOGAB is
significantly better than that of the other methods.

Table 10.5 The p-values for t-test comparing mean value of $\mathcal{BI}$ index of MOGAB with that of
other algorithms for real datasets

Algorithm	p-values		
	Yeast	Human	Leukemia
SGAB	5.73E-12	1.25E-12	9.96E-18
CC	0.006414	1.25E-07	0.000273
RWB	2.35E-69	1.12E-37	4.77E-39
Bimax	1.75E-19	3.98E-17	1.21E-12
OPSM	0.046793	0.896341	2.48E-50
ISA	1.97E-29	1.34E-07	1.55E-97
BiVisu	1.18E-61	3.53E-108	1.47E-42

10.6.8 Biological Significance Test

The biological relevance of the biclusters can be verified based on the GO annotation
database (http://db.yeastgenome.org/cgi-bin/GO/goTermFinder). This is used to test
the functional enrichment of a group of genes in terms of three structured, controlled
vocabularies (ontologies), viz., biological processes, molecular functions and biological components. The p-value of a statistical significance test is used to find the
probability of getting values of a test statistic that are at least equal to the observed
test statistic. The degree of functional enrichment (p-values) is computed using a
cumulative hypergeometric distribution that measures the probability of finding the
number of genes involved in a given GO term within a bicluster. From a given GO
category, the probability p of getting k or more genes within a cluster of size n can
be defined as [352, 398]

$$p = 1 - \sum_{i=0}^{k-1} \frac{\binom{f}{i}\binom{g-f}{n-i}}{\binom{g}{n}}, \tag{10.32}$$

where f and g denote the total number of genes within a category and within the genome, respectively. This signifies how well the genes in the bicluster match with the different GO categories. If the majority of genes in a bicluster have the same biological function, then it is unlikely that this takes place by chance, and the p-value of the category will be close to 0.

The biological significance test has been conducted at 1% significance level. For different algorithms, the number of biclusters for which the most significant GO terms have p-value less than 0.01 (1% significance level) are as follows: MOGAB, 32; SGAB, 19; CC, 10; RWB, 7; Bimax, 14; OPSM, 6; ISA, 6; and BiVisu, 17. The top five most significant biclusters produced by each algorithm on Yeast data have been examined. Table 10.6 reports the GO terms along with their p-values for each of these biclusters. For example, the most significant five biclusters produced by MOGAB are mostly enriched with the GO terms cytosolic part (p-value = $1.4E -45$), ribosomal subunit (p-value = $1.6E - 45$), translation (p-value = $3.8E - 41$), RNA metabolic process (p-value = $8.4E - 25$) and DNA metabolic process (p-value = $3.1E - 21$). It is evident from the table that MOGAB beats all other algorithms in terms of the p-values for the top five functionally enriched biclusters. Also, note that the significant GO terms have very low p-value (much less than 0.01), which indicates that the biclusters are biologically significant.

Table 10.6 Result of biological significance test: The top five functionally enriched significant biclusters produced by each algorithm for Yeast data. Corresponding GO terms and the p-values are reported

Biclusters	MOGAB	SGAB	CC	RWB	Bimax	OPSM	ISA	BiVisu
Bicluster 1	cytosolic part GO:0044445 pval: 1.4E-45	cytosolic part GO:0044445 (1.4E-45)	cytosolic part GO:0044445 (4.2E-45)	ribosome biogenesis and assembly GO:0042254 (9.3E-09)	ribonucleo-protein complex GO:0030529 (9.4E-11)	intracellular membrane-bound organelle GO:0043231 (2.8E-09)	cytosolic part GO:0044445 (3.6E-44)	ribonucleo-protein complex GO:0030529 (1.4E-20)
Bicluster 2	ribosomal subunit GO:0033279 (1.6E-45)	ribosome GO:0005840 (1.5E-25)	translation GO:0006412 (1.5E-21)	RNA metabolic process GO:0016070 (4.9E-08)	cytosolic part GO:0044445 (1.3E-10)	protein modification process GO:0006464 (2.8E-08)	sulfur metabolic process GO:0006790 (6.9E-10)	ribosome biogenesis and assembly GO:0042254 (9.5E-20)
Bicluster 3	translation GO:0006412 (3.8E-41)	translation GO:0006412 (7.4E-24)	ribosome biogenesis and assembly GO:0042254 (1.9E-15)	MAPKKK cascade GO:0000165 (2.5E-06)	sulfur metabolic process GO:0006790 (4.2E-10)	biopolymer modification GO:0043412 (3.1E-07)	macromolecule biosynthetic process GO:0009059 (2.9E-05)	RNA metabolic process GO:0016070 (5.8E-18)
Bicluster 4	RNA metabolic process GO:0016070 (8.4E-25)	chromosome GO:0005694 (2.3E-13)	ribonucleo-protein complex biogenesis and assembly GO:0022613 (2.5E-12)	RNA processing GO:0006396 (2.6E-06)	chromosome GO:0005694 (1.1E-09)	carbohydrate metabolic process GO:0005975 (1.4E-06)	nucleic acid binding GO:0003676 (7.3E-04)	RNA processing GO:0006396 (4.5E-16)
Bicluster 5	DNA metabolic process GO:0006259 (3.1E-21)	RNA metabolic process GO:0016070 (1.3E-11)	mitochondrial part GO:0044429 (9.1E-12)	response to osmotic stress GO:0006970 (3.9E-06)	cellular bud GO:0005933 (2.4E-09)	M phase of meiotic cell cycle GO:0051327 (3.2E-05)	establishment of cellular localization GO:0051649 (7.8E-04)	ribonucleo-protein complex biogenesis and assembly GO:0022613 (3.3E-15)

10.7 Incorporating Fuzziness in MOGAB

In this section we discuss how fuzziness can be incorporated into MOGAB [304] in order to handle overlapping biclusters in a better way. The fuzzified MOGAB algorithm (FuzzMOGAB) uses a similar encoding policy to that in MOGAB. Also, the selection, crossover and mutation operations are kept same. The main difference is in the fitness computation part, which is motivated by fuzzy the K-medoids clustering algorithm [262]. First, the fuzzy K-medoids algorithm is described. Thereafter, the fitness computation of and the process of obtaining final solutions for FuzzMO-GAB have been discussed. Finally a comparative study on the performance of MO-GAB and FuzzMOGAB has been made for the three real-life microarray datasets mentioned above.

10.7.1 Fuzzy K-Medoids Clustering

The fuzzy K-medoids [262] algorithm is the extension of the well-known fuzzy C-means [62] algorithm replacing cluster means with cluster medoids. A medoid is defined as follows: Let $Y = \{y_1, y_2, \ldots, y_v\}$ be a set of v objects. The medoid of Y is an object $O \in Y$ such that the sum of distances from O to other objects in Y is minimum, i.e., the following criterion is minimized:

$$\mathscr{D}(O, Y) = \sum_{i=1}^{v} D(O, y_i). \tag{10.33}$$

Here $D(O, y_i)$ denotes the distance measure between O and y_i.

The aim of the fuzzy K-medoids algorithm is to cluster the dataset $X = \{x_1, x_2, \ldots, x_n\}$ into K partitions so that the following criterion is minimized:

$$J_m(U, Z : X) = \sum_{k=1}^{n} \sum_{i=1}^{K} u_{ik}^m D(z_i, x_k), \tag{10.34}$$

where m is the fuzzy exponent. $U = [u_{ik}]$ denotes the $K \times n$ fuzzy partition matrix and u_{ik} (between 0 and 1) denotes the membership degree of kth object for the ith cluster. $Z = \{z_1, z_2, \ldots, z_K\}$ represents the set of cluster medoids. For probabilistic clustering, $\sum_{i=1}^{K} u_{ik} = 1, \forall k \in \{1, 2, \ldots, n\}$.

The fuzzy K-medoids algorithm involves iteratively estimating the partition matrix followed by computation of new cluster medoids. It starts with random initial K medoids, and then at every iteration it finds the fuzzy membership of each object in every cluster using the following equation [262]:

$$u_{ik} = \frac{1}{\sum_{j=1}^{K} \left(\frac{D(z_i,x_k)}{D(z_j,x_k)}\right)^{\frac{1}{m-1}}}, \quad \text{for } 1 \le i \le K; \ 1 \le k \le n. \tag{10.35}$$

Note that while computing u_{ik} using Equation 10.35, if $D(z_j,x_k)$ is equal to zero for some j, then u_{ik} is set to zero for all $i = 1,\ldots,K$, $i \ne j$, while u_{jk} is set equal to 1.

Based on the membership values, the cluster medoids are recomputed as follows: The medoid z_i of the ith cluster will be $z_i = x_p$, where

$$p = \arg\min_{1 \le j \le n} \sum_{k=1}^{n} u_{ik}^m D(x_j,x_k). \tag{10.36}$$

The algorithm terminates when there is no significant improvement in the J_m value (Equation 10.34). Finally, each object is assigned to the cluster for which it has maximum membership.

10.7.2 Fitness Computation in FuzzMOGAB

Given a valid string (i.e., the string contains no repetition of gene or condition indices, and $2 \le M \le \lceil \sqrt{\mathscr{G}} \rceil$, $2 \le N \le \lceil \sqrt{\mathscr{C}} \rceil$), first all the gene clusters and the condition clusters encoded in it are extracted. Thereafter, the fuzzy membership μ_{ik} for each gene g_k in the ith gene cluster gc_i is computed as follows:

$$\mu_{ik} = \frac{1}{\sum_{j=1}^{M} \left(\frac{D(gc_i,g_k)}{D(gc_j,g_k)}\right)^{\frac{1}{m-1}}}, \quad \text{for } 1 \le i \le M; \ 1 \le k \le \mathscr{G}. \tag{10.37}$$

Here $D(.,.)$ is a distance function. Euclidean distance measure has been adopted. Subsequently, the gene index gc_i representing the ith gene cluster is replaced by q_i, where,

$$q_i = \arg\min_{1 \le j \le \mathscr{G}} \sum_{k=1}^{\mathscr{G}} \mu_{ik}^m D(g_j,g_k). \tag{10.38}$$

Also the membership matrix for the genes is recomputed.

The fuzzy membership η_{ik} for each condition c_k in the ith condition cluster cc_i is computed as

$$\eta_{ik} = \frac{1}{\sum_{j=1}^{N} \left(\frac{D(cc_i,c_k)}{D(cc_j,c_k)}\right)^{\frac{1}{m-1}}}, \quad \text{for } 1 \le i \le N; \ 1 \le k \le \mathscr{C}. \tag{10.39}$$

Next, the condition index cc_i representing the ith condition cluster is replaced by r_i, where

$$r_i = \arg\min_{1 \le j \le \mathscr{C}} \sum_{k=1}^{\mathscr{C}} \eta_{ik}^m D(c_j, c_k). \tag{10.40}$$

Finally, the membership matrix for the conditions is recomputed. The chromosome is also updated using new gene and condition indices representing the gene clusters and condition clusters, respectively.

As the two objective functions, the fuzzified versions of mean squared residue and row variance are used. The fuzzy volume of bicluster $B(I,J)$ corresponding to the $< gc_x, cc_y >$ pair is defined as

$$fvol(I,J) = \sum_{i=1}^{\mathscr{G}} \sum_{j=1}^{\mathscr{C}} \mu_{xi}^m \eta_{yj}^m. \tag{10.41}$$

Here I is a fuzzy set corresponding to the fuzzy gene cluster centered at gc_x. It consists of all genes g_j with membership degree μ_{xi}, $1 \le i \le \mathscr{G}$. Similarly, J is a fuzzy set corresponding to the fuzzy condition cluster centered at cc_y. It consists of all conditions c_j with membership degree η_{yj}, $1 \le j \le \mathscr{C}$.

The residue of an element a_{ij} of the fuzzy bicluster $B(I,J)$ is defined as

$$fr_{ij} = a_{ij} - a_{iJ} - a_{Ij} + a_{IJ}, \tag{10.42}$$

where

$$a_{iJ} = \frac{\sum_{j=1}^{\mathscr{C}} \eta_{yj}^m a_{ij}}{\sum_{j=1}^{\mathscr{C}} \eta_{yj}^m},$$

$$a_{Ij} = \frac{\sum_{i=1}^{\mathscr{G}} \mu_{xj}^m a_{ij}}{\sum_{i=1}^{\mathscr{G}} \mu_{xj}^m}, \quad \text{and}$$

$$a_{IJ} = \frac{\sum_{i=1}^{\mathscr{G}} \sum_{j=1}^{\mathscr{C}} \mu_{xi}^m \eta_{yj}^m a_{ij}}{fvol(I,J)}.$$

The fuzzy mean squared residue ($FMSR(I,J)$) of the fuzzy bicluster $B = (I,J)$ is defined as

$$FMSR(I,J) = \frac{1}{fvol(I,J)} \sum_{i=1}^{\mathscr{G}} \sum_{j=1}^{\mathscr{C}} \mu_{xi}^m \eta_{yj}^m fr_{ij}^2. \tag{10.43}$$

The fuzzy row variance of $B(I,J)$ is computed as

$$fvar(I,J) = \frac{1}{fvol(I,J)} \sum_{i=1}^{\mathscr{G}} \sum_{j=1}^{\mathscr{C}} \mu_{xi}^m \eta_{yj}^m (a_{ij} - a_{iJ})^2. \tag{10.44}$$

All the fuzzy biclusters represented by each gene-cluster condition-cluster pair $< gc_x, cc_y >$, $1 \le x \le M$, $1 \le y \le N$, are extracted and the two objective functions $FMSR$ and $fvar$ are computed for each of them as mentioned above. Note that $FMSR$ is to be minimized whereas $fvar$ is to be maximized to have good quality large and non-trivial biclusters. As FuzzMOGAB is also posed as a minimization algorithm, the two objective functions (f_1 and f_2) are taken as follows:

$$f_1(I,J) = \frac{FMSR(I,J)}{\delta},$$ (10.45)

and

$$f_2(I,J) = \frac{1}{1 + fvar(I,J)}.$$ (10.46)

The denominator of f_2 is chosen in such a way as to avoid the accidental divide by zero condition when the fuzzy row variance becomes 0. Note that both f_1 and f_2 are to be minimized to obtain highly coherent and non-trivial biclusters. For each encoded fuzzy bicluster, the fitness vector $\overline{f} = \{f_1, f_2\}$ is computed. The fitness vector of a string is then the average of the fitness vectors of all fuzzy biclusters encoded in it.

Note that due to the randomness of the genetic operators, invalid strings with repeated gene and/or condition indices may arise at any point in the algorithm. The invalid strings are given fitness vector $\overline{f} = \{X, X\}$, where X is an arbitrary large number. Thus the invalid strings will be automatically out of the competition in subsequent generations.

10.7.3 Obtaining Final Biclusters in FuzzMOGAB

The FuzzMOGAB algorithm has been executed for a fixed number of generations with a fixed population size. When the final generation completes, the fuzzy biclusters are extracted from the non-dominated strings. For a gene-cluster condition-cluster pair $< gc_x, cc_y >$, the fuzzy bicluster $B(I,J)$ can be interpreted as a regular bicluster $B'(I',J')$, where,

$$I' = \{i | \mu_{xi} >= \alpha_g\}, \quad 1 \le i \le \mathscr{G},$$ (10.47)

and

$$J' = \{j | \eta_{yj} >= \alpha_c\}, \quad 1 \le j \le \mathscr{C}.$$ (10.48)

Here α_g and α_c denote two threshold parameters on membership degrees. The threshold α_g is taken as $1/M$ and α_c is taken as $1/N$, where M and N are the number of gene clusters and condition clusters encoded in the string, respectively. After extracting the regular biclusters from the fuzzy ones, all the δ-biclusters are returned by FuzzMOGAB.

10.7.4 Comparison of MOGAB and FuzzMOGAB

Both MOGAB and FuzzMOGAB have been executed for 100 generations with population size 50, crossover probability 0.8 and mutation probability 0.1 on the three real-life datasets, viz., Yeast, Human and Leukemia. Table 10.7 reports the comparative $\mathscr{BI}$ scores for the 100 best biclusters (in terms of $\mathscr{BI}$ index scores) found from all the three datasets as produced by the MOGAB and FuzzMOGAB algorithms. It is evident from the table that FuzzMOGAB outperforms MOGAB in terms of both average and minimum $\mathscr{BI}$ index scores for all the datasets. Moreover, the standard deviations of the $\mathscr{BI}$ index scores for FuzzMOGAB are smaller than those for MOGAB, which indicates that all the biclusters produced by FuzzMOGAB are more equally interesting than those produced by MOGAB. These results establish that incorporation of fuzziness into MOGAB leads to improved performance of the algorithm.

Table 10.7 Comparison of the $\mathscr{BI}$ index values of MOGAB and FuzzMOGAB for Yeast, Human and Leukemia datasets. Bracketed values represent the standard deviations

Algorithms	Yeast		Human		Leukemia	
	Average	Minimum	Average	Minimum	Average	Minimum
MOGAB	0.3329 (0.061)	0.2123	0.4710 (0.059)	0.2283	0.1698 (0.069)	0.0553
FuzzMOGAB	0.3105 (0.054)	0.2055	0.4518 (0.046)	0.2174	0.1548 (0.062)	0.0486

10.8 Summary

Biclustering is a method for simultaneous clustering of both genes and conditions of a microarray gene expression matrix. Unlike clustering, biclustering methods try to capture local modules, i.e., sets of genes that are coregulated and coexpressed in a subset of conditions. In recent times, there has been tremendous growth in biclustering research and a large number of algorithms have been proposed. In this chapter, first we have discussed different issues and definitions regarding the biclustering problem and presented a comprehensive review of biclustering algorithms of different categories along with their pros and cons.

Thereafter, the biclustering problem in gene expression data has been posed as a multiobjective optimization problem. In this regard, a recently proposed multiobjective biclustering algorithm (MOGAB) that simultaneously optimizes the mean squared residue and row variance of the biclusters in order to discover coherent and nontrivial biclusters is discussed in detail. MOGAB uses a variable chromosome length representation and utilizes NSGA-II as the underlying optimization strategy. The performance of MOGAB has been demonstrated both qualitatively (visually) and quantitatively (using a match score and the $\mathscr{BI}$ index) on a simulated dataset and three well-known real-life microarray datasets. Also, a comparative study with other well-known biclustering methods has been made to establish its effectiveness. Subsequently, biological significance tests based on Gene Ontology have been carried out to show that MOGAB is able to identify biologically significant biclusters.

Finally, it has been discussed how fuzziness can be incorporated into MOGAB in order to handle overlapping biclusters in a better way. In this regard a fuzzy version of MOGAB, called FuzzMOGAB, has been described. The performance comparison between MOGAB and its fuzzy version reveals that incorporation of fuzziness leads to better performance of the algorithm.

References

1. *IRS Data Users Handbook.* NRSA, Hyderabad, India, Rep. IRS/NRSA/NDC/HB-01/86, 1986.
2. J. P. Adrahams and M. Breg. Prediction of RNA secondary structure including pseudoknotting by computer simulation. *Nucleic Acids Research*, 18:3035–3044, 1990.
3. C. C. Aggarwal and P. S. Yu. Outlier detection for high dimensional data. In *SIGMOD '01: Proceedings of the 2001 ACM SIGMOD International Conference on Management of Data*, pages 37–46, New York, NY, USA, 2001. ACM.
4. R. Agrawal, T. Imielinski, and A. N. Swami. Mining association rules between sets of items in large databases. In P. Buneman and S. Jajodia, editors, *Proceedings of the 1993 ACM SIGMOD International Conference on Management of Data*, pages 207–216, Washington, DC, May 1993.
5. R. Agrawal and R. Srikant. Fast algorithms for mining association rules in large databases. In *VLDB '94: Proceedings of the 20th International Conference on Very Large Data Bases*, pages 487–499, San Francisco, CA, USA, 1994. Morgan Kaufmann.
6. J. S. Aguilar-Ruiz. Shifting and scaling patterns from gene expression data. *Bioinformatics*, 21(20):3840–3845, 2005.
7. W. Ahmad and A. Khokhar. cHawk: A highly efficient biclustering algorithm using bigraph crossing minimization. In *Proc. 2nd International Workshop on Data Mining and Bioinformatics*, Vienna, Austria.
8. T. Akutsu, S. Miyano, and S. Kuhara. Identification of genetic networks from a small number of gene expression patterns under the Boolean network model. In *Proceedings of the Pacific Symposium on Biocomputing*, volume 99, pages 17–28, 1999.
9. F. Al-Shahrour, R. Diaz-Uriarte, and J. Dopazo. FatiGO: A Web tool for finding significant associations of gene ontology terms with groups of genes. *Bioinformatics*, 20(4):578–580, 2004.
10. D. Alahakoon, S. K. Halgamuge, and B. Srinivasan. Dynamic self organizing maps with controlled growth for knowledge discovery. *IEEE Transactions on Neural Networks*, 11:601–614, 2000.
11. E. Alba, J. Garcia-Nieto, L. Jourdan, and E-G. Talbi. Gene selection in cancer classification using PSO/SVM and GA/SVM hybrid algorithms. In *Proc. IEEE Cong. Evol. Comput.*, pages 284–290, 2007.
12. B. Alberts, A. Johnson, J. Lewis, M. Raff, K. Roberts, and P. Walters. *The Shape and Structure of Proteins.* Garland Science, New York and London, fourth edition, 2002.
13. A. A. Alizadeh, M. B. Eisen, R. Davis, C. Ma, I. Lossos, A. Rosenwald, J. Boldrick, R. Warnke, R. Levy, W. Wilson, M. Grever, J. Byrd, D. Botstein, P. O. Brown, and L. M.

Straudt. Distinct types of diffuse large B-cell lymphomas identified by gene expression profiling. *Nature*, 403:503–511, 2000.

14. U. Alon, N. Barkai, D. Notterman, K. Gish, S. Ybarra, D. Mack, and A. J. Levine. Broad patterns of gene expression revealed by clustering analysis of tumor and normal colon tissues probed by oligonucleotide arrays. In *Proc. Nat. Academy of Sciences*, volume 96, pages 6745–6750, USA, 1999.

15. M. R. Anderberg. *Cluster Analysis for Application*. Academic Press, 1973.

16. L. N. Andersen, J. Larsen, L. K. Hansen, and M. Hintz-Madsen. Adaptive regularization of neural classifiers. In *Proc. IEEE Workshop on Neural Networks for Signal Processing VII*, pages 24–33, New York, USA, 1997.

17. T. W. Anderson. *An Introduction to Multivariate Statistical Analysis*. Wiley, New York, 1958.

18. S. Ando and H. Iba. Quantitative modeling of gene regulatory network: Identifying the network by means of genetic algorithms. *Genome Informatics*, 11:278–280, 2000.

19. S. Ando and H. Iba. Inference of gene regulatory model by genetic algorithms. In *Proceedings of the Congress on Evolutionary Computation*, volume 1, pages 712–719, 2001.

20. S. Ando, E. Sakamoto, and H. Iba. Evolutionary modeling and inference of gene network. *Information Sciences – Informatics and Computer Science: An International Journal*, 145(3–4):237–259, 2002.

21. H. C. Andrews. *Mathematical Techniques in Pattern Recognition*. Wiley Interscience, New York, 1972.

22. E. Angeleri, B. Apolloni, D. de Falco, and L. Grandi. DNA fragment assembly using neural prediction techniques. *Int. J. Neural Syst.*, 9(6):523–544, 1999.

23. F. Angiulli, E. Cesario, and C. Pizzuti. Gene expression biclustering using random walk strategies. In *Proc. 7th Int. Conf. on Data Warehousing and Knowledge Discovery (DAWAK'05)*, Copenhagen, Denmark, 2005.

24. F. Angiulli, E. Cesario, and C. Pizzuti. Random walk biclustering for microarray data. *Information Sciences*, 178(6):1479–1497, 2008.

25. M. Ankerst, M. Breunig, H.-P. Kriegel, and J. Sander. Optics: Ordering points to identify the clustering structure. In *ACM SIGMOD Int. Conf. Management of Data (SIGMOD'99)*, pages 49–60, Philadelphia, 1999.

26. C. Anselmi, G. Bocchinfuso, P. De Santis, M. Savino, and A. Scipioni. A theoretical model for the prediction of sequence-dependent nucleosome thermodynamic stability. *Journal of Biophysics*, 79(2):601–613, 2000.

27. M. Ashburner, C. A. Ball, J. A. Blake, D. Botstein, H. Butler, J. M. Cherry, A. P. Davis, K. Dolinski, S. S. Dwight, J. T. Eppig, M. A. Harris, D. P. Hill, L. Issel-Tarver, A. Kasarskis, S. Lewis, J. C. Matese, J. E. Richardson, M. Ringwald, G. M. Rubin, and G. Sherlock. Gene Ontology: Tool for the unification of biology. *Nature Genetics*, 25(1):25–29, May 2000.

28. J. Atkinson-Abutridy, C. Mellish, and S. Aitken. Combining information extraction with genetic algorithms for text mining. *IEEE Intelligent Systems*, 19(3):22–30, 2004.

29. T. Back, F. Hoffmeister, and H-P. Schwefel. A survey of evolution strategies. In *Proc. of the Fourth International Conference on Genetic Algorithms*, pages 2–9. Morgan Kaufmann, 1991.

30. A. Bagchi, S. Bandyopadhyay, and U. Maulik. Determination of molecular structure for drug design using variable string length genetic algorithm. In *Workshop on Soft Computing, High Performance Computing (HiPC): New Frontiers in High-Performance Computing*, pages 145–154, Hyderabad, 2003.

31. T. M. Bakheet and A. J. Doig. Properties and identification of human protein drug targets. *Bioinformatics*, 25(4):451–457, January 2009.

32. J. Bala, K. De Jong, J. Huang, H. Vafaie, and H. Wechsler. Using learning to facilitate the evolution of features for recognizing visual concepts. *Evolutionary Computation*, 4(3):297–311, 1996.

33. P. Baldi and P. F. Baisnee. Sequence analysis by additive scales: DNA structure for sequences and repeats of all lengths. *Bioinformatics*, 16:865–889, 2000.

34. J. F. Baldwin. Knowledge from data using fuzzy methods. *Pattern Recognition Letters*, 17:593–600, 1996.

35. S. Bandyopadhyay, A. Bagchi, and U. Maulik. Active site driven ligand design: An evolutionary approach. *Journal of Bioinformatics and Computational Biology*, 3(5):1053–1070, 2005.

36. S. Bandyopadhyay and U. Maulik. Non-parametric genetic clustering: Comparison of validity indices. *IEEE Trans. Systems, Man and Cybernetics Part C*, 31(1):120–125, 2001.

37. S. Bandyopadhyay and U. Maulik. Efficient prototype reordering in nearest neighbor classification. *Pattern Recognition*, 35(12):2791–2799, December 2002.

38. S. Bandyopadhyay and U. Maulik. An evolutionary technique based on k-means algorithm for optimal clustering in R^N. *Information Science*, 146:221–237, 2002.

39. S. Bandyopadhyay and U. Maulik. Genetic clustering for automatic evolution of clusters and application to image classification. *Pattern Recognition*, 35(6):1197–1208, 2002.

40. S. Bandyopadhyay, U. Maulik, L. B. Holder, and D. J. Cook. *Advanced Methods for Knowledge Discovery from Complex Data (Advanced Information and Knowledge Processing)*. Springer-Verlag, London, 2005.

41. S. Bandyopadhyay, U. Maulik, and A. Mukhopadhyay. Multiobjective genetic clustering for pixel classification in remote sensing imagery. *IEEE Trans. Geosci. Remote Sens.*, 45(5):1506–1511, 2007.

42. S. Bandyopadhyay, U. Maulik, and M. K. Pakhira. Clustering using simulated annealing with probabilistic redistribution. *Int. J. Pattern Recog. and Artificial Intelligence*, 15(2):269–285, 2001.

43. S. Bandyopadhyay, U. Maulik, and J. T. L. Wang. *Analysis of Biological Data: A Soft Computing Approach*. World Scientific, 2007.

44. S. Bandyopadhyay, A. Mukhopadhyay, and U. Maulik. An improved algorithm for clustering gene expression data. *Bioinformatics*, 23(21):2859–2865, 2007.

45. S. Bandyopadhyay, C. A. Murthy, and S. K. Pal. Pattern classification using genetic algorithms. *Pattern Recognition Letters*, 16:801–808, 1995.

46. S. Bandyopadhyay, C. A. Murthy, and S. K. Pal. Pattern classification using genetic algorithms: Determination of H. *Pattern Recognition Letters*, 19(13):1171–1181, 1998.

47. S. Bandyopadhyay, C. A. Murthy, and S. K. Pal. Theoretical performance of genetic pattern classifier. *Journal of the Franklin Institute*, 336:387–422, 1999.

48. S. Bandyopadhyay and S. K. Pal. Pattern classification with genetic algorithms: Incorporation of chromosome differentiation. *Pattern Recognition Letters*, 18:119–131, 1997.

49. S. Bandyopadhyay and S. K. Pal. Pixel classification using variable string genetic algorithms with chromosome differentiation. *IEEE Trans. Geosci. Remote Sens.*, 39(2):303–308, 2001.

50. S. Bandyopadhyay and S. K. Pal. *Classification and Learning Using Genetic Algorithms: Applications in Bioinformatics and Web Intelligence*. Natural Computing Series. Springer, 2007.

51. S. Bandyopadhyay, S. K. Pal, and U. Maulik. Incorporating chromosome differentiation in genetic algorithms. *Information Sciences*, 104(3-4):293–319, 1998.

52. S. Bandyopadhyay, S. Saha, U. Maulik, and K. Deb. A simulated annealing-based multiobjective optimization algorithm: AMOSA. *IEEE Transactions on Evolutionary Computation*, 12(3):269–283, 2008.

53. S. Barkow, S. Bleuler, A. Prelic, P. Zimmermann, and E. Zitzler. BicAT: A biclustering analysis toolbox. *Bioinformatics*, 22(10):1282–1283, 2006.

54. V. Batenburg, A. P. Gultyaev, and C. W. A. Pleij. An APL-programmed genetic algorithm for the prediction of RNA secondary structure. *Journal of Theoretical Biology*, 174(3):269–280, 1995.

55. A. Baykasogcaronlu. Goal programming using multiple objective tabu search. *Journal of the Operational Research Society*, 52(12):1359–1369, December 2001.

56. M. J. Bayley, G. Jones, P. Willett, and M. P. Williamson. Genfold: A genetic algorithm for folding protein structures using NMR restraints. *Protein Science*, 7(2):491–499, 1998.

57. Y. Bazi and F. Melgani. Semisupervised PSO-SVM regression for biophysical parameter estimation. *IEEE Trans. Geosci. Remote Sens.*, 45(6):1887–1895, 2007.

58. M. L. Beckers, L. M. Buydens, J. A. Pikkemaat, and C. Altona. Application of a genetic algorithm in the conformational analysis of methylene-acetal-linked thymine dimers in DNA: Comparison with distance geometry calculations. *Journal of Biomol NMR*, 9(1):25–34, 1997.

59. A. Ben-Dor, B. Chor, R. Karp, and Z. Yakhini. Discovering local structure in gene expression data: The order-preserving sub-matrix problem. In *Proc. 6th Annual International Conference on Computational Biology*, volume 1-58113-498-3, pages 49–57, 2002.

60. A. Ben-Dor, R. Shamir, and Z. Yakhini. Clustering gene expression patterns. *J. Computational Biology*, 6(3–4):281–297, 1999.

61. A. Ben-Hur and G. Isabelle. Detecting stable clusters using principal component analysis. *Methods Mol. Biol.*, 224:159–182.

62. J. C. Bezdek. *Pattern Recognition with Fuzzy Objective Function Algorithms*. Plenum, New York, 1981.

63. J. C. Bezdek, R. Ehrlich, and W. Full. FCM: Fuzzy c-means algorithm. *Computers and Geoscience*, 1984.

64. J. C. Bezdek and R. J. Hathaway. VAT: A tool for visual assessment of (cluster) tendency. In *Proceedings of International Joint Conference on Neural Networks*, volume 3, pages 2225–2230, 2002.

65. J. C. Bezdek and N. R. Pal. Some new indexes of cluster validity. *IEEE Transactions on Systems, Man and Cybernetics*, 28:301–315, 1998.

66. D. Bhandari, C. A. Murthy, and S. K. Pal. Genetic algorithm with elitist model and its convergence. *Int. J. Pattern Recognition and Artificial Intelligence*, 10:731–747, 1996.

67. P. J. Bickel and K. A. Doksum. *Mathematical Statistics: Basic Ideas and Selected Topics*. Holden-Day, San Francisco, 1977.

68. Joseph P. Bigus. *Data Mining with Neural Networks: Solving Business Problems from Application Development to Decision Support*. McGraw-Hill, 1996.

69. S. Bleuler, A. Prelic, and E. Zitzler. An EA framework for biclustering of gene expression data. In *Proc. IEEE Congress on Evolutionary Computation (CEC-04)*, pages 166–173, 2004.

70. S. Bogdanovic and B. Langlands. *Systems Biology: The Future of Integrated Drug Discovery*. Strategic Management Report Series. PJB Publications Ltd., 2004.

71. P. A. N. Bosman and D. Thierens. The balance between proximity and diversity in multi-objective evolutionary algorithms. *IEEE Transactions Evolutionary Computation*, 7(2):174–188, 2003.

72. F. Bovolo, L. Bruzzone, and L. Marconcini. A novel approach to unsupervised change detection based on a semisupervised SVM and a similarity measure. *IEEE Trans. Geosci. Remote Sens.*, 46(7):2070–2082, July 2008.

73. J. Branke, K. Deb, H. Dierolf, and M. Osswald. Finding knees in multi-objective optimization. In *Proceedings of the Eighth International Conference on Parallel Problem Solving from Naure*, pages 722–731, Berlin, Germany, 2004. Springer-Verlag.

74. F. Z. Brill, D. E. Brown, and W. N. Martin. Genetic algorithms for feature selection for counterpropagation networks. Technical Report IPC-TR-90-004, Charlottesville, VA, USA, 1990.

75. L. Bruzzone, M. Chi, and M. Marconcini. A novel transductive SVM for semisupervised classification of remote-sensing images. *IEEE Trans. Geosci. Remote Sens.*, 44(11):3363–3373, November 2006.

76. K. Bryan, P. Cunningham, and N. Bolshakova. Biclustering of expression data using simulated annealing. In *Proc. 18th IEEE Symposium on Computer-Based Medical Systems, (CBMS-05)*, pages 383–388, Dublin, Ireland, 2005.

77. R. Bueno, A. J. M. Traina, and C. Traina. Accelerating approximate similarity queries using genetic algorithms. In *Proc. ACM Symposium on Applied Computing (SAC-05)*, pages 617–622, New York, NY, USA, 2005. ACM.

78. S. Busygin, G. Jacobsen, E. Krämer, and C. Ag. Double conjugated clustering applied to leukemia microarray data. In *Proc. 2nd SIAM ICDM, Workshop on Clustering High Dimensional Data*, 2002.

79. R. Caballero, M. Laguna, R. Marti, and J. Molina. Multiobjective clustering with metaheuristic optimization technology. Technical report, Leeds School of Business in the University of Colorado at Boulder, CO, 2006. http://leeds-faculty.colorado.edu/laguna/articles/mcmot.pdf.

80. L. Campbell, Y. Li, and M. Tipping. An efficient feature selection algorithm for classification of gene expression data. In *Proc. NIPS Workshop on Machine Learning Techniques for Bioinformatics*, Vancouver, Canada, 2001.

81. G. Camps-Valls, L. Gomez-Chova, J. Munoz-Mari, J. L. Rojo-Alvarez, and M. Martinez-Ramon. Kernel-based framework for multitemporal and multisource remote sensing data classification and change detection. *IEEE Trans. Geosci. Remote Sens.*, 46(6):1822–1835, June 2008.

82. E. Cantú-Paz. Feature subset selection by estimation of distribution algorithms. In *Proc. Genetic and Evolutionary Computation Conference (GECCO '02)*, pages 303–310, San Francisco, CA, USA, 2002. Morgan Kaufmann.

83. P. Carmona-Saez, R. D. Pascual-Marqui, F. Tirado, J. M. Carazo, and A. Pascual-Montano. Biclustering of gene expression data by non-smooth non-negative matrix factorization. *BMC Bioinformatics*, 7, 2006.

84. C. A. Del Carpio. A parallel genetic algorithm for polypeptide three dimensional structure prediction: A transputer implementation. *Journal of Chemical Information and Computer Sciences*, 36(2):258–269, 1996.

85. H. C. Causton, J. Quackenbush, and A. Brazma. *Microarray gene expressions data analysis: A beginner's guide*. Blackwell Pub., April 2003.

86. A. Celikyilmaz and I. B. Türkşen. Validation criteria for enhanced fuzzy clustering. *Pattern Recognition Letters*, 29(2):97–108, 2008.

87. A. Chakraborty. Biclustering of gene expression data by simulated annealing. In *Proc. Eighth International Conference on High-Performance Computing in Asia-Pacific Region (HPCASIA'05)*, 2005.

88. A. Chakraborty and H. Maka. Biclustering of gene expression data using genetic algorithm. In *Proc. IEEE Symp. Computational Intelligence in Bioinformatics and Computational Biology (CIBCB-05)*, 2005.

89. K. C. C. Chan and W.-H. Au. Mining fuzzy association rules. In *Proc. Sixth International Conference on Information and Knowledge Management (CIKM-97)*, pages 209–215. ACM, 1997.

90. J. Chanussot, J. A. Benediktsson, and M. Fauvel. Classification of remote sensing images from urban areas using a fuzzy possibilistic model. *IEEE Geosci. Remote Sens. Lett.*, 3(1):40–44, January 2006.

91. C. Chen, L. H. Wang, C. Kao, M. Ouhyoung, and W. Chen. Molecular binding in structure-based drug design: A case study of the population-based annealing genetic algorithms. *IEEE International Conference on Tools with Artificial Intelligence*, pages 328–335, 1998.

92. J. Chen and B. Yuan. Detecting functional modules in the yeast protein-protein interaction network. *Bioinformatics*, 22(18):2283–2290, July 2006.

93. J. Cheng and R. Greiner. Learning Bayesian belief network classifiers: Algorithms and system. In *Proc. 14th Biennial Conference of the Canadian Society on Computational Studies of Intelligence*, pages 141–151, London, UK, 2001. Springer-Verlag.

94. Y. Cheng and G. M. Church. Biclustering of gene expression data. In *Proc. Int. Conf. on Intelligent Systems for Molecular Biology (ISMB'00)*, pages 93–103, 2000.

95. D. A. Chiang, L. R. Chow, and Y. F. Wang. Mining time series data by a fuzzy linguistic summary system. *Fuzzy Sets and Systems*, 112:419–432, 2000.

96. S. Chiba, K. Sugawara, and T. Watanabe. Classification and function estimation of protein by using data compression and genetic algorithms. In *Proceedings of the Congress on Evolutionary Computation*, volume 2, pages 839–844, 2001.

97. P. Chiu, A. Girgensohn, W. Polak, E. Rieffel, and L. Wilcox. A genetic algorithm for video segmentation and summarization. In *Proc. IEEE Int. Conf. Multimedia and Expo (ICME 2000)*, volume 3, pages 1329–1332, 2000.

98. C. R. Cho, M. Labow, M. Reinhardt, J. van Oostrum, and M. C. Peitsch. The application of systems biology to drug discovery. *Current Opinion in Chemical Biology*, 10(4):294–302, July 2006.

99. H. Cho, I.S. Dhilon, Y. Guan, and S. Sra. Minimum sum-squared residue co-clustering of gene expression data. In *Proc. 4th SIAM Int. Conf. on Data Mining*, 2004.

100. R. J. Cho, M. J. Campbell, E. A. Winzeler, L. Steinmetz, A. Conway, L. Wodica, and T. G. Wolfsberg et al. A genome-wide transcriptional analysis of mitotic cell cycle. *Mol. Cell.*, 2:65–73, 1998.

101. P. Y. Chou and G. D. Fasman. Prediction of the secondary structure of proteins from their amino acid sequence. *Adv. Enzymol. Relat. Areas Mol. Biol.*, 47:45–148, 1978.

102. S. Chu, J. DeRisi, M. Eisen, J. Mulholland, D. Botstein, P. O. Brown, and I. Herskowitz. The transcriptional program of sporulation in budding yeast. *Science*, 282:699–705, October 1998.

103. H-Y. Chuang, E. Lee, Y.-T. Liu, D. Lee, and T. Ideker. Network-based classification of breast cancer metastasis. *Molecular Systems Biology*, 3, October 2007.

104. M. Clerc. *Particle Swarm Optimization*. Wiley, 2006.

105. G. P. Coelho, F. O. França, and F. J. Zuben. A multi-objective multipopulation approach for biclustering. In *Proc. 7th International Conference on Artificial Immune Systems (ICARIS-08)*, pages 71–82, Berlin, Heidelberg, 2008. Springer-Verlag.

106. C. A. Coello Coello. A comprehensive survey of evolutionary-based multiobjective optimization techniques. *Knowledge and Information Systems*, 1(3):129–156, 1999.

107. C.A. Coello Coello. Evolutionary multiobjective optimization: A historical view of the field. *IEEE Comput. Intell. Mag.*, 1(1):28–36, 2006.

108. C.A. Coello Coello, G. B. Lamont, and D. A. van Veldhuizen. *Evolutionary Algorithms for Solving Multi-Objective Problems*. Genetic and Evolutionary Computation. Springer, Berlin–Heidelberg, Germany, 2nd edition, 2007.

109. B. Comes and A. Kelemen. Probabilistic neural network classification for microarray data. In *Proceedings of the International Joint Conference on Neural Networks, 2003*, volume 3, pages 1714–1717, July 2003.

110. B. Contreras-Moreira, P. W. Fitzjohn, M. Offman, G. R. Smith, and P. A. Bates. Novel use of a genetic algorithm for protein structure prediction: Searching template and sequence alignment space. *Proteins*, 53(6):424–429, 2003.

111. L. Cooper, D. Corne, and M. Crabbe. Use of a novel hill-climbing genetic algorithm in protein folding simulations. *Computational Biology and Chemistry*, 27(6):575–580, 2003.

112. T. H. Cormen, C. E. Leiserson, R. L. Rivest, and C. Stein. *Introduction to Algorithms*. The MIT Press, 2nd edition, September 2001.

113. D. W. Corne, N. R. Jerram, J. D. Knowles, and M. J. Oates. PESA-II: Region-based selection in evolutionary multiobjective optimization. In L. Spector, E. D. Goodman, A. Wu, W. B. Langdon, H.-M. Voigt, M. Gen, S. Sen, M. Dorigo, S. Pezeshk, M. H. Garzon, and E. Burke, editors, *Proc. Genetic and Evolutionary Computation Conference (GECCO-2001)*, pages 283–290, San Francisco, California, USA, 2001. Morgan Kaufmann.

114. D. W. Corne, J. D. Knowles, and M. J. Oates. The Pareto envelope-based selection algorithm for multiobjective optimization. In *Proc. Parallel Problem Solving from Nature VI Conference*, pages 839–848. Springer, 2000.

115. K. Crammer and Y. Singer. On the algorithmic implementation of multiclass kernel-based vector machines. *J. Machine Learning Research*, 2:265–292, 2001.

116. I. Das. On characterizing the 'knee' of the Pareto curve based on normal-boundary intersection. *Structural Optimization*, 18(2-3):107–115, 1999.

117. R. Das and D. Whitley. The only challenging problems are deceptive: Global search by solving order-1 hyperplane. In R. K. Belew and L. B. Booker, editors, *Proceedings of the 4th International Conference on Genetic Algorithms*, pages 166–173, San Mateo, 1991. Morgan Kaufmann.

118. D. L. Davies and D. W. Bouldin. A cluster separation measure. *IEEE Transactions on Pattern Analysis and Machine Intelligence*, 1:224–227, 1979.

119. L. Davis. *Genetic Algorithms and Simulated Annealing*. Morgan Kaufmann, San Francisco, CA, USA, 1987.

120. L. Davis, editor. *Handbook of Genetic Algorithms*. Van Nostrand Reinhold, New York, 1991.

121. T. E. Davis and J. C. Principe. A simulated annealing-like convergence theory for the simple genetic algorithm. In R. K. Belew and J. B. Booker, editors, *Proc. 4th Int. Conf. Genetic Algorithms*, pages 174–181. Morgan Kaufmann, San Mateo, 1991.

122. J. E. Dayhoff. *Neural Network Architectures: An Introduction*. Van Nostrand Reinhold, New York, 1990.

123. K. A. De Jong and W. M. Spears. Learning concept classification rules using genetic algorithms. In *Proc. 12th international joint conference on Artificial intelligence (IJCAI-91)*, pages 651–656, San Francisco, CA, USA, 1991. Morgan Kaufmann.

124. M. C. P. de Souto, I. G. Costa, D. S. A. de Araujo, T. B. Ludermir, and A. Schliep. Clustering cancer gene expression data: a comparative study. *BMC Bioinformatics*, 9(497), November 2009.

125. K. Deb. *Multi-objective Optimization Using Evolutionary Algorithms*. John Wiley and Sons, England, 2001.

126. K. Deb. Multi-objective evolutionary algorithms: Introducing bias among Pareto-optimal solutions. In *Advances in Evolutionary Computing: Theory and Applications*, pages 263–292, London, UK, 2003. Springer-Verlag.

127. K. Deb, S. Agrawal, A. Pratap, and T. Meyarivan. A fast elitist non-dominated sorting genetic algorithm for multi-objective optimization: NSGA-II. In *Proceedings of the Parallel Problem Solving from Nature VI Conference, Springer. Lecture Notes in Computer Science No. 1917*, pages 849–858. Paris, France, 2000.

128. K. Deb, M. Mohan, and S. Mishra. Towards a quick computation of well-spread Pareto-optimal solutions. In *Proceedings of the 2nd International Conference on Evolutionary Multi-Criterion Optimization (EMO-03)*, pages 222–236, Berlin, Heidelberg, 2003. Springer-Verlag.

129. K. Deb, M. Mohan, and S. Mishra. Evaluating the ε-domination based multi-objective evolutionary algorithm for a quick computation of Pareto-optimal solutions. *Evolutionary Computation*, 13(4):501–525, 2005.

130. K. Deb, A. Pratap, S. Agrawal, and T. Meyarivan. A fast and elitist multiobjective genetic algorithm: NSGA-II. *IEEE Transactions on Evolutionary Computation*, 6:182–197, 2002.

131. M. Delattre and P. Hansen. Bicriterion cluster analysis. *IEEE Transactions on Pattern Analysis and Machine Intelligence*, 2(4):277–291, 1980.

132. D. Dembele and P. Kastner. Fuzzy c-means method for clustering microarray data. *Bioinformatics*, 19(8):973–980, 2003.

133. A. P. Dempster, N. M. Laird, and D. B. Rubin. Maximum likelihood from incomplete data via the EM algorithm. *Journal of the Royal Statistical Society*, 39:1–38, 1977.

134. G. Desjardins. A genetic algorithm for text mining. In *Proc. Sixth Int. Conf. Data Mining, Text Mining and Their Business Applications*, pages 133–142, 2005.

135. P. A. Devijver and J. Kittler. *Pattern Recognition: A Statistical Approach*. Prentice-Hall, London, 1982.

136. S. Dharan and A. S. Nair. Biclustering of gene expression data using reactive greedy randomized adaptive search procedure. *BMC Bioinformatics*, 10(Suppl. 1), 2009.

137. P. Dimaggio, S. McAllister, C. Floudas, X. J. Feng, J. Rabinowitz, and H. Rabitz. Biclustering via optimal re-ordering of data matrices in systems biology: Rigorous methods and comparative studies. *BMC Bioinformatics*, 9(1), 2008.

138. Y. Ding and C. E. Lawrence. A statistical sampling algorithm for RNA secondary structure prediction. *Nucl. Acids Res.*, 31(24):7280–7301, December 2003.

139. I. Dinu, J. Potter, T. Mueller, Q. Liu, A. Adewale, G. Jhangri, G. Einecke, K. Famulski, P. Halloran, and Y. Yasui. Improving gene set analysis of microarray data by SAM-GS. *BMC Bioinformatics*, 8(242), 2007.

140. F. Divina and J. S. Aguilar-Ruiz. A multi-objective approach to discover biclusters in microarray data. In *Proc. 9th Annual Conference on Genetic and Evolutionary Computation (GECCO-07)*, pages 385–392, New York, NY, USA, 2007. ACM.

141. F. Divina and J. S. Aguilarruiz. Biclustering of expression data with evolutionary computation. *IEEE Transactions on Knowledge and Data Engineering*, 18:590–602, 2006.

142. DNA microarray, http://en.wikipedia.org/wiki/DNA_microarray .

143. H. Dopazo, J. Santoyo, and J. Dopazo. Phylogenomics and the number of characters required for obtaining an accurate phylogeny of eukaryote model species. *Bioinformatics*, 20(1):116–121, 2004.

144. Georg Dorffner. Neural networks for time series processing. *Neural Network World*, 6(4):447–468, 1996.

145. K. B. Duan, J. C. Rajapakse, H. Wang, and F. Azuaje. Multiple SVM-RFE for gene selection in cancer classification with expression data. *IEEE Transactions on Nanobioscience*, 4(3):228–234, September 2005.

146. R. O. Duda and P. E. Hart. *Pattern Classification and Scene Analysis*. Wiley, New York, 1973.

147. J. C. Dunn. A fuzzy relative of the ISODATA process and its use in detecting compact well-separated clusters. *J. Cybernetics*, 3:32–57, 1973.

148. J. C. Dunn. Well separated clusters and optimal fuzzy partitions. *J. Cyberns.*, 4:95–104, 1974.

149. Y. J. Edwards and A. Cottage. Bioinformatics methods to predict protein structure and function. a practical approach. *Molecular Biotechnology*, 23(2):139–166, February 2003.

150. L. Ein-Dor, O. Zuk, and E. Domany. Thousands of samples are needed to generate a robust gene list for predicting outcome in cancer. *Proc. National Academy of Sciences, USA*, 103:5923–5928, 2006.

151. M. B. Eisen, P. T. Spellman, P. O. Brown, and D. Botstein. Cluster analysis and display of genome-wide expression patterns. In *Proc. Nat. Academy of Sciences*, pages 14863–14868, USA, 1998.

152. M. Ester, H.-P. Kriegel, J. Sander, and X. Xu. A density-based algorithm for discovering clusters in large spatial databases with noise. In *2nd Int. Conf. on Knowledge Discovery and Data Mining*, pages 226–231, 1996.

153. M. Ester, H.-P. Kriegel, and X. Xu. Knowledge discovery in large spatial databases: Focusing techniques for efficient class identification. In *Proc. 4th International Symposium on Large Spatial Databases (SSD'95)*, pages 67–82, Portland, Maine, 1995.

154. M. Fauvel, J. Chanussot, and J. A. Benediktsson. Decision fusion for the classification of urban remote sensing images. *IEEE Trans. Geosci. Remote Sens.*, 44(10):2828–2838, October 2006.

155. U. M. Fayyad, G. Piatetsky-Shapiro, P. Smyth, and R. Uthurusamy, editors. *Advances in Knowledge Discovery and Data Mining*. MIT Press, Menlo Park, 1996.

156. L. Fei and L. Juan. Biclustering of gene expression data with a new hybrid multi-objective evolutionary algorithm of NSGA-II and EDA. In *Proc. Int. Conf. Bioinformatics and Biomedical Engineering (ICBBE 2008)*, May 2008.

157. X. Fei, S. Lu, H. F. Pop, and L. R. Liang. GFBA: A biclustering algorithm for discovering value-coherent biclusters. In *Proc. Int. Symp. Bioinformatics Research and Applications (ISBRA 2007)*, May 2007.

158. L. Feng and Z. Huaibei. Biclustering of gene expression data based on bucketing technique. pages 359–362, July 2007.

159. G. A. Ferguson and Y. Takane. *Statistical Analysis in Psychology and Education*. McGraw-Hill Ryerson Limited, sixth edition, 2005.

160. M. Filippone, F. Masulli, and S. Rovetta. Unsupervised gene selection and clustering using simulated annealing. In I. Bloch, A. Petrosino, and A. Tettamanzi, editors, *Proc. WILF 2005*, volume 3849 of *Lecture Notes in Computer Science*, pages 229–235. Springer, 2005.

161. Ian W. Flockhart. GA-MINER: Parallel data mining with hirarchical genetic algorithms – final report. Technical Report EPCC-AIKMS-GA-MINER-REPORT 1.0, Edinburgh, UK, 1995.

162. Ian W. Flockhart and Nicholas J. Radcliffe. A genetic algorithm-based approach to data mining. In Evangelos Simoudis, Jia Wei Han, and Usama Fayyad, editors, *Proceedings of the Second International Conference on Knowledge Discovery and Data Mining (KDD-96)*, pages 299–302, Portland, Oregon, USA, August 2-4 1996. AAAI Press.

163. C. M. Fonseca and P. J. Fleming. Genetic algorithms for multiobjective optimization: Formulation, discussion and generalization. In *Proc. Fifth International Conference on Genetic Algorithms*, pages 416–423. Morgan Kaufmann, 1993.

164. H. Frigui. Adaptive image retrieval using the fuzzy integral. In *Proceedings of NAFIPS 99*, pages 575–579, New York, USA, 1999. IEEE Press.

165. K. S. Fu. *Syntactic Pattern Recognition and Applications*. Academic Press, London, 1982.

166. K. Fukunaga. *Introduction to Statistical Pattern Recognition*. Academic Press, New York, 1990.

167. T. S. Furey, N. Cristianini, N. Duffy, D. W. Bednarski, M. Schummer, and D. Haussler. Support vector machine classification and validation of cancer tissue samples using microarray expression data. *Bioinformatics*, 16(10):906–914, October 2000.

168. V. Ganti, J. Gehrke, and R. Ramakrishnan. CACTUS-clustering categorical data using summeries. In *Proc. ACM SIGKDD*, 1999.

169. C. Garcia-Martinez, O. Cordon, and F. Herrera. A taxonomy and an empirical analysis of multiple objective ant colony optimization algorithms for the bi-criteria TSP. *European Journal of Operational Research*, 180(1):116–148, July 2007.

170. J. Garnier, J. F. Gibrat, and B. Robson. GOR method for predicting protein secondary structure from amino acid sequence. *Methods Enzymol.*, 266:540–553, 1996.

171. C. Gaspin and T. Schiex. Genetic algorithms for genetic mapping. In *Selected Papers from the Third European Conference on Artificial Evolution*, pages 145–156, London, UK, 1998. Springer-Verlag.

172. E. S. Gelsema and L. Kanal, editors. *Pattern Recognition in Practice II*. North Holland, Amsterdam, 1986.

173. G. Getz, E. Levine, and E. Domany. Coupled two-way cluster analysis of gene microarray data. In *Proc. National Academy of Sciences*, pages 12079–12084, USA, 2000.

174. N. Ghoggali and F. Melgani. Genetic SVM approach to semisupervised multitemporal classification. *IEEE Geosci. Remote Sens. Lett.*, 5(2):212–216, 2008.

175. D. Gibson, J. Kelinberg, and P. Raghavan. Clustering categorical data: An approach based on dynamical systems. In *Proc. VLDB*, pages 222–236, 2000.

176. G. Gibson and S. V. Muse. *The Evolution of the Genome*. Elsevier, second edition, 2004.

177. C. Lee Giles, Steve Lawrence, and Ah Chung Tsoi. Noisy time series prediction using a recurrent neural network and grammatical inference. *Machine Learning*, 44(1/2):161–183, July/August 2001.

178. K. A. Giuliano, J. R. Haskins, and D. L. Taylor. Advances in high content screening for drug discovery. *ASSAY and Drug Development Technologies*, 1(4):565–577, August 2003.

179. J. J. Goeman and P. Bühlmann. Analyzing gene expression data in terms of gene sets: methodological issues. *Bioinformatics*, 23(8):980–987, April 2007.

180. B. Goethals. *Efficient Frequent Pattern Mining*. PhD thesis, University of Limburg, Belgium, 2002.

181. D. E. Goldberg. *Genetic Algorithms in Search, Optimization and Machine Learning*. Addison-Wesley, New York, 1989.

182. D. E. Goldberg and J. Richardson. Genetic algorithms with sharing for multimodal function optimization. In *Proceedings of the Second International Conference on Genetic Algorithms and Their Application*, pages 41–49, Hillsdale, NJ, USA, 1987. L. Erlbaum Associates Inc.

183. D. E. Goldberg and P. Segrest. Finite Markov chain analysis of genetic algorithms. In *Proceedings of the Second International Conference on Genetic Algorithms and Their Application*, pages 1–8, Hillsdale, NJ, USA, 1987. L. Erlbaum Associates Inc.

184. T. R. Golub, D. K. Slonim, P. Tamayo, C. Huard, M. Gaasenbeek, J. P. Mesirov, H. Coller, M. L. Loh, J. R. Downing, M .A. Caligiuri, C. D. Bloomfield, and E. S. Lander. Molecular classification of cancer: Class discovery and class prediction by gene expression monitoring. *Science*, 286:531–537, 1999.

185. R. C. Gonzalez and M. G. Thomason. *Syntactic Pattern Recognition: An Introduction*. Addison-Wesley, Reading, 1978.

186. D. P. Greene and S. F. Smith. Competition-based induction of decision models from examples. *Mach. Learn.*, 13(2-3):229–257, 1993.

187. L. Groll and J. Jakel. A new convergence proof of fuzzy c-means. *IEEE Trans. Fuzzy Systems*, 13(5):717–720, 2005.

188. J. Gu and J. S. Liu. Bayesian biclustering of gene expression data. *BMC Bioinformatics*, 9(Suppl. 1), 2008.

189. S. Guha, R. Rastogi, and K. Shim. ROCK: A robust clustering algorithms for categorical atrributes. In *Proc. IEEE Int. Conf. on Data Eng. (ICDE-99)*, 1999.

190. A. P. Gultyaev, V. Batenburg, and C. W. A. Pleij. The computer simulation of RNA folding pathways using a genetic algorithm. *Journal of Molecular Biology*, 250:37–51, 1995.

191. J. Gunnels, P. Cull, and J. L. Holloway. Genetic algorithms and simulated annealing for gene mapping. In *Proceedings of the First IEEE Conference on Evolutionary Computation*, pages 385–390, USA, 1994. Lawrence Erlbaum Associates, Inc.

192. D. Gusfield. *Algorithms on Strings, Trees, and Sequences: Computer Science and Computational Biology*. Cambridge University Press, January 1997.

193. I. Guyon, J. Weston, S. Barnhill, and V. Vapnik. Gene selection for cancer classification using support vector machines. *Machine Learning*, 46(1-3):389–422, 2002.

194. M. Halkidi, Y. Batistakis, and M. Vazirgiannis. On clustering validation techniques. *J. Intelligent Information Systems*, 17(2/3):107–145, 2001.

195. R. E. Hammah and J. H. Curran. Validity measures for the fuzzy cluster analysis of orientations. *IEEE Transactions on Pattern Analysis and Machine Intelligence*, pages 1467–1472, 2000.

196. K. Hammond, R. Burke, C. Martin, and S. Lytinen. FAQ finer: A case-based approach to knowledge navigation. In *Working Notes of the AAAI Spring Symposium: Information Gathering from Heterogeneous, Distributed Environments*, pages 69–73, Stanford University, 1995. AAAI Press.

197. J. Han and M. Kamber. *Data Mining: Concepts and Techniques*. Morgan Kaufmann, San Francisco, USA, 2000.

198. L. Han and H. Yan. Fuzzy biclustering for DNA microarray data analysis. In *Proc. IEEE Int. Conf. Fuzzy Systems, FUZZ-IEEE 2008. (IEEE World Congress on Computational Intelligence)*, pages 1132–1138, June 2008.

199. J. Handl and J. Knowles. Multiobjective clustering around medoids. In *Proc. IEEE Cong. Evol. Comput.*, volume 1, pages 632–639, 2005.

200. J. Handl and J. Knowles. An evolutionary approach to multiobjective clustering. *IEEE Trans. Evol. Comput.*, 11(1):56–76, 2006.

201. J. Handl and J. Knowles. Feature subset selection in unsupervised learning via multiobjective optimization. *International Journal of Computational Intelligence Research*, 2(3):217–238, 2006.

202. J. Handl and J. Knowles. *Multiobjective Clustering and Cluster Validation*, volume 16 of *Comput. Intell.*, pages 21–47. Springer, 2006.

203. J. Handl, J. Knowles, and D. B. Kell. Computational cluster validation in post-genomic data analysis. *Bioinformatics*, 21(15):3201–3212, 2005.

204. T. Hanne. On the convergence of multiobjective evolutionary algorithms. *European Journal of Operational Research*, 117:553–564, 1999.

205. J. A. Hartigan. *Clustering Algorithms*. Wiley, 1975.

206. J.A. Hartigan. Direct clustering of a data matrix. *J. American Statistical Association*, 67(337):123–129, 1972.

207. E. Hartuv and R. Shamir. A clustering algorithm based on graph connectivity. *Information Processing Letters*, 76(200):175–181, 2000.

208. S. Haykin. *Neural Networks: A Comprehensive Foundation*. Macmillan College Publishing Company, New York, 1994.

209. D. O. Hebb. *The Organization of Behaviour*. Wiley, New York, 1949.

210. S. Helbig and D. Pateva. On several concepts for ε-efficiency. *OR Spectrum*, 16(3):179–186, 1994.

211. R. Herwig, A. Poustka, C. Mueller, H. Lehrach, and J. O'Brien. Large-scale clustering of cDNA fingerprinting data. *Genome Research*, 9(11):1093–1105, 1999.

212. J. Hipp, U. Güntzer, and G. Nakhaeizadeh. Algorithms for association rule mining – a general survey and comparison. *SIGKDD Explorations*, 2(1):58–64, July 2000.

213. S. Hochreiter and K. Obermayer. Sequence classification for protein analysis. In *Snowbird Workshop*, Snowbird, Utah, 2005. Computational and Biological Learning Society.

214. J. Holland. *Adaptation in Natural and Artificial Systems*. University of Michigan Press, 1975.

215. M. Hollander and D. A. Wolfe. *Nonparametric Statistical Methods*. Second edition, 1999.

216. J. Horn. Finite Markov chain analysis of genetic algorithms with niching. In *Proceedings of the 5th International Conference on Genetic Algorithms*, pages 110–117, San Francisco, CA, USA, 1993. Morgan Kaufmann.

217. J. Horn and N. Nafpliotis. Multiobjective optimization using niched Pareto genetic algorithm. Technical Report IlliGAL Report 93005, University of Illinois at Urbana-Champaign, Urbana, Illinois, USA, 1993.

218. T. Hoya and A. Constantidines. An heuristic pattern correction scheme for GRNNs and its application to speech recognition. In *Proc. IEEE Signal Processing Society Workshop*, pages 351–359, 1998.

219. C.-W. Hsu and C.-J. Lin. A comparison of methods for multi-class support vector machines. *IEEE Transactions on Neural Networks*, 13(2):415–425, 2002.

220. J. Hu and X. Yang-Li. Association rules mining using multi-objective coevolutionary algorithm. pages 405–408, Los Alamitos, CA, USA, 2007. IEEE Computer Society.

221. Y.-J. Hu, S. B. Sandmeyer, C. McLaughlin, and D. F. Kibler. Combinatorial motif analysis and hypothesis generation on a genomic scale. *Bioinformatics*, 16(3):222–232, 2000.

222. Z. Huang. Clustering large data sets with mixed numeric and categorical values. In *First Pacific-Asia Conference on Knowledge Discovery and Data Mining*, Singapore, 1997. World Scientific.

223. Z. Huang. Extensions to the k-means algorithm for clustering large data sets with categorical values. *Data Mining Knowledge Discovery*, 2(3):283–304, 1998.

224. Z. Huang and M. K. Ng. A fuzzy k-modes algorithm for clustering categorical data. *IEEE Transactions on Fuzzy Systems*, 7(4), 1999.

225. M. Hulin. Analysis of schema distribution. In R. K. Belew and L. B. Booker, editors, *Proceedings of the 4th International Conference on Genetic Algorithms*, pages 190–196, San Mateo, 1991. Morgan Kaufmann.

226. Michael Hüsken and Peter Stagge. Recurrent neural networks for time series classification. *Neurocomputing*, 50(C), 2003.

227. H. T. Huynh, J.-J. Kim, and Y. Won. DNA microarray classification with compact single hidden-layer feed forward neural networks. In *FBIT '07: Proceedings of the 2007 Frontiers in the Convergence of Bioscience and Information Technologies*, pages 193–198, Washington, DC, USA, 2007. IEEE Computer Society.

228. T. Ideker, O. Ozier, B. Schwikowski, and A. F. Siegel. Discovering regulatory and signalling circuits in molecular interaction networks. *Bioinformatics*, 18(suppl. 1):S233–240, July 2002.

229. J. Ihmels, S. Bergmann, and N. Barkai. Defining transcription modules using large-scale gene expression data. *Bioinformatics*, 20:1993–2003, 2004.

230. J. Ihmels, G. Friedlander, S. Bergmann, O. Sarig, Y. Ziv, and N. Barkai. Revealing modular organization in the yeast transcriptional network. *Nature Genetics*, 31:370–377, 2002.

231. H. Iijima and Y. Naito. Incremental prediction of the side-chain conformation of proteins by a genetic algorithm. In *Proceedings of the IEEE Conference on Evolutionary Computation*, volume 1, pages 362–367, 1994.

232. T. Imielinski and H. Mannila. A database perspective on knowledge discovery. *Communications of the ACM*, 39(11):58–64, 1996.

233. T. Imielinski, A. Virmani, and A. Abdulghani. A discovery board application programming interface and query language for database mining. In *Proceedings of KDD 96*, pages 20–26, Portland, Oregon, 1996.

234. W. H. Inmon. The data warehouse and data mining. *Communications of the ACM*, 39(11):49–50, November 1996.

235. I. Inza, P. Larrañaga, R. Etxeberria, and B. Sierra. Feature subset selection by Bayesian network-based optimization. *Artificial Intelligence*, 123(1-2):157–184, 2000.

236. V. R. Iyer, M. B. Eisen, D. T. Ross, G. Schuler, T. Moore, J.C.F. Lee, J. M. Trent, L. M. Staudt, Jr. J. Hudson, M. S. Boguski, D. Lashkari, D. Shalon, D. Botstein, and P. O. Brown. The transcriptional program in the response of the human fibroblasts to serum. *Science*, 283:83–87, 1999.

237. A. K. Jain and R. C. Dubes. *Algorithms for Clustering Data*. Prentice-Hall, Englewood Cliffs, NJ, 1988.

238. A. K. Jain and R. C. Dubes. Data clustering: A review. *ACM Computing Surveys*, 31, 1999.

239. J. R. Jang, C. Sun, and E. Mizutani. *Neuro-Fuzzy and Soft Computing: A Computational Approach to Learning and Machine Intelligence*. Pearson Education, 1996.

240. A. Jaszkiewicz. Do multiple-objective metaheuristics deliver on their promises? A computational experiment on the set-covering problem. *IEEE Transactions Evolutionary Computation*, 7(2):133–143, 2003.

241. D. Jiang, C. Tang, and A. Zhang. Cluster analysis for gene expression data: A survey. *IEEE Trans. on Knowl. and Data Eng.*, 16(11):1370–1386, 2004.

242. M. Karzynski, A. Mateos, J. Herrero, and J. Dopazo. Using a genetic algorithm and a perceptron for feature selection and supervised class learning in DNA microarray data. *Artificial Intelligence Review*, 20(1-2):39–51, 2003.

243. L. Kaufman and P. J. Rousseeuw. *Finding Groups in Data: An Introduction to Cluster Analysis*. John Wiley & Sons, NY, US, 1990.

244. A. Kel, A. Ptitsyn, V. Babenko, S. Meier-Ewert, and H. Lehrach. A genetic algorithm for designing gene family-specific oligonucleotide sets used for hybridization: The G protein-coupled receptor protein superfamily. *Bioinformatics*, 14(3):259–270, 1998.

245. J. D. Kelly and L. Davis. Hybridizing the genetic algorithm and the K nearest neighbors classification algorithm. In R. K. Belew and L. B. Booker, editors, *Proc. fourth International Conference on Genetic Algorithms*, pages 377–383, San Mateo, CA, USA, 1991. Morgan Kaufmann.

246. J. Khan, J. S. Wei, M. Ringner, L. H. Saal, M. Ladanyi, F. Westermann, F. Berthold, M. Schwab, C. R. Antonescu, C. Peterson, and P. S. Meltzer. Classification and diagnostic prediction of cancers using gene expression profiling and artificial neural networks. *Nature Medicine*, 7(6):673–679, June 2001.

247. O. Khayat, H. R. Shahdoosti, and A. J. Motlagh. A hybrid GA & back propagation approach for gene selection and classification of microarray data. In *Proc. 7th WSEAS International Conference on Artificial Intelligence, Knowledge Engineering and Databases (AIKED-08)*, pages 56–61, Cambridge, UK, 2008. World Scientific and Engineering Academy and Society (WSEAS).

248. M. Khimasia and P. Coveney. Protein structure prediction as a hard optimization problem: The genetic algorithm approach. *Molecular Simulation*, 19:205–226, 1997.

249. S. Y. Kim, J. W. Lee, and J. S. Bae. Effect of data normalization on fuzzy clustering of DNA microarray data. *BMC Bioinformatics*, 7(134), 2006.

250. Y. Kim, W. N. Street, and F. Menczer. Evolutionary model selection in unsupervised learning. *Intell. Data Anal.*, 6(6):531–556, 2002.

251. Y. B. Kim and J. Gao. Unsupervised gene selection for high dimensional data. In *Proc. Sixth IEEE Symposium on BionInformatics and BioEngineering (BIBE-06)*, pages 227–234, Washington, DC, USA, 2006. IEEE Computer Society.

252. S. Kirkpatrik, C. D. Gelatt, and M. P. Vecchi. Optimization by simulated annealing. *Science*, 220:671–680, 1983.

253. Y. Kluger, R. Basri, J. T. Chang, and M. Gerstein. Spectral biclustering of microarray data: Coclustering genes and conditions. *Genome Research*, 13(4):703–716, April 2003.

254. E. M. Knorr, R. T. Ng, and V. Tucakov. Distance-based outliers: Algorithms and applications. *VLDB Journal: Very Large Data Bases*, 8(3–4):237–253, 2000.

255. J. D. Knowles and D. W. Corne. The Pareto Archived Evolution Strategy: A new baseline algorithm for Pareto multiobjective optimisation. In *Proc. IEEE Cong. Evol. Comput.*, pages 98–105, Piscataway, NJ, 1999. IEEE Press.

256. S. Knudsen. Promoter2.0: For the recognition of PolII promoter sequences. *Bioinformatics*, 15:356–361, 1999.

257. A. Koenig. Interactive visualization and analysis of hierarchical projections for data mining. *IEEE Transactions on Neural Networks*, 11:615–624, 2000.

258. J. R. Koza. *Genetic Programming: On the Programming of Computers by Means of Natural Selection (Complex Adaptive Systems)*. The MIT Press, December 1992.

259. N. Krasnogor, D. Pelta, P. M. Lopez, P. Mocciola, and E. Canal. Genetic algorithms for the protein folding problem: A critical view. In C. Fyfe and E. Alpaydin, editors, *Proceedings of the Engineering Intelligent Systems*, pages 353–360. ICSC Academic Press, 1998.

260. K. Krishna and M. N. Murty. Genetic K-means algorithm. *IEEE Transactions on Systems, Man and Cybernetics Part B*, 29(3):433–439, 1999.

261. R. Krishnapuram, A. Joshi, O. Nasraoui, and L. Yi. Low complexity fuzzy relational clustering algorithms for Web mining. *IEEE Transactions on Fuzzy Systems*, 9:595–607, 2001.

262. R. Krishnapuram, A. Joshi, and L. Yi. A fuzzy relative of the k-medoids algorithm with application to document and snippet clustering. In *Proc. IEEE Intl. Conf. Fuzzy Systems - FUZZ-IEEE 99*, pages 1281–1286, Seoul, South Korea, 1999.

263. R. Krovi. Genetic algorithms for clustering: A preliminary investigation. In *Proc. 25th Hawaii International Conference on System Science*, volume 4, pages 540–544, 1992.

264. K. Praveen Kumar, S. Sharath, G. Rio D'Souza, and K. C. Sekaran. Memetic NSGA: A multi-objective genetic algorithm for classification of microarray data. In *Proc. 15th International Conference on Advanced Computing and Communications (ADCOM-07)*, pages 75–80, Washington, DC, USA, 2007. IEEE Computer Society.

265. F. Kursawe. A variant of evolution strategies for vector optimization. In *Proc. 1st Workshop on Parallel Problem Solving from Nature (PPSN-I)*, pages 193–197, London, UK, 1991. Springer-Verlag.

266. S. H. Kwon. Cluster validity index for fuzzy clustering. *Electronic Lett.*, 34(22):2176–2177, 1998.

267. C. Lai, M. Reinders, L. V. Veer, and L. Wessels. A comparison of univariate and multivariate gene selection techniques for classification of cancer datasets. *BMC Bioinformatics*, 7(1), 2006.

268. P. Larranaga and J. A. Lozano. *Estimation of Distribution Algorithms: A New Tool for Evolutionary Computation*. Kluwer Academic Publisher, MA, 2001.

269. M. Laumanns, L. Thiele, K. Deb, and E. Zitzler. Combining convergence and diversity in evolutionary multi-objective optimization. *Evolutionary Computation*, 10(3):263–282, 2002.

270. C. Lawrence, S. Altschul, M. Boguski, J. Liu, A. Neuwald, and J. Wootton. Detecting subtle sequence signals: A Gibbs sampling strategy for multiple alignment. *Science*, 262:208–214, 1993.

271. B. Lazareva-Ulitsky, K. Diemer, and P. D. Thomas. On the quality of tree-based protein classification. *Bioinformatics*, 21(9):1876–1890, 2005.

272. L. Lazzeroni and A. Owen. Plaid models for gene expression data. *Statistica Sinica*, 12:61–86, 2002.

273. S.-K. Lee, Y.-H. Kim, and B. R. Moon. Finding the optimal gene order in displaying microarray data. In *GECCO*, pages 2215–2226, 2003.

274. A. M. Lesk. *Introduction to Bioinformatics*. Oxford University Press, Oxford, 2002.

275. V. G. Levitsky and A. V. Katokhin. Recognition of eukaryotic promoters using a genetic algorithm based on iterative discriminant analysis. *In Silico Biology*, 3: 0008(1-2):81–87, 2003.

276. S. Li and M. Tan. Gene selection and tissue classification based on support vector machine and genetic algorithm. pages 192–195, July 2007.

277. G. E. Liepins and M. D. Vose. Deceptiveness and genetic algorithm dynamics. In G. J. E. Rawlins, editor, *Foundations of Genetic Algorithms*, pages 36–50, 1991.

278. Yinghua Lin and George A. Cunningham III. A new approach to fuzzy-neural system modeling. *IEEE Trans. on Fuzzy Systems*, 3(2):190–198, May 1995.

279. B. Liu, C. Wan, and L. Wang. An efficient semi-unsupervised gene selection method via spectral biclustering. *IEEE Transactions on NanoBioscience*, 5(2):110–114, 2006.

280. H. Liu and H. Motoda. *Feature Selection for Knowledge Discovery and Data Mining*. Kluwer Academic Publishers, Norwell, MA, USA, 1998.

281. J. Liu, Z. Li, X. Hu, and Y. Chen. Biclustering of microarray data with MOPSO based on crowding distance. *BMC Bioinformatics*, 10(Suppl 4), 2008.

282. D. J. Lockhart and E. A. Winzeler. Genomics, gene expreesion and DNA arrays. *Nature*, 405:827–836, 2000.

283. H. J. Lu, R. Setiono, and H. Liu. Effective data mining using neural networks. *IEEE Transactions on Knowledge and Data Engineering*, 15:14–25, 2003.

284. Y. Lu, S. Lu, F. Fotouhi, Y. Deng, and S. J. Brown. FGKA: a fast genetic k-means clustering algorithm. In *Proc. 2004 ACM Symposium on Applied Computing (SAC-04)*, pages 622–623. ACM, 2004.

285. Y. Lu, S. Lu, F. Fotouhi, Y. Deng, and S. J. Brown. Incremental genetic k-means algorithm and its application in gene expression data analysis. *BMC Bioinformatics*, 5(1), 2004.

286. A. V. Lukashin and R. Fuchs. Analysis of temporal gene expression profiles: clustering by simulated annealing and determining the optimal number of clusters. *Bioinformatics*, 17(5):405–419, 2001.

287. B. Ma, J. Tromp, and M. Li. PatternHunter: Faster and more sensitive homology search. *Bioinformatics*, 18(3):440–445, March 2002.

288. P. C. H. Ma, K. C. C. Chan, Y. Xin, and D. K. Y. Chiu. An evolutionary clustering algorithm for gene expression microarray data analysis. *IEEE Transactions on Evolutionary Computation*, 10(3):296–314, June 2006.

289. Q. Ma and J. T. L. Wang. Application of Bayesian neural networks to protein sequence classification. In *ACM SIGKDD Int. Conf. on Knowledge Discovery and Data Mining*, pages 305–309, Boston, MA, USA, 2000.

290. S. C. Madeira and A. L. Oliveira. An efficient biclustering algorithm for finding genes with similar patterns in time-series expression data. In *Proc. 5th Asia Pacific Bioinformatics Conference (APBC-2007), Series in Advances in Bioinformatics and Computational Biology*, volume 5, pages 67–80, Hong Kong, January 2007. Imperial College Press.

291. M. A. Mahfouz and M. A. Ismail. BIDENS: Iterative density based biclustering algorithm with application to gene expression analysis. *Proc. World Academy of Science, Engineering and Technology*, 37:342–348, 2009.

292. E. G. Mansoori, M. J. Zolghadri, and S. D. Katebi. SGERD: A steady-state genetic algorithm for extracting fuzzy classification rules from data. *IEEE Trans. Fuzzy Systems*, 16(4):1061–1071, August 2008.

293. A. Mateos, J. Herrero, J. Tamames, and J. Dopazo. Supervised neural networks for clustering conditions in DNA array data after reducing noise by clustering gene expression profiles. In *Microarray Data Analysis II*, pages 91–103. Kluwer Academic Publishers, 2002.

294. D. H. Mathews. Using an RNA secondary structure partition function to determine confidence in base pairs predicted by free energy minimization. *RNA*, 10(8):1178–1190, August 2004.

295. D. H. Mathews, M. D. Disney, J. L. Childs, S. J. Schroeder, M. Zuker, and D. H. Turner. Incorporating chemical modification constraints into a dynamic programming algorithm for prediction of RNA secondary structure. *Proceedings of the National Academy of Sciences of the United States of America*, 101(19):7287–7292, May 2004.

296. H. Matsuda. Protein phylogenetic inference using maximum likelihood with a genetic algorithm. In *Proc. Pacific Symposium on Biocomputing*, pages 512–523, London, 1996. World Scientific.

297. C. A. Mattson, A. A. Mullur, and A. Messac. Smart Pareto filter: Obtaining a minimal representation of multiobjective design space. *Engineering Optimization*, 36(6):721–740, 2004.

298. U. Maulik and S. Bandyopadhyay. Genetic algorithm based clustering technique. *Pattern Recognition*, 33:1455–1465, 2000.

299. U. Maulik and S. Bandyopadhyay. Performance evaluation of some clustering algorithms and validity indices. *IEEE Trans. Pattern Anal. Mach. Intell.*, 24(12):1650–1654, 2002.

300. U. Maulik and S. Bandyopadhyay. Fuzzy partitioning using a real-coded variable-length genetic algorithm for pixel classification. *IEEE Trans. Geosci. Remote Sens.*, 41(5):1075–1081, 2003.

301. U. Maulik and A. Mukhopadhyay. Simulated annealing based automatic fuzzy clustering combined with ANN classification for analyzing microarray data. *Computers and Operations Research*, 37(8):1369–1380, 2009.

302. U. Maulik, A. Mukhopadhyay, and S. Bandyopadhyay. Combining Pareto-optimal clusters using supervised learning for identifying co-expressed genes. *BMC Bioinformatics*, 10(27), 2009.

303. U. Maulik, A. Mukhopadhyay, and S. Bandyopadhyay. Finding multiple coherent biclusters in microarray data using variable string length multiobjective genetic algorithm. *IEEE Transactions on Information Technology in Biomedicine*, 13(6):969–976, 2009.

304. U. Maulik, A. Mukhopadhyay, S. Bandyopadhyay, M. Q. Zhang, and X. Zhang. Multiobjective fuzzy biclustering in microarray data: Method and a new performance measure. In

Proc. IEEE World Congress on Computational Intelligence (WCCI 2008)/IEEE Congress on Evolutionary Computation (CEC 2008), pages 383–388, Hong Kong, 2008.

305. S. Medasani and R. Krishnapuram. A fuzzy approach to complex linguistic query based image retrieval. In *Proc. of NAFIPS 99*, pages 590–594, 1999.

306. P. Meksangsouy and N. Chaiyaratana. DNA fragment assembly using an ant colony system algorithm. In *Proc. IEEE Congress on Evolutionary Computation (2003)*, 2003.

307. F. Melgani and L. Bruzzone. Classification of hyperspectral remote sensing images with support vector machines. *IEEE Trans. Geosci. Remote Sens.*, 42(8):1778–1790, Aug 2004.

308. Z. Michalewicz. *Genetic Algorithms + Data Structures = Evolution Programs*. Springer-Verlag, New York, 1992.

309. M. S. Mohamad, S. Omatu, S. Deris, and M. Yoshioka. A recursive genetic algorithm to automatically select genes for cancer classification. In J. M. Corchado, J. F. de Paz, M. Rocha, and F. F. Riverola, editors, *IWPACBB*, volume 49 of *Advances in Soft Computing*, pages 166–174. Springer, 2008.

310. S. Mohamed, D. Rubin, and T. Marwala. Multi-class protein sequence classification using fuzzy ARTMAP. In *Proc. IEEE International Conference on Systems, Man and Cybernetics, 2006. SMC '06*, volume 2, pages 1676–1681, October 2006.

311. M. Morita, R. Sabourin, F. Bortolozzi, and C. Y. Suen. Unsupervised feature selection using multi-objective genetic algorithms for handwritten word recognition. In *Proc. Seventh International Conference on Document Analysis and Recognition (ICDAR-03)*, pages 666–671, Washington, DC, USA, 2003. IEEE Computer Society.

312. G. M. Morris, D. S. Goodsell, R. S. Halliday, R. Huey, W. E. Hart, R. K. Belew, and A. J. Olsoni. Automated docking using a Lamarckian genetic algorithm and an empirical binding free energy function. *Journal of Computational Chemistry*, 19(14):1639–1662, 1998.

313. A. Mukhopadhyay and U. Maulik. Towards improving fuzzy clustering using support vector machine: Application to gene expression data. *Pattern Recognition*, 42(11):2744–2763, 2009.

314. A. Mukhopadhyay and U. Maulik. Unsupervised pixel classification in satellite imagery using multiobjective fuzzy clustering combined with SVM classifier. *IEEE Transactions on Geoscience and Remote Sensing*, 47(4):1132–1138, 2009.

315. A. Mukhopadhyay, U. Maulik, and S. Bandyopadhyay. Multi-objective genetic algorithm based fuzzy clustering of categorical attributes. *IEEE Transactions on Evolutionary Computation*, 13(5):991–1005.

316. A. Mukhopadhyay, U. Maulik, and S. Bandyopadhyay. *Multiobjective Evolutionary Approach to Fuzzy Clustering of Microarray Data*, chapter 13, pages 303–326. World Scientific, 2007.

317. A. Mukhopadhyay, U. Maulik, and S. Bandyopadhyay. Unsupervised cancer classification through SVM-boosted multiobjective fuzzy clustering with majority voting ensemble, In *Proc. IEEE Congress on Evolutionary Computation (CEC-09)*, pages 255–261, May 2009.

318. A. Mukhopadhyay, U. Maulik, and S. Bandyopadhyay. On biclustering of gene expression data. *Current Bioinformatics*, 5(3):204–216, 2010.

319. T. M. Murali and S. Kasif. Extracting conserved gene expression motifs from gene expression data. In *Proc. Pacific Symposium on Biocomputing*, volume 8, pages 77–88, 2003.

320. H. Murao, H. Tamaki, and S. Kitamura. A coevolutionary approach to adapt the genotype-phenotype map in genetic algorithms. In *Proc. IEEE Congress on Evolutionary Computation (CEC-02)*, pages 1612–1617, Washington, DC, USA, 2002. IEEE Computer Society.

321. C. A. Murthy, D. Bhandari, and S. K. Pal. ε-optimal stopping time for genetic algorithms with elitist model. *Fundamenta Informaticae*, 35:91–111, 1998.

322. C. A. Murthy and N. Chowdhury. In search of optimal clusters using genetic algorithms. *Pattern Recognition Letters*, 17(8):825–832, 1996.

323. A. E. Nassar and R. E. Talaat. Strategies for dealing with metabolite elucidation in drug discovery and development. *Drug Discovery Today*, 9(7):317–327, April 2004.

324. A. Y. Ng, M. I. Jordan, and Y. Weiss. On spectral clustering: Analysis and an algorithm. In *Advances in Neural Information Processing Systems 14*, pages 849–856. MIT Press, 2001.

325. R. T. Ng and J. Han. CLARANS: A method for clustering objects for spatial data mining. *IEEE Transactions on Knowledge and Data Engineering*, 14(5):1003–1016, 2002.

326. E. Noda, A. A. Freitas, and H. S. Lopes. Discovering interesting prediction rules with a genetic algorithm. In P. J. Angeline, Z. Michalewicz, M. Schoenauer, X. Yao, and A. Zalzala, editors, *Proc. IEEE Congress on Evolutionary Computation (CEC-99)*, volume 2, pages 1322–1329, Mayflower Hotel, Washington, DC, USA, 1999. IEEE Press.

327. C. Notredame and D. G. Higgins. SAGA: Sequence alignment by genetic algorithm. *Nucleic Acids Research*, 24(8):1515–1524, 1996.

328. C. Notredame, E. A. O'Brien, and D. G. Higgins. RAGA: RNA sequence alignment by genetic algorithm. *Nucleic Acids Research*, 25(22):4570–4580, 1997.

329. H. Ogata, Y. Akiyama, and M. Kanehisa. A genetic algorithm based molecular modeling technique for RNA stem-loop structures. *Nucleic Acids Research*, 23(3):419–426, 1995.

330. A. Oliver, N. Monmarché, and G. Venturini. Interactive design of Web sites with a genetic algorithm. In *Proceedings of the IADIS International Conference WWW/Internet*, pages 355–362, Lisbon, Portugal, November 13–15 2002.

331. A. Oliver, O. Regragui, N. Monmarché, and G. Venturini. Genetic and interactive optimization of Web sites. In *Eleventh International World Wide Web Conference*, Honolulu, Hawaii, 7-11 May 2002.

332. I. Ono, H. Fujiki, M. Ootsuka, N. Nakashima, N. Ono, and S. Tate. Global optimization of protein 3-dimensional structures in NMR by a genetic algorithm. In *Proceedings of the Congress on Evolutionary Computation*, volume 1, pages 303–308, 2002.

333. M. K. Pakhira, S. Bandyopadhyay, and U. Maulik. A study of some fuzzy cluster validity indices, genetic clustering and application to pixel classification. *Fuzzy Sets and Systems*, 155:191–214, 2005.

334. M.K. Pakhira, S. Bandyopadhyay, and U. Maulik. Validity index for crisp and fuzzy clusters. *Pattern Recognition*, 37:487–501, 2004.

335. N. R. Pal and J. C. Bezdek. On cluster validity for the fuzzy c-means model. *IEEE Transactions on Fuzzy Systems*, 3:370–379, 1995.

336. S. K. Pal, S. Bandyopadhyay, and C. A. Murthy. Genetic algorithms for generation of class boundaries. *IEEE Transactions on Systems, Man and Cybernetics*, 28(6):816–828, 1998.

337. S. K. Pal, S. Bandyopadhyay, and S. S. Ray. Evolutionary computation in bioinformatics: A review. *IEEE Trans. Systems, Man, and Cybernetics, Part C: Applications and Reviews*, 36(5):601–615, September 2006.

338. S. K. Pal and P. Mitra. Multispectral image segmentation using the rough set initialized EM algorithm. *IEEE Transactions on Geoscience and Remote Sensing*, 11:2495–2501, 2002.

339. H. Pan, J. Zhu, and D. Han. Genetic algorithms applied to multi-class clustering for gene expression data. *Genomics Proteomics Bioinformatics*, 1(4):279–287, November 2003.

340. Y. H. Pao. *Adaptive Pattern Recognition and Neural Networks*. Addison-Wesley, New York, 1989.

341. R. V. Parbhane, S. Unniraman, S. S. Tambe, V. Nagaraja, and B. D. Kulkarni. Optimum DNA curvature using a hybrid approach involving an artificial neural network and genetic algorithm. *Journal of Biomolecular Structure and Dynamics*, 17(4):665–672, 2000.

342. R. J. Parsons, S. Forrest, and C. Burks. Genetic algorithms, operators, and DNA fragment assembly. *Machine Learning*, 21(1-2):11–33, 1995.

343. R. J. Parsons and M. E. Johnson. DNA fragment assembly and genetic algorithms. New results and puzzling insights. *International Conference on Intelligent Systems in Molecular Biology*, pages 277–284, AAAI Press, Menlo Park, CA, 1995.

344. A. W. P. Patton, III and E. Goldman. A standard GA approach to native protein conformation prediction. In *Proceedings of the International Conference on Genetic Algorithms*, volume Morgan Kaufmann, pages 574–581, 1995.

345. L. Pauling, R. B. Corey, and H. R. Branson. The structure of proteins; two hydrogen-bonded helical configurations of the polypeptide chain. *Proc. National Academy of Sciences*, 37(4):205–211, 1951.

346. T. Pavlidis. *Structural Pattern Recognition*. Springer, 1977.

347. W. Pedrycz. Fuzzy set technology in knowledge discovery. *Fuzzy Sets and Systems*, 98:279–290, 1998.

348. W. Pedrycz and M. Reformat. Genetic optimization with fuzzy coding. In F. Herrera and J. L. Verdegay, editors, *Genetic Algorithms and Soft Computing: Studies in Fuzziness and Soft Computing*, volume 8, pages 51–67. Physica-Verlag, 1996.

349. F. Picarougne, C. Fruchet, A. Oliver, N. Monmarché, and G. Venturini. Recherche d'information sur internet par algorithme génétique. In *Actes des quatrièmes journals nationales de la ROADEF*, pages 247–248, Paris, France, 2002.

350. F. Picarougne, C. Fruchet, A. Oliver, N. Monmarché, and G. Venturini. Web searching considered as a genetic optimization problem. In *Local Search Two Day Workshop*, London, UK, 16-17 April 2002.

351. F. Picarougne, N. Monmarché, A. Oliver, and G. Venturini. Web mining with a genetic algorithm. In *Eleventh International World Wide Web Conference*, Honolulu, Hawaii, 7-11 May 2002.

352. A. Prelic, S. Bleuler, P. Zimmermann, A. Wille, P. Buhlmann, W. Gruissem, L. Hennig, L. Thiele, and E. Zitzler. A systematic comparison and evaluation of biclustering methods for gene expression data. *Bioinformatics*, 22(9):1122–1129, 2006.

353. F. Provost and V. Kolluri. A survey of methods for scaling up inductive algorithms. *Data Mining and Knowledge Discovery*, 3(2):131–169, 1999.

354. N. Qian and T. J. Sejnowski. Predicting the secondary structure of globular proteins using neural network models. *Journal Molecular Biology*, 202(4):865–884, 1988.

355. Z. S. Qin. Clustering microarray gene expression data using weighted Chinese restaurant process. *Bioinformatics*, 22(16):1988–1997, 2006.

356. J. R. Quinlan. *C4.5: Programs for Machine Learning*. Morgan Kaufmann, San Mateo, California, 1993.

357. A. A. Rabow and H. A. Scheraga. Improved genetic algorithm for the protein folding problem by use of a Cartesian combination operator. *Protein Science*, 5:1800–1815, 1996.

358. N. J. Radcliffe. Forma analysis and random respectful recombination. In R. K. Belew and L. B. Booker, editors, *Proceedings of the 4th International Conference on Genetic Algorithms*, pages 222–229, San Mateo, 1991. Morgan Kaufmann.

359. S. Ramaswamy, R. Rastogi, and K. Shim. Efficient algorithms for mining outliers from large data sets. *SIGMOD Rec.*, 29(2):427–438, 2000.

360. F. Rapaport, A. Zinovyev, M. Dutreix, E. Barillot, and J. P. Vert. Classification of microarray data using gene networks. *BMC Bioinformatics*, 8(35), February 2007.

361. M. L. Raymer, W. F. Punch, E. D. Goodman, and L. A. Kuhn. Genetic programming for improved data mining: An application to the biochemistry of protein interactions. In John R. Koza, David E. Goldberg, David B. Fogel, and Rick L. Riolo, editors, *Genetic Programming 1996: Proceedings of the First Annual Conference*, pages 375–380, Stanford University, CA, USA, 1996. MIT Press.

362. M. Raza, I. Gondal, D. Green, and R. L. Coppel. Feature selection and classification of gene expression profile in hereditary breast cancer. pages 315–320, December 2004.

363. R. J. Reece. *Analysis of Genes and Genomes*. John Wiley & Sons, 2004.

364. M. Reyes Sierra and C. A. Coello Coello. Improving PSO-based multi-objective optimization using crowding, mutation and ε-dominance. In *Proceedings of the 3rd international conference on Evolutionary multi-criterion optimization (EMO-05)*, pages 505–519. Springer, 2005.

365. M. Reyes-Sierra and C. A. Coello Coello. Multi-objective particle swarm optimizers: A survey of the state-of-the-art. *International Journal of Computational Intelligence Research*, 2(3):287–308, 2006.

366. P. Reymonda, H. Webera, M. Damonda, and E. E. Farmera. Differential gene expression in response to mechanical wounding and insect feeding in arabidopsis. *Plant Cell*, 12:707–720, 2000.

367. S. K. Riis and A. Krogh. Improving prediction of protein secondary structure using structured neural networks and multiple sequence alignments. *Journal of Computational Biology*, 3:163–183, 1996.

368. C. D. Rosin, R. S. Halliday, W. E. Hart, and R. K. Belew. A comparison of global and local search methods in drug docking. In *Proceedings of the International Conference on Genetic Algorithms*, pages 221–228, 1997.

369. P.J. Rousseeuw. Silhouettes: A graphical aid to the interpretation and validation of cluster analysis. *J. Comp. App. Math*, 20:53–65, 1987.

370. X. Ruan, J. Wang, H. Li, and X. Li. A method for cancer classification using ensemble neural networks with gene expression profile. In *Proc. 2nd International Conference on Bioinformatics and Biomedical Engineering, 2008. ICBBE 2008*, pages 342–346, May 2008.

371. G. Rudolph. On a multi-objective evolutionary algorithm and its convergence to the Pareto set. In *Proceedings of the 5th IEEE Conference on Evolutionary Computation*, pages 511–516. IEEE Press, 1998.

372. G. Rudolph and A. Agapie. Convergence properties of some multi-objective evolutionary algorithms. In *Proceedings of the IEEE Congress on Evolutionary Computation (CEC 2000)*, pages 1010–1016. IEEE Press, 2000.

373. M. Saggar, A. K. Agrawal, and A. Lad. Optimization of association rule mining using improved genetic algorithms. In *Proc. IEEE Int. Conf. Systems, Man and Cybernetics, 2004*, volume 4, pages 3725–3729, October 2004.

374. A. Salamov and V. Solovyev. Prediction of protein secondary structure by combining nearest-neighbor algorithms and multiple sequence alignments. *Journal of Molecular Biology*, 247:11–15, 1995.

375. S. Salzberg and S. Cost. Predicting protein secondary structure with a nearest-neighbor algorithm. *Journal of Molecular Biology*, 227:371–374, 1992.

376. J. Sander, M. Ester, H.-P. Kriegel, and X. Xu. Density-based clustering in spatial databases: The algorithm GDBSCAN and its applications. *Data Mining and Knowledge Discovery*, 2(2):169–194, 1998.

377. J. D. Schaffer. Multiple objective optimization with vector evaluated genetic algorithms. In *Proceedings of the 1st International Conference on Genetic Algorithms*, pages 93–100, Hillsdale, NJ, USA, 1985. L. Erlbaum Associates Inc.

378. S. Schulze-Kremer and U. Tiedemann. Parameterizing genetic algorithms for protein folding simulation. *System Sciences, Hawaii International Conference on Biotechnology Computing*, 5:345–354, 1994.

379. S. Z. Selim and M. A. Ismail. K-means type algorithms: A generalized convergence theorem and characterization of local optimality. *IEEE Transactions on Pattern Analysis and Machine Intelligence*, 6:81–87, 1984.

380. J. Setubal and J. Meidanis. *Introduction to Computational Molecular Biology*. International Thomson Publishing, Boston, MA, 1999.

381. W. Shannon, R. Culverhouse, and J. Duncan. Analyzing microarray data using cluster analysis. *Pharmacogenomics*, 4(1):41–51, 2003.

382. B. A. Shapiro and J. Navetta. A massively parallel genetic algorithm for RNA secondary structure prediction. *Journal of Supercomputing*, 8:195–207, 1994.

383. B. A. Shapiro and J. C. Wu. An annealing mutation operator in the genetic algorithms for RNA folding. *Computer Applications in the Biosciences*, 12:171–180, 1996.

384. R. Sharan, M.-K. Adi, and R. Shamir. CLICK and EXPANDER: A system for clustering and visualizing gene expression data. *Bioinformatics*, 19:1787–1799, 2003.

385. W. M. Shen, K. Ong, B. Mitbander, and C. Zaniolo. Metaqueries for data mining. In U. M. Fayyad, G. Piatetsky-Shapiro, P. Smyth, and R. Uthurusamy, editors, *Advances in Knowledge Discovery and Data Mining*, pages 375–398. AAAI Press, 1996.

386. Q. Sheng, Y. Moreau, and B. D. Moor. Biclustering microarray data by Gibbs sampling. *Bioinformatics*, 19:196–205, 2003.

387. P. D. Shenoy, K. G. Srinivasa, K. R. Venugopal, and L. M. Patnaik. Dynamic association rule mining using genetic algorithms. *Intell. Data Anal.*, 9(5):439–453, 2005.

388. X-J. Shi and H. Lei. A genetic algorithm-based approach for classification rule discovery. In *Proc. Int. Conf. Information Management, Innovation Management and Industrial Engineering (ICIII-08)*, volume 1, pages 175–178, December 2008.

389. A. Skourikhine. Phylogenetic tree reconstruction using self-adaptive genetic algorithm. In *IEEE International Symposium on Bioinformatics and Biomedical Engineering*, pages 129–134, 2000.

390. L. Smith. *Chaos: A Very Short Introduction*. Oxford University Press, USA, March 2007.

391. H. Spath. *Cluster Analysis Algorithms*. Ellis Horwood, Chichester, UK, 1989.

392. D. F. Specht. A general regression neural network. *IEEE Transcations on Neural Networks*, 2:568–576, 1991.

393. N. Srinivas and K. Deb. Multiobjective optimization using nondominated sorting in genetic algorithms. *Evolutionary Computation*, 2(3):221–248, 1994.

394. G. Stehr, H. Graeb, and K. Antreich. Performance trade-off analysis of analog circuits by normal-boundary intersection. In *Proceedings of 40th Design Automation Conference*, pages 958–963, Anaheim, CA, 2003. IEEE Press.

395. J. Suzuki. A Markov chain analysis on simple genetic algorithms. *IEEE Trans. Systems, Man and Cybernetics*, 25(4):655–659, April 1995.

396. P. Tamayo, D. Slonim, J. Mesirov, Q. Zhu, S. Kitareewan, E. Dmitrovsky, E.S. Lander, and T.R. Golub. Interpreting patterns of gene expression with self-organizing maps: Methods and application to hematopoietic differentiation. In *Proc. Nat. Academy of Sciences*, volume 96, pages 2907–2912, USA, 1999.

397. P.-N. Tan, V. Kumar, and J. Srivastava. Selecting the right interestingness measure for association patterns. In *Proc. Eighth ACM SIGKDD International Conference on Knowledge Discovery and Data Mining*, pages 32–41, New York, USA, 2002. ACM Press.

398. A. Tanay, R. Sharan, and R. Shamir. Discovering statistically significant biclusters in gene expression data. *Bioinformatics*, 18:S136–S144, 2002.

399. C. Tang, L. Zhang, I. Zhang, and M. Ramanathan. Interrelated two-way clustering: An unsupervised approach for gene expression data analysis. In *Proc. Second IEEE Int. Symp. Bioinformatics and Bioengineering*, pages 41–48, 2001.

400. D. K. Tasoulis, V. P. Plagianakos, and M. N. Vrahatis. Unsupervised clustering of bioinformatics data. In *European Symposium on Intelligent Technologies, Hybrid Systems and their implementation on Smart Adaptive Systems*, pages 47–53, 2004.

401. S. Tavazoie, J.D. Hughes, M.J. Campbell, R.J. Cho, and G.M. Church. Systematic determination of genetic network architecture. *Nature Genet.*, 22:281–285, 1999.

402. C. M. Taylor and A. Agah. Data mining and genetic algorithms: Finding hidden meaning in biological and biomedical data. In *Computational Intelligence in Biomedicine and Bioinformatics*, pages 49–68. 2008.

403. B. A. Tchagang and A. H. Tewfik. DNA microarray data analysis: A novel biclustering algorithm approach. *EURASIP J. Appl. Signal Process.*, 2006.

404. A. Teller and M. Veloso. Program evolution for data mining. *The International Journal of Expert Systems*, 8(3):213–236, 1995.

405. L. Teng and L.-W. Chan. Biclustering gene expression profiles by alternately sorting with weighted correlated coefficient. In *Proc. IEEE International Workshop on Machine Learning for Signal Processing*, pages 289–294, 2006.

406. R. Tibshirani, G. Walther, and T. Hastie. Estimating the number of clusters in a dataset via the gap statistic. *Journal of the Royal Statistical Society: Series B (Statistical Methodology)*, 63(2):411–423, 2001.

407. W.-C. Tjhi and C. Lihui. Flexible fuzzy co-clustering with feature-cluster weighting. In *Proc. 9th Int. Conf. Control, Automation, Robotics and Vision (ICARCV 2006)*, December 2006.

408. J. T. Tou and R. C. Gonzalez. *Pattern Recognition Principles*. Addison-Wesley, Reading, 1974.

409. H. K. Tsai, J. M. Yang, and C. Y. Kao. Applying genetic algorithms to finding the optimal order in displaying the microarray data. In *Proceedings of the Genetic and Evolutionary Computation Conference (GECCO)*, pages 610–617, 2002.

410. G. E. Tsekouras, D. Papageorgiou, S. Kotsiantis, C. Kalloniatis, and P. Pintelas. Fuzzy clustering of categorical attributes and its use in analyzing cultural data. *Int. J. Computational Intelligence*, 1(2):147–151, 2004.

411. M. Tyagi, F. Bovolo, A. K. Mehra, S. Chaudhuri, and L. Bruzzone. A context-sensitive clustering technique based on graph-cut initialization and expectation-maximization algorithm. *IEEE Geosci. Remote Sens. Lett.*, 5(1):21–25, January 2008.

412. R. Unger and J. Moult. Genetic algorithms for protein folding simulations. *Journal of Molecular Biology*, 231(1):75–81, 1993.

413. R. Unger and J. Moult. A genetic algorithms for three dimensional protein folding simulations. In *Proceedings of the International Conference on Genetic Algorithms*, pages 581–588. Morgan Kaufmann, 1993.

414. R. Unger and J. Moult. On the applicability of genetic algorithms to protein folding. In *Proceedings of the Hawaii International Conference on System Sciences*, volume 1, pages 715–725, 1993.

415. P. J. M. van Laarhoven and E. H. L. Aarts. *Simulated Annealing: Theory and Applications*. Kluwer Academic Publisher, 1987.

416. W. N. van Wieringen, D. Kun, R. Hampel, and A-L. Boulesteix. Survival prediction using gene expression data: A review and comparison. *Computational Statistics and Data Analysis*, 53(5):1590–1603, 2009.

417. V. Vapnik. *Statistical Learning Theory*. Wiley, New York, USA, 1998.

418. M. Villalobos-Arias, C. A. Coello Coello, and O. Hernndez-Lerma. Asymptotic convergence of metaheuristics for multiobjective optimization problems. *Soft Computing*, 10(11):1001–1005, 2006.

419. M. D. Vose and G. E. Liepins. Punctuated equilibria in genetic search. *Complex Systems*, 5:31–44, 1991.

420. M. D. Vose and A. H. Wright. The Walsh transform and the theory of simple genetic algorithm. In S. K. Pal and P. P. Wang, editors, *Genetic Algorithms for Pattern Recognition*, Boca Raton, 1996. CRC Press.

421. William W. M. Rand. Objective criteria for the evaluation of clustering methods. *Journal of the American Statistical Association*, 66(336):846–850, 1971.

422. P. P. Wakabi-Waiswa and V. Baryamureeba. Extraction of interesting association rules using genetic algorithms. *International Journal of Computing and ICT Research*, 2(1):26–33, June 2008.

423. D. Wang, N. K. Lee, and T. S. Dillon. Extraction and optimization of fuzzy protein sequences classification rules using GRBF neural networks. *Neural Information Processing - Letters and Reviews*, 1(1):53–57, 2003.

424. J. T. L. Wang, Q. Ma, D. Shasha, and C. H. Wu. New techniques for extracting features from protein sequences. *IBM Systems Journal*, 40(2):426–441, 2001.

425. L. Wang, J. Zhu, and H. Zou. Hybrid huberized support vector machines for microarray classification. In *Proceedings of the 24th International Conference on Machine Learning (ICML'07)*, pages 983–990, New York, NY, USA, 2007. ACM.

426. S. Y. Wang and K. Tai. Structural topology design optimization using genetic algorithms with a bit-array representation. *Computer Methods in Applied Mechanics and Engineering*, 194:3749–3770, 2005.

427. W. Wanga and Y. Zhanga. On fuzzy cluster validity indices. *Fuzzy Sets and Systems*, 158(19):2095–2117, 2007.
428. M. A. Wani. Incremental hybrid approach for microarray classification. In *Proceedings of the 2008 Seventh International Conference on Machine Learning and Applications, ICMLA'08*, pages 514–520, Washington, DC, USA, 2008. IEEE Computer Society.
429. B. Waske and J. A. Benediktsson. Fusion of support vector machines for classification of multisensor data. *IEEE Trans. Geosci. Remote Sens.*, 45(12):3858–3866, 2007.
430. M. Waterman. RNA structure prediction. In *Methods in Enzymology*, volume 164. Academic Press, USA, 1988.
431. G. I. Webb. Efficient search for association rules. In *Proc. Sixth ACM SIGKDD International Conference on Knowledge Discovery and Data Mining*, pages 99–107. The Association for Computing Machinery, 2000.
432. Q. Wei and G. Chen. Mining generalized association rules with fuzzy taxonomic structures. In *Proceedings of NAFIPS 99*, pages 477–481, New York, USA, 1999. IEEE Press.
433. X. Wen, S. Fuhrman, G. S. Michaels, D. B. Carr, S. Smith, J. L. Barker, and R. Somogyi. Large-scale temporal gene expression mapping of central nervous system development. In *Proc. Nat. Academy of Sciences*, volume 95, pages 334–339, USA, 1998.
434. W. Wetcharaporn, N. Chaiyaratana, and S. Tongsima. DNA fragment assembly by ant colony and nearest neighbour heuristics. In *ICAISC*, pages 1008–1017, 2006.
435. L. D. Whitley. Fundamental principles of deception in genetic search. In G. J. E. Rawlins, editor, *Foundations of Genetic Algorithms*, pages 221–241, 1991.
436. K. C. Wiese and E. Glen. A permutation-based genetic algorithm for the RNA folding problem: A critical look at selection strategies, crossover operators, and representation issues. *Biosystems*, 72(1-2):29–41, 2003.
437. P. Willet. Genetic algorithms in molecular recognition and design. *Trends in Biotechnology*, 13(12):516–521, 1995.
438. M. P. Windham. Cluster validity for the fuzzy c-means clustering algorithm. *IEEE Transactions on Pattern Analysis and Machine Intelligence*, 4:357–363, 1982.
439. A. S. Wu and R. K. Lindsay. A survey of intron research in genetics. In *Proceedings of the 4th Conference on Parallel Problem Solving from Nature*, Lecture Notes in Computer Science, pages 101–110, Berlin Heidelberg New York, 1996. Springer.
440. Y. Xiao and D. Williams. Genetic algorithms for docking of Actinomycin D and Deoxyguanosine molecules with comparison to the crystal structure of Actinomycin D-Deoxyguanosine complex. *Journal of Physical Chemistry*, 98:7191–7200, 1994.
441. B. Xie, S. Chen, and F. Liu. Biclustering of gene expression data using PSO-GA hybrid. In *Proc. Int. Conf. Bioinformatics and Biomedical Engineering (ICBBE 2007)*, pages 302–305, 2007.
442. X. L. Xie and G. Beni. A validity measure for fuzzy clustering. *IEEE Transactions on Pattern Analysis and Machine Intelligence*, 13:841–847, 1991.
443. Y. Xu, V. Olman, and D. Xu. Minimum spanning trees for gene expression data clustering. *Genome Informatics*, 12:24–33, 2001.
444. F. Xue, A. C. Sanderson, and R. J. Graves. Multi-objective differential evolution - algorithm, convergence analysis, and applications. *Proc. IEEE Congress on Evolutionary Computation, CEC-2005*, 1:743–750, September 2005.
445. R. R. Yager. Database discovery using fuzzy sets. *International Journal of Intelligent Systems*, 11:691–712, 1996.
446. J. Yang, W. Wang, H. Wang, and P. Yu. Enhanced biclustering on expression data. In *Proc. 3rd IEEE Conf. Bioinformatics and Bioengineering (BIBE'03)*, pages 321–327, 2003.
447. K. Y. Yeung and R. E. Bumgarner. Multiclass classification of microarray data with repeated measurements: Application to cancer. *Genome Biology*, 4, 2003.
448. K. Y. Yeung, D. R. Haynor, and W. L. Ruzzo. Validating clustering for gene expression data. *Bioinformatics*, 17(4):309–318, 2001.

449. K. Y. Yeung and W. L. Ruzzo. An empirical study on principal component analysis for clustering gene expression data. *Bioinformatics*, 17(9):763–774, 2001.

450. K. Y. Yip, D. W. Cheung, M. K. Ng, and K.-H. Cheung. Identifying projected clusters from gene expression profiles. *J. Biomedical Informatics*, 37:345–357, October 2004.

451. T. Yokoyama, T. Watanabe, A. Taneda, and T. Shimizu. A Web server for multiple sequence alignment using genetic algorithm. *Genome Informatics*, 12:382–383, 2001.

452. S. Yoon, C. Nardini, L. Benini, and G. De Micheli. Discovering coherent biclusters from gene expression data using zero-suppressed binary decision diagrams. *IEEE/ACM Trans. Comput. Biol. Bioinformatics*, 2(4):339–354, 2005.

453. N. Yosef, Z. Yakhini, A. Tsalenko, V. Kristensen, A-L. Borresen-Dale, E. Ruppin, and R. Sharan. A supervised approach for identifying discriminating genotype patterns and its application to breast cancer data. *Bioinformatics*, 23 ECCB:e91–e98, 2006.

454. L. Zadeh. Fuzzy sets. *Information and Control*, 8:338–353, 1965.

455. L. A. Zadeh. Calculus of fuzzy restrictions. In L. A. Zadeh et al., editors, *Fuzzy Sets and Their Application to Cognitive and Decision Process*. Academic Press, 1975.

456. L. A. Zadeh. Fuzzy logic and approximate reasoning. *Synthese*, 30:407–428, 1977.

457. L. A. Zadeh. Soft computing and fuzzy logic. *IEEE Softw.*, 11(6):48–56, 1994.

458. L. A. Zadeh, K. S. Fu, K. Tanaka, and M. Shimura, editors. *Fuzzy Sets and Their Applications to Cognitive and Decision Process*. Academic Press, London, 1975.

459. C. Zhang and A. K. C. Wong. A genetic algorithm for multiple molecular sequence alignment. *Bioinformatics*, 13:565–581, 1997.

460. C. Zhang and A. K. C. Wong. A technique of genetic algorithm and sequence synthesis for multiple molecular sequence alignment. *IEEE International Conference on Systems, Man, and Cybernetics*, 3:2442–2447, 1998.

461. J. Zhang, B. Liu, X. Jiang, H. Zhao, M. Fan, Z. Fan, J. J. Lee, T. Jiang, T. Jiang, and S. W. Song. A systems biology-based gene expression classifier of glioblastoma predicts survival with solid tumors. *PLoS ONE*, 4(7):e6274, July 2009.

462. P. Zhang, X. Wang, and P. X. Song. Clustering categorical data based on distance vectors. *The Journal of the American Statistical Association*, 101(473):355–367, 2006.

463. T. Zhang, R. Ramakrishnan, and M. Livny. BIRCH: A new data clustering algorithm and its applications. *Data Mining and Knowledge Discovery*, 1(2):141–182, 1997.

464. L. Zhao and M. J. Zaki. MicroCluster: Efficient deterministic biclustering of microarray data. *IEEE Intelligent Systems*, 20(6):40–49, 2005.

465. Li. Zhuo, J. Zheng, X. Li, F. Wang, B. Ai, and J. Qian. A genetic algorithm based wrapper feature selection method for classification of hyperspectral images using support vector machine. In *Int. Arch. Photogrammetry, Remote Sensing and Spatial Information Sciences*, volume XXXVII (B7), pages 397–402, Beijing, 2008.

466. E. Zitzler, M. Laumanns, and L. Thiele. SPEA2: Improving the Strength Pareto Evolutionary Algorithm. Technical Report 103, Universität Zürich, Switzerland, 2001.

467. E. Zitzler and L. Thiele. An evolutionary algorithm for multiobjective optimization: The strength Pareto approach. Technical Report 43, Universität Zürich, Switzerland, 1998.

468. M. Zuker and P. Stiegler. Optimal computer folding of large RNA sequences using thermodynamics and auxiliary information. *Nucleic Acids Research*, 9:133–148, 1981.

Index

279